AF357836

Cell Differentiation, Oxidative Stress, and Oxygen Radicals—in Honor of Prof. Michael Breitenbach

Cell Differentiation, Oxidative Stress, and Oxygen Radicals—in Honor of Prof. Michael Breitenbach

Guest Editors

Mark Rinnerthaler
Markus Ralser

Basel • Beijing • Wuhan • Barcelona • Belgrade • Novi Sad • Cluj • Manchester

Guest Editors

Mark Rinnerthaler
Department of Biosciences
and Medical Biology
University of Salzburg
Salzburg
Austria

Markus Ralser
Biochemie
Charité –Universitätsmedizin Berlin
Berlin
Germany

Editorial Office
MDPI AG
Grosspeteranlage 5
4052 Basel, Switzerland

This is a reprint of the Special Issue, published open access by the journal *Biomolecules* (ISSN 2218-273X), freely accessible at: www.mdpi.com/journal/biomolecules/special_issues/Honor_Michael_Breitenbach.

For citation purposes, cite each article independently as indicated on the article page online and using the guide below:

Lastname, A.A.; Lastname, B.B. Article Title. *Journal Name* **Year**, *Volume Number*, Page Range.

ISBN 978-3-7258-2616-2 (Hbk)
ISBN 978-3-7258-2615-5 (PDF)
https://doi.org/10.3390/books978-3-7258-2615-5

Contents

Preface

This volume is dedicated to exploring the intricacies of cell differentiation, oxidative stress, and oxygen radicals, which are all topics that have been foundational to Professor Michael Breitenbach's distinguished career. Through this collection, we aim to present cutting-edge research on cellular responses to oxidative environments and the resulting implications for health and disease. Motivated by a commitment to honor Breitenbach's pioneering contributions, this work is intended for molecular biologists, clinicians, and students.

Grateful acknowledgment goes to the contributing authors and colleagues, whose collaboration and insights made this volume possible.

Mark Rinnerthaler and Markus Ralser
Guest Editors

Editorial

Editorial on the Research Topic 'Cell Differentiation, Oxidative Stress, and Oxygen Radicals—In Honor of Prof. Michael Breitenbach'

Markus Ralser [1,2,3] and **Mark Rinnerthaler** [4,*]

1   Department of Biochemistry, Charité Universitätsmedizin Berlin, 10117 Berlin, Germany; markus.ralser@charite.de
2   The Wellcome Centre for Human Genetics, Nuffield Department of Medicine, University of Oxford, Oxford OX3 7BN, UK
3   Max Planck Institute for Molecular Genetics, 14195 Berlin, Germany
4   Department of Biosciences and Medical Biology, Paris-Lodron University Salzburg, 5020 Salzburg, Austria
*   Correspondence: mark.rinnerthaler@plus.ac.at

Citation: Ralser, M.; Rinnerthaler, M. Editorial on the Research Topic 'Cell Differentiation, Oxidative Stress, and Oxygen Radicals—In Honor of Prof. Michael Breitenbach'. *Biomolecules* **2024**, *14*, 920. https://doi.org/10.3390/biom14080920

Received: 28 May 2024
Accepted: 18 July 2024
Published: 29 July 2024

This Special Issue of *Biomolecules* is dedicated to the life and work of our mentor and outstanding scientist, Michael Breitenbach, and marks his 80th birthday, which he celebrated in 2023. As a pioneer in yeast research and a significant contributor to biochemistry and molecular biology, Michael Breitenbach has left an impressive mark on the history of science. As a leitmotif for this Special Issue, we have curated a collection of topics that align with Michael's diverse scientific interests. Throughout his career, he demonstrated an insatiable curiosity and a willingness to explore a wide array of subjects in great detail and at a depth that escaped others. We aim for this diversity to be reflected in the eclectic range of contributions within this Special Issue.

Born in Mauer (south of Vienna, Austria) in 1943, Michael commenced his studies in chemistry at the University of Vienna in 1962. His formative scientific experiences took root in the laboratory of Professor Otto Hoffmann-Ostenhof, where he studied the process of myo-inositol phosphorylation in the erythrocytes of chickens. This process influenced him profoundly, prompting him to immediately switch to other model systems. As a postdoc, he worked on the structure of bacterial ferredoxin in the research group of Klaus Gersonde (1972–1974). Subsequently, in 1974, he established a working group at Otto Hoffmann-Ostenhof's institute, focusing on yeast sporulation. Later, as a full professor at the University of Salzburg, Austria, he dedicated himself primarily to unraveling aging processes, using yeast cells as a model system. Yeast has remained a central theme in his research to this day, but he has made seminal contributions to other fields as well.

Michael's work was pivotal in clarifying several key biological processes. His contributions range from unraveling sporulation-specific metabolic processes, such as inositol phospholipid and dityrosine synthesis, to the study of nonsense suppressor tRNAs, the GTPase Ras2, research on ribosomes (especially the ribosomal protein RPL10), research on apoptosis, the oxidative stress signaling pathways (especially NADPH oxidases), asymmetric inheritance of organelles, and the role of mitochondria in the aging process. Beyond yeast, he made significant contributions to research on allergens, notably cloning and characterizing the allergen Betv1a, the major Birch pollen allergen, for the first time. Without claiming to be exhaustive, publications have also been produced under his aegis that deal with tumor migration or investigate skin diseases, such as epidermolysis bullosa.

Among the several characteristics that make Michael a special mentor to us, two stand out. The first is his seemingly infinite knowledge, not only about molecular biology, but also about a much broader range of natural sciences, history, culture, and society, to name a few. It is hard to imagine any conversation to which Michael would not be able to add profound insight, typically to a level of detail that would be missed by others. His second unique trait is his unwavering ability to remember what truly matters. In situations where the logic of a scientist collides with our society's rules, laws, and bureaucracy, his mantra has guided us: "Rules cannot be more important than the people for whom they are made".

This diversity of scientific interest is also deliberately reflected in this Special Issue, which is contributed to exclusively by long-standing colleagues and collaborators.

Carlo (Bruschi), Valentina (Tosato), and Jason (Sims) have repeatedly worked with Michael [1], addressing questions such as whether aging or oxidative stress can affect genomic stability. An article entitled "**Timing of Chromosome DNA Integration throughout the Yeast Cell Cycle**" [2] was written for this issue, which shows that chromosomal integration of DNA elements is highly cell cycle-specific and can lead to an increase in oxygen radicals.

Nikolaus (Netzer) has been researching cellular changes triggered by hypoxia with Michael [3]. In the article "**Oxidative Stress Reaction to Hypobaric–Hyperoxic Civilian Flight Conditions**" [4], it was shown that oxygen supplementation, according to the recommendations of the Federal Aviation Administration, does not lead to any change in oxidative stress markers in a flight simulation.

Campbell (Gourlay) and Michael have developed a strong interest in the interaction network of ROS (reactive oxygen species) and RAS (rat sarcoma)signaling and their influence on the actin cytoskeleton [5]. In the publication "**Elevated Levels of Mislocalized, Constitutive Ras Signalling Can Drive Quiescence by Uncoupling Cell-Cycle Regulation from Metabolic Homeostasis**" [6], it is shown that the overexpression of constitutively active Ras2 alleles drives yeast cells into quiescence, a process that is mainly initiated by the loss of metabolic control.

Andrey (Kozlov) has been in contact with Michael for quite some time, and has successfully contributed to a Special Issue edited by Michael entitled "Oxidative Stress and Oxygen Radicals" [7]. The article "**Effect of mitoTEMPO on Redox Reactions in Different Body Compartments Upon Endotoxemia in Rats**" [8] was written for the anniversary volume, in which the influence of the mitochondria-targeted antioxidant mitoTEMPO on immune cells and liver cells after an injection of lipopolysaccharides was investigated.

Jing (Li) and Gianni (Liti) have co-authored publications with Michael that deal with mutations and their influence on the nuclear genome and mtDNA, as well as on a population basis [9]. In their paper, "**Spontaneous Mutation Rates and Spectra of Respiratory-Deficient Yeast**" [10], a comprehensive genome-wide screening was used to determine mutation rates and spectra that occur as compensatory reactions in non-respiring yeast strains.

Johannes (Grillari) and Markus (Schosserer), as well as Michael, were part of the National Research Network NFN S93 funded by the Austrian Science Fund (FWF) from 2005 to 2010. Over the years, the three have proven to be not only competent cooperation partners, but also good friends, which has been expressed in many joint publications [11], among other things. For this anthology, senopathies, which are responsible for many aging phenomena, were examined in more detail in the article titled "**Senopathies—Diseases Associated with Cellular Senescence**" [12].

An extremely exciting and personal article was written by Heimo (Breiteneder) and Dietrich (Kraft). The article, "**The History and Science of the Major Birch Pollen Allergen Bet v 1**" [13], describes, in detail, how an ingenious team of scientists isolated, cloned, and characterized the first (pollen) allergen [14].

Shaoping (Li) and Michael share a deep interest in traditional Chinese medicine [15], which has resulted in numerous discussions and joint project proposals. In the article "**Quality Evaluation of Ophiopogon japonicus from Two Authentic Geographical Origins in China Based on Physicochemical and Pharmacological Properties of Their Polysaccharides**" [16], the polysaccharide composition of the dwarf lilyturf was determined.

During his studies in Vienna, Michael not only attended biochemistry, but also seminars in philosophy. A former student and good friend, Gregor (Greslehner) [17], wrote a philosophical article entitled "**Molecular Biology—Pleonasm or Denotation for a Discipline of Its Own? Reflections on the Origins of Molecular Biology and Its Situation Today**" [18], which rounds out Michael's broad interests.

The two editors, Markus (Ralser) and Mark (Rinnerthaler), supported by the long-term cooperation partners Nana (Grüning) and Nikolaus (Bresgen), would also like to pay tribute to the birthday boy with one publication each. As we have both worked with Michael for many years (and still do) and have produced countless publications together (e.g., [19,20]), we do not want to list common interfaces with Michael (of which there are many). One paper, "**Monogenic Disorders of ROS Production and the Primary Anti-Oxidative Defense**" [21], discusses how a breakdown of cellular redox homeostasis can lead to the onset of human disease, while the other, "**The Janus-Faced Role of Lipid Droplets in Aging: Insights from the Cellular Perspective**" [22], discusses how protein and lipid damage can be detoxified with lipid droplets.

Funding: MRi was supported by the Austrian Science Fund FWF grant P33511 to MRi.

Conflicts of Interest: The authors declare no conflict of interest.

References

1. Tosato, V.; West, N.; Zrimec, J.; Nikitin, D.V.; Del Sal, G.; Marano, R.; Breitenbach, M.; Bruschi, C.V. Bridge-Induced Translocation between NUP145 and TOP2 Yeast Genes Models the Genetic Fusion between the Human Orthologs Associated with Acute Myeloid Leukemia. *Front. Oncol.* **2017**, *7*, 231. [CrossRef] [PubMed]
2. Tosato, V.; Rossi, B.; Sims, J.; Bruschi, C.V. Timing of Chromosome DNA Integration throughout the Yeast Cell Cycle. *Biomolecules* **2023**, *13*, 614. [CrossRef] [PubMed]
3. Netzer, N.C.; Breitenbach, M. Metabolic changes through hypoxia in humans and in yeast as a comparable cell model. *Sleep Breath.* **2010**, *14*, 221–225. [CrossRef] [PubMed]
4. Netzer, N.C.; Jaekel, H.; Popp, R.; Gostner, J.M.; Decker, M.; Eisendle, F.; Turner, R.; Netzer, P.; Patzelt, C.; Steurer, C.; et al. Oxidative Stress Reaction to Hypobaric–Hyperoxic Civilian Flight Conditions. *Biomolecules* **2024**, *14*, 481. [CrossRef] [PubMed]
5. Leadsham, J.E.; Sanders, G.; Giannaki, S.; Bastow, E.L.; Hutton, R.; Naeimi, W.R.; Breitenbach, M.; Gourlay, C.W. Loss of cytochrome c oxidase promotes RAS-dependent ROS production from the ER resident NADPH oxidase, Yno1p, in yeast. *Cell Metab.* **2013**, *18*, 279–286. [CrossRef] [PubMed]
6. Piper-Brown, E.; Dresel, F.; Badr, E.; Gourlay, C.W. Elevated Levels of Mislocalised, Constitutive Ras Signalling Can Drive Quiescence by Uncoupling Cell-Cycle Regulation from Metabolic Homeostasis. *Biomolecules* **2023**, *13*, 1619. [CrossRef] [PubMed]
7. Weidinger, A.; Kozlov, A.V. Biological Activities of Reactive Oxygen and Nitrogen Species: Oxidative Stress Signal Transduction. *Biomolecules* **2015**, *5*, 472–484. [CrossRef] [PubMed]
8. Weidinger, A.; Meszaros, A.T.; Dumitrescu, S.; Kozlov, A.V. Effect of mitoTEMPO on Redox Reactions in Different Body Compartments upon Endotoxemia in Rats. *Biomolecules* **2023**, *13*, 794. [CrossRef] [PubMed]
9. Li, J.; Rinnerthaler, M.; Hartl, J.; Weber, M.; Karl, T.; Breitenbach-Koller, H.; Mulleder, M.; Vowinckel, J.; Marx, H.; Sauer, M.; et al. Slow Growth and Increased Spontaneous Mutation Frequency in Respiratory Deficient afo1(-) Yeast Suppressed by a Dominant Mutation in ATP3. *G3* **2020**, *10*, 4637–4648. [CrossRef]
10. Tu, X.; Wang, F.; Liti, G.; Breitenbach, M.; Yue, J.X.; Li, J. Spontaneous Mutation Rates and Spectra of Respiratory-Deficient Yeast. *Biomolecules* **2023**, *13*, 501. [CrossRef]
11. Schosserer, M.; Grillari, J.; Breitenbach, M. The Dual Role of Cellular Senescence in Developing Tumors and Their Response to Cancer Therapy. *Front. Oncol.* **2017**, *7*, 278. [CrossRef] [PubMed]
12. Lushchak, O.; Schosserer, M.; Grillari, J. Senopathies—Diseases associated with cellular senescence. *Biomolecules* **2023**, *13*, 966. [CrossRef] [PubMed]
13. Breiteneder, H.; Kraft, D. The history and science of the major birch pollen allergen Bet v 1. *Biomolecules* **2023**, *13*, 1151. [CrossRef] [PubMed]
14. Breiteneder, H.; Pettenburger, K.; Bito, A.; Valenta, R.; Kraft, D.; Rumpold, H.; Scheiner, O.; Breitenbach, M. The gene coding for the major birch pollen allergen Betv1, is highly homologous to a pea disease resistance response gene. *EMBO J.* **1989**, *8*, 1935–1938. [CrossRef] [PubMed]
15. Wang, L.Y.; Cheong, K.L.; Wu, D.T.; Meng, L.Z.; Zhao, J.; Li, S.P. Fermentation optimization for the production of bioactive polysaccharides from Cordyceps sinensis fungus UM01. *Int. J. Biol. Macromol.* **2015**, *79*, 180–185. [CrossRef] [PubMed]
16. Chen, Z.; Zhu, B.; Peng, X.; Li, S.; Zhao, J. Quality evaluation of Ophiopogon japonicus from two authentic geographical origins in China based on physicochemical and pharmacological properties of their polysaccharides. *Biomolecules* **2022**, *12*, 1491. [CrossRef] [PubMed]
17. Greslehner, G.P. Not by structures alone: Can the immune system recognize microbial functions? *Stud. Hist. Philos. Biol. Biomed. Sci.* **2020**, *84*, 101336. [CrossRef] [PubMed]

18. Greslehner, G.P. "Molecular Biology"—Pleonasm or Denotation for a Discipline of Its Own? Reflections on the Origins of Molecular Biology and Its Situation Today. *Biomolecules* **2023**, *13*, 1511. [CrossRef] [PubMed]
19. Rinnerthaler, M.; Buttner, S.; Laun, P.; Heeren, G.; Felder, T.K.; Klinger, H.; Weinberger, M.; Stolze, K.; Grousl, T.; Hasek, J.; et al. Yno1p/Aim14p, a NADPH-oxidase ortholog, controls extramitochondrial reactive oxygen species generation, apoptosis, and actin cable formation in yeast. *Proc. Natl. Acad. Sci. USA* **2012**, *109*, 8658–8663. [CrossRef]
20. Stincone, A.; Prigione, A.; Cramer, T.; Wamelink, M.M.; Campbell, K.; Cheung, E.; Olin-Sandoval, V.; Gruning, N.M.; Kruger, A.; Tauqeer Alam, M.; et al. The return of metabolism: Biochemistry and physiology of the pentose phosphate pathway. *Biol. Rev. Camb. Philos. Soc.* **2015**, *90*, 927–963. [CrossRef]
21. Grüning, N.M.; Ralser, M. Monogenic Disorders of ROS Production and the Primary Anti-Oxidative Defense. *Biomolecules* **2024**, *14*, 206. [CrossRef] [PubMed]
22. Bresgen, N.; Kovacs, M.; Lahnsteiner, A.; Felder, T.K.; Rinnerthaler, M. The janus-faced role of lipid droplets in aging: Insights from the cellular perspective. *Biomolecules* **2023**, *13*, 912. [CrossRef] [PubMed]

Article

Oxidative Stress Reaction to Hypobaric–Hyperoxic Civilian Flight Conditions

Nikolaus C. Netzer [1,2,3,4,*], Heidelinde Jaekel [5], Roland Popp [6], Johanna M. Gostner [5], Michael Decker [7], Frederik Eisendle [1], Rachel Turner [1], Petra Netzer [2], Carsten Patzelt [4], Christian Steurer [4], Marco Cavalli [4], Florian Forstner [4] and Stephan Pramsohler [2,3] on behalf of the Hypoxiflight Study Group

[1] Institute of Mountain Emergency Medicine, Eurac Research, Noi Park Campus, Via Hypatia 2, 39100 Bozen, Italy; frederik.eisele@eurac.edu (F.E.); rachel.turner@eurac.edu (R.T.)
[2] Hermann Buhl Institute for Hypoxia and Sleep Medicine Research, Department Psychology and Sport Science, University Innsbruck, 6020 Innsbruck, Austria; petra.netzer@gmx.net (P.N.); s.pramsohler@gmx.net (S.P.)
[3] Division Sports Medicine and Rehabilitation, Department Internal Medicine, University Hospitals, 89070 Ulm, Germany
[4] Terra X Cube, Eurac Research, 39100 Bozen, Italy; carsten.patzelt@eurac.edu (C.P.); christian.steurer@eurac.edu (C.S.); florian.forstner@eurac.edu (F.F.)
[5] Institute of Medical Biochemistry, Medical University of Innsbruck, 6020 Innsbruck, Austria; heidelinde.jaekel@i-med.ac.at (H.J.); johanna.gostner@i-med.ac.at (J.M.G.)
[6] Sleep Medicine Work Group, Department Psychiatry and Psychotherapy, University Hospitals, University Regensburg, 93053 Regensburg, Germany; roland.popp@medbo.de
[7] Institute for Aerospace Physiology, Department Physiology, Medical School, Case Western Reserve University, Cleveland, OH 44120, USA; mjd6@case.edu
* Correspondence: nikolaus.netzer@eurac.edu

Citation: Netzer, N.C.; Jaekel, H.; Popp, R.; Gostner, J.M.; Decker, M.; Eisendle, F.; Turner, R.; Netzer, P.; Patzelt, C.; Steurer, C.; et al. Oxidative Stress Reaction to Hypobaric–Hyperoxic Civilian Flight Conditions. *Biomolecules* **2024**, *14*, 481. https://doi.org/10.3390/biom14040481

Academic Editors: Mark Rinnerthaler, Markus Ralser and Viswanathan Natarajan

Received: 21 December 2023
Revised: 27 March 2024
Accepted: 7 April 2024
Published: 15 April 2024

Abstract: Background: In military flight operations, during flights, fighter pilots constantly work under hyperoxic breathing conditions with supplemental oxygen in varying hypobaric environments. These conditions are suspected to cause oxidative stress to neuronal organ tissues. For civilian flight operations, the Federal Aviation Administration (FAA) also recommends supplemental oxygen for flying under hypobaric conditions equivalent to higher than 3048 m altitude, and has made it mandatory for conditions equivalent to more than 3657 m altitude. Aim: We hypothesized that hypobaric–hyperoxic civilian commercial and private flight conditions with supplemental oxygen in a flight simulation in a hypobaric chamber at 2500 m and 4500 m equivalent altitude would cause significant oxidative stress in healthy individuals. Methods: Twelve healthy, COVID-19-vaccinated (third portion of vaccination 15 months before study onset) subjects (six male, six female, mean age 35.7 years) from a larger cohort were selected to perform a 3 h flight simulation in a hypobaric chamber with increasing supplemental oxygen levels (35%, 50%, 60%, and 100% fraction of inspired oxygen, FiO_2, via venturi valve-equipped face mask), switching back and forth between simulated altitudes of 2500 m and 4500 m. Arterial blood pressure and oxygen saturation were constantly measured via radial catheter and blood samples for blood gases taken from the catheter at each altitude and oxygen level. Additional blood samples from the arterial catheter at baseline and 60% oxygen at both altitudes were centrifuged inside the chamber and the serum was frozen instantly at $-21\ °C$ for later analysis of the oxidative stress markers malondialdehyde low-density lipoprotein (M-LDL) and glutathione-peroxidase 1 (GPX1) via the ELISA test. Results: Eleven subjects finished the study without adverse events. Whereas the partial pressure of oxygen (PO_2) levels increased in the mean with increasing oxygen levels from baseline 96.2 mm mercury (mmHg) to 160.9 mmHg at 2500 m altitude and 60% FiO_2 and 113.2 mmHg at 4500 m altitude and 60% FiO_2, there was no significant increase in both oxidative markers from baseline to 60% FiO_2 at these simulated altitudes. Some individuals had a slight increase, whereas some showed no increase at all or even a slight decrease. A moderate correlation (Pearson correlation coefficient 0.55) existed between subject age and glutathione peroxidase levels at 60% FiO_2 at 4500 m altitude. Conclusion: Supplemental oxygen of 60% FiO_2 in a flight simulation, compared to flying in cabin pressure levels equivalent to 2500 m–4500 m altitude, does not lead to a significant increase or decrease in the oxidative stress markers M-LDL and GPX1 in the serum of arterial blood.

Keywords: high altitude; hyperoxia; flight conditions; oxidative stress; glutathione-peroxidase

1. Introduction

In flight operations with fighter jets and in space crafts, inspired levels of hypoxic and hyperoxic air for pilots and astronauts vary, largely due to the instability of the oxygen delivery system and pressure losses in the flight cabin. This can have an influence on the blood flow and oxygenation of the brain and on cognitive performance [1]. The change of oxygenation of arterial blood can also change peripheral arterial blood pressure and blood flow [2]. The rapid changes in oxygenation of neuronal cells and the oxidative stress can lead to the production of inflammatory neuronal cell markers [3]. Hyperoxia is suspected to cause brain damage and possible white matter densities, proven at least in animal research [4,5]. However, in humans also, hyperoxic flight operations for space flights showed microstructural changes in neuronal tissues as a predisposition for white matter densities [6].

Additionally, in humans, repeated hypobaric hyperoxia increases inflammatory markers and might lead to symptoms of chronic fatigue [7].

Previous investigations with different FiO_2 (fraction of inspired oxygen) levels of supplemental oxygen in simulated flight operations showed a dose-dependent response to higher FiO_2 levels concerning cerebral blood flow and brain oxygenation; a statistical difference in response could be seen at FiO_2 levels from 60% on [8]. Despite this reduction in blood flow, it seems that higher oxygenation enhances cognitive function and high-density EEG (electroencephalogram) activity. For the in-flight performance of flight crews, supplemental oxygen and high FiO_2 could be an advantage [9]. This is the main reason why not only is supplemental oxygen standard in military flight operations, but also, the United States Federal Aviation Administration (FAA) recommends supplemental oxygen in non-pressurized civilian airplanes at altitudes from 3750 m and makes it mandatory from altitudes of 4200 m [10]. Even in pressurized airplanes, supplemental oxygen must be ready for pilots and flight crews [11].

Magnetic resonance imaging (MRI) scans have shown in studies with hypoxic conditions in humans a reduction of cerebral blood flow in several brain regions, although from a physiologic standpoint, one might assume that hypoxia increases blood flow and hyperoxia decreases blood flow [12].

Hypoxia-related accidents in military flight operations have been recorded and published for a long period of time [13–15], but they also affect civilian flight operations and, according to post-accident investigations, have led to major crashes like the one with famous golfer Payne Stewart on board, as well as to a Greek airplane crash with 121 fatalities [16,17]. Therefore, supplemental oxygen for flight operations makes sense not only in military flights, but also for all civilian flight operations. Some researchers even lobby a reduction of hypoxia in civilian flights for crew and passengers from an altitude level of 2400 m in the pressurized cabin down to 1500 m equivalent altitude [18].

Supplemental oxygen in civilian flight operations for flight crews is usually delivered at FiO_2 levels of 35–60% via face masks. The question is if supplemental oxygen and the achievement of a hyperoxic arterial blood gas level lead to a possibly harmful level of oxidative stress for neuronal cells.

Therefore, we aimed to investigate, in a crossover physiology study with a flight simulation with healthy humans in a hypobaric chamber at the most common cabin altitude levels in civilian flight operations, if the delivery of supplemental oxygen leads to a significant increase of the oxidative stress markers malondialdehyde-modified low-density lipoprotein (M-LDL) and glutathione peroxidase 1 (GPX1) in arterial blood. We hypothesized that supplemental oxygen with an FiO_2 level of 60% would lead to a significant increase of the two oxidative stress markers versus baseline.

2. Methods

2.1. Subjects

Subjects were recruited via social media and Blackboard announcement. The inclusion criteria were an age between 20 and 55 years, an overall completely healthy physical status without any need for medications, non-smoking, non-pregnancy, and a COVID-19 vaccination status with three completed COVID-19 vaccinations at least 15 months before the study onset.

The initial cohort from this recruitment consisted of 50 volunteers, of which 12 males and 12 females were willing to participate and were scheduled for this study. All 24 persons underwent an extensive clinical exam by an internal medicine physician, including a lung function test (NDD Easy One, NDD Zuerich, Zuerich, Switzerland) to exclude any ongoing respiratory disease or actual infection. All subjects had an FVC (forced vital capacity) and FEV1 (forced expiratory volume in one second) > 100% of the normal values based on their age and body size.

Via coin tossing, while keeping an equal sex distribution, the 24 subjects were divided into two groups. Group one performed the flight simulation with a radial catheter and group two performed the flight simulation parallel without a radial catheter for blood sampling, standing by as reserve in case of problems with the puncture of the arteria radialis when placing the radial catheter in a subject of group one. The mean age of the 12 subjects from group one was 35.7 years and the standard deviation (SD) was 12.35 years.

The study protocol, according to the standards of the Declaration of Helsinki including amendments, and the subject information were reviewed and accepted by the ethical review board of the province of Bozen/South Tyrol (106-2020) and also by the ethical review board of the University of Innsbruck, Austria (42-2019).

2.2. Study Site and Flight Simulation

Terra X Cube is one of the world's largest civilian hypobaric chambers, with a size of 137 m^2 and a room height of 6 m (822 cubic meter volume). It is located in Bozen, Italy, at an altitude of 322 m above sea level and is incorporated in Eurac Research. The maximal reachable altitude is 9000 m; the maximal rise in altitude is normally 6 m/s. For the Hypoxiflight study, we stretched the rising speed to 10 m/s and used a speed of 50 m/s for the decrease of altitude. The room temperature was kept at a comfortable 22 °C at all times. One subject from group one and one subject from standby group two were in the chamber at the same time for three hours. The study was performed on six consecutive days with two shifts per day: one from 9–12 a.m. (9.00–12.00) and one from 2–5 p.m. (14.00–17.00).

The radial catheter (BD Flow-Switch, Becton Dickinson, Heidelberg, Germany) was placed in the arteria radialis of the non-prominent arm of the subject in a supine position under ultrasound control by an anesthesiologist before the start of the trial and was connected to a patient monitor (Phillips IntelliVue, Phillips Healthcare, Amsterdam, The Netherlands). The arterial blood pressure (BP), arterial oxygen saturation (SaO_2), and pulse frequency were constantly recorded from that time point on. Baseline arterial blood samples (one for blood gases, one for centrifugation to gain serum) were drawn from a three-way valve before the first venturi valve-equipped face mask was put on the subject with 35% FiO_2 and 8 L/min O_2 flow.

Since PO_2 (partial oxygen pressure) analyzed from arterial blood gases is considered the gold standard for the measurement of arterial blood oxygenation, we concentrated on this parameter for the assessment of the oxidative level of the subjects at the different steps. Blood gas analysis was performed via a Roche Opti cartridge system (Opti CCA TS, Opti Medical, Atlanta, GA, USA) on site inside the chamber. The analyzer was calibrated for the specific altitude each day at each step.

After baseline measurements, the altitude was increased and decreased back and forth between 2500 m and 4500 m, respectively, and the equivalent barometric pressure was achieved according to a specific protocol mimicking a standard civilian flight operation (for example, a flight over a mountain range like the Alps or the Rocky Mountains) or after cabin pressure loss in a pressurized airline jet. The flight simulation protocol is shown in Figure 1. After each return from 4500 m to 2500 m altitude, respectively, with the equivalent air pressure in the Terra X Cube, the face masks were changed on the subject with the specific venturi valves for the FiO_2 steps of 35%, 50%, 60%, and 100%. Blood gas analysis was performed at each FiO_2 step and at each altitude of the protocol.

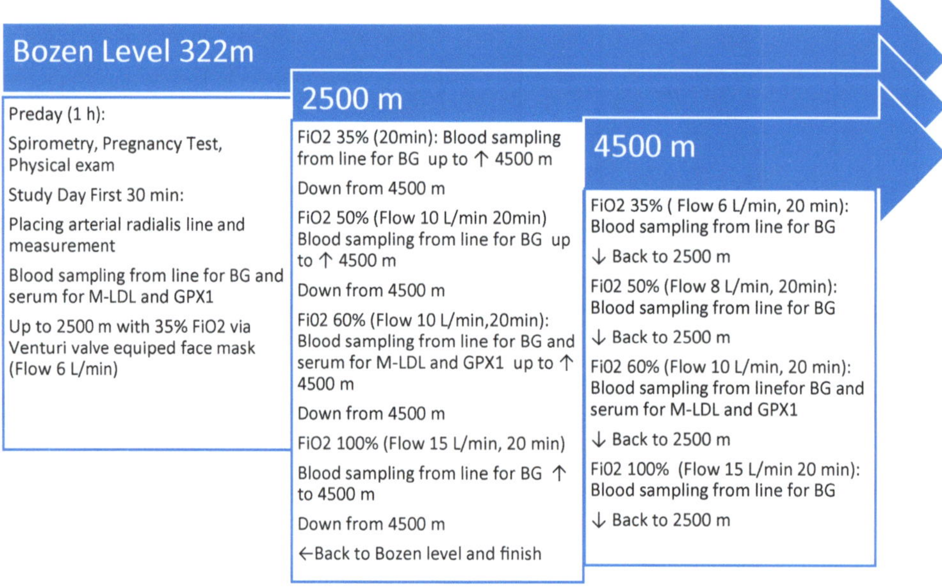

Figure 1. Study protocol time course with altitudes, supplemental oxygen levels, and blood sampling. FiO_2 = fraction of inspired oxygen, BG = blood gases, Flow = airflow of inspired oxygen, M-LDL = malondialdehyde-modified low-density lipoprotein, GPX1 = glutathione peroxidase 1, L = liter, min = minutes.

2.3. Assessment of Oxidative Stress

According to the leading literature, we have chosen malondialdehyde-modified low-density lipoprotein (M-LDL) and glutathione peroxidase 1 (GPX1) as oxidative stress markers for our analysis, and the proof of the hypothesis that there is a correlation between arterial hyperoxia and oxidative stress in the serum from arterial blood in our human subjects [19–22]. According to the German Ministry of Science, these are also the two markers that express oxidative stress the best on a feasible level of cost and analysis efforts [23].

Additional arterial blood for centrifugation inside the hypobaric chamber was drawn from the radial catheter at baseline and at FiO_2 of 60% at both altitudes. The serum samples were transported outside the chamber via an airlock immediately after centrifugation and frozen at $-21\,^{\circ}C$ for further analysis at the biochemistry laboratory of the Medical University of Innsbruck. For the transport from Bozen to Innsbruck (driving time 1 h 15 min), after the collection of all serum samples at the end of the study, the serum was kept in a Styrofoam box with dry ice. In the biochemistry laboratory, the biomarkers oxidized-LDL/MDA adducts and GPX1 were analyzed via ELISA, ox-LDL/MDA adduct ELISA K7810 (Immundiagnostik, Bensheim, Germany), and Human Glutathione Peroxidase 1 ELISA ab193767 (Abcam plc, Cambridge, UK). Each test was repeated twice to prove reproducibility.

2.4. Statistical Analyses

An a priori power analysis was performed based on our hypothesis with an estimated high correlation and a Pearson correlation coefficient of 0.8 between PO_2 values and oxidative stress biomarker ELISA results. At a significance level of p 0.05 (5%) and a power of 0.8 (80%), the necessary number of subjects was nine [24].

All values here are presented in means and standard deviation. The significance of difference between two values was calculated via Student's *t*-test for paired variables; a significant difference between variables is presented as $p \leq 0.05$. The correlation between parameters was calculated according to Pearson and expressed with the Pearson correlation coefficient. All statistical work was performed with the Excel 2021 (Microsoft, Seattle, WA, USA) statistics program.

3. Results

Eleven of the twelve selected subjects and the twelve standby subjects finished the flight simulation in the hypobaric chamber according to the stepwise altitude increase and decrease and supplemental oxygen increase without any problems. One subject (female, 48 years) dropped out after baseline and 35% FiO_2 supplemental oxygen at simulated 2500 m due to subjectively felt stress breathing under the face mask and subsequent hyperventilation, leading to dizziness. After taking the facemask off, she calmed down to a completely normal respiration and all feelings of dizziness disappeared, but she did not want to continue with the protocol.

Arterial PO_2 measured from the blood gases increased from baseline (322 m altitude, 9761 kPa air pressure) in the mean 96.2 mmHg (SD 9.9 mmHg) to 160.9 mmHg (mean, SD 20.5 mmHg) at 2500 m simulated altitude with 60% FiO_2. At the equivalent 4500 m and 60% FiO_2, the mean PO_2 was at 113.2 mmHg (SD 5.7 mmHg). All individual PO_2 values from baseline to 35%, 50%, 60%, and 100% FiO_2 at 2500 m and 4500 m simulated altitude are presented in Figure 2.

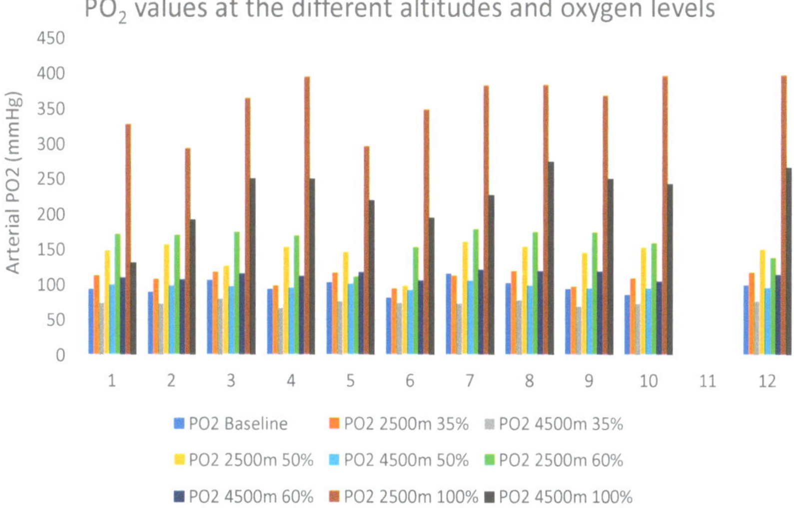

Figure 2. Arterial blood gas from radial line partial pressure level of oxygen in the single subjects at single points of measurement. PO_2 = partial pressure of oxygen in arterial blood, 35–100% = fraction of inspired oxygen through venturi valve-equipped face mask at the specific hypobaric altitude level, X-Axis: subject number.

Mean malondialdehyde-modified LDL (M-LDL) increased in the mean from 49.0 ng/mL (SD 58.02 ng/mL) at baseline to 65.5 ng/mL (SD 69.5 ng/mL) at 2500 m 60% FiO_2 and 73.2 ng/mL (SD 99.2 ng/mL) at 4500 m 60% FiO_2. One subject (age 38) started with a much higher base value of M-LDL (189.9 ng/mL, reason unknown; the subject had no signs of infection and lung function better than the suggested normal values) and further increased to 222.2 ng/mL at 2500 m 60% FiO_2 and 310.9 ng/mL at 4500 m 60% FiO_2. If this subject had been left out from the calculation, the mean M-LDL for the other 10 subjects would have fallen between 2500 m 60% FiO_2 and 4500 m 60% FiO_2 from 43.1 ng/mL to 39.3 ng/mL with much lower standard deviations. With and without this subject, the difference between the baseline value and value at 2500 m 60%FiO_2 or 4500 m 60% FiO_2 did not reach significance (n.s.). Figure 3 presents the M-LDL means and SD at the different protocol time points of measure.

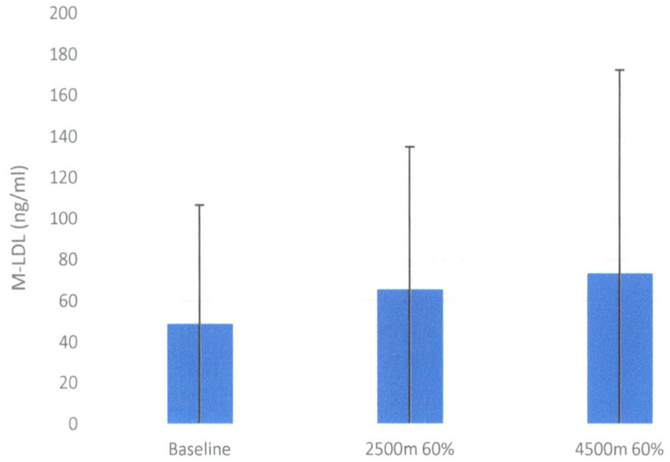

Figure 3. Mean values of malondialdehyde-modified low-density lipoprotein (M-LDL) at baseline and at different altitudes with 60% inspired oxygen. Data are represented as mean ± SD.

Glutathione peroxidase 1 (GPX1) in the mean decreased from baseline 33.5 g/mL (SD 29.5 ng/mL) to 25.9 ng/mL (SD 20.3 ng/mL) at 2500 m 60% FiO_2 and also slightly from baseline to 29.6 ng/mL (SD 25.9 ng/mL) at 4500 m 60% FiO_2. Likewise, with M-LDL, there was no significant difference between GPX1 at baseline and one of the altitudes with 60% supplemental O_2 or in the different values between altitudes. Figure 4 shows the single GPX1 values for each subject.

There was no correlation between M-LDL and arterial PO_2 values at baseline or at one of the altitudes with supplemental 60% oxygen. The Pearson correlation coefficient for M-LDL at 2500 m and 60% FiO_2 was 0.02 and at 4500 m and 60% FiO_2 0.11. For GPX1, there was a slight negative correlation with PO_2 values at 2500 m and 60% FiO_2 (the Pearson correlation coefficient was -30.0) and a moderate negative correlation with PO_2 values at 4500 m and 60% FiO_2, with a Peason correlation coefficient of -0.38. Figure 5 shows the correlation curve for GPX1 and arterial PO_2 at 4500 m and 60% FiO_2.

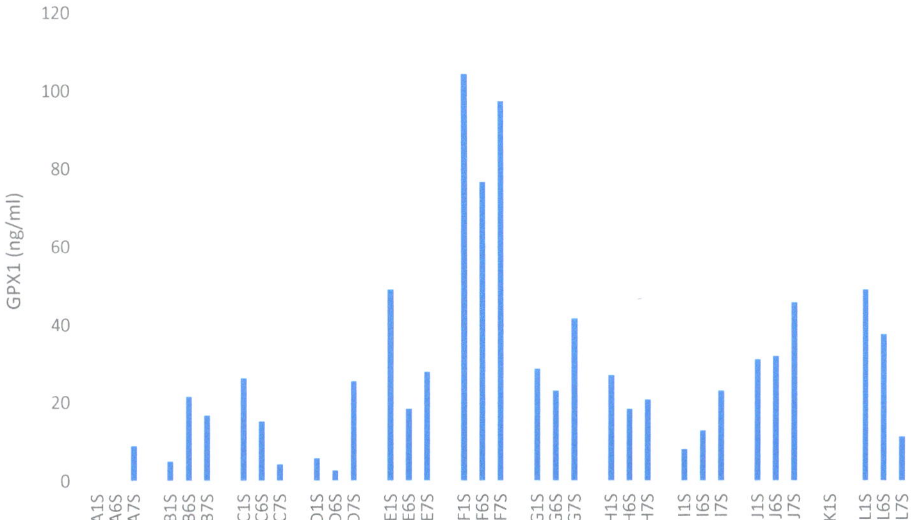

Figure 4. Glutathione peroxidase values of the single subjects at the different points of measurement. GPX1 = glutathione peroxidase 1, A–L = subjects (subjects are marked with letters instead of numbers; the order is the same as in Figure 2), 1S = baseline at 322 m without supplemental oxygen, 6S = 2500 m equivalent altitude with 60% FiO_2, 7S = 4500 m equivalent altitude with 60% FiO_2.

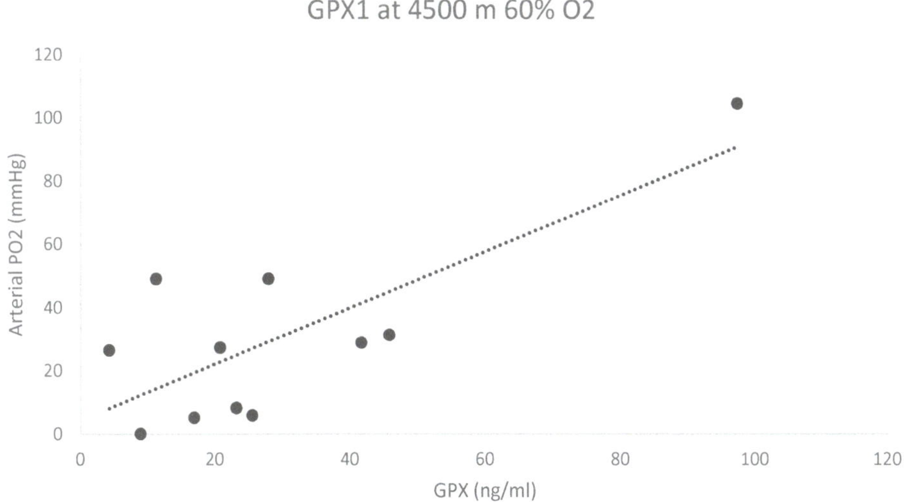

Figure 5. Correlation point-cloud graph of glutathione peroxidase 1 (GPX) values in correlation with arterial PO_2 levels. 4500 m 60% = hypobaric pressure equivalent to 4500 m altitude and 60% FiO_2 through venturi valve-equipped face mask (correlation coefficient 0.38).

There is a moderate correlation between subject age and GPX1 at 4500 m and 60% FiO_2, with a correlation coefficient of 0.55.

4. Discussion

To the best of our knowledge, this is the first hypobaric–hyperoxic flight simulation with specific commonly occurring altitudes and recommended maximal supplemental oxygen levels for civilian airflight operations. Our results show that the recommended oxidative stress biomarkers [19–23] malondialdehyde-modified low-density lipoprotein (M-LDL) and glutathione peroxidase 1 (GPX1) do not significantly increase in a simulated three-hour flight with changing altitudes changing the barometric pressure in a hypobaric chamber. This leads to the assumption that a several-hours flight with supplemental oxygen through a face mask might not cause a higher level of oxidative stress resulting in neuronal cell damage.

The baseline levels of the two measured oxidative stress biomarkers in the serum from arterial blood showed a high inter-individual variation, which led to large standard deviations from the mean values, and they did not change much in the single subjects with the altitude at 60% FiO_2. We do not know why the baseline levels were so individually different in our healthy subject population with clinically excluded ongoing infection or major stress factors. Age did not play a major role here, at least not concerning the baseline values. However, at 4500 m and 60% supplemental oxygen, there was a moderate correlation between age and GPX1 values. The higher the age, the higher the level of this oxidative stress biomarker.

Our results confirm, in some way, the outcomes of former studies concerning military flight operations from members of our work group [7], who showed that flight operators with higher inflammatory biomarkers have a higher tendency to develop signs of chronic fatigue, but that these higher levels of inflammatory biomarkers cannot be, or are only in part, causally related to the number of performed flights and level of hypobaric hyperoxia. In fighter pilots in hypobaric hypoxic conditions, it seems that glutathione and three other oxidative stress markers increased with a higher tendency of oxidative stress elevation in experienced fighter pilots with more flight hours [25]. This could lead to two assumptions. One is that oxidative stress adds up over a longer period of flight hours, in accordance with results that showed more inflammatory biomarkers and greater chronic fatigue with total flight time in our previous studies [7]. The other is that hyperoxia might indeed suppress oxidative stress to a certain level.

With a non-significant increase of M-LDL and GPX1 at lower and higher altitudes at hypobaric air pressure, respectively, our results show that supplemental oxygen up to 60% FiO_2 is safe under the aspect of the non-development of oxidative stress, although there is a moderate correlation between arterial PO_2 levels and GPX1 values. The positive aspects of supplemental oxygen, as recommended by authorities for flight operations in airplanes without pressurized cabins or pressure loss in normally pressurized cabins, seem to make sense, especially since most studies show a significant cognitive decline of pilots in hypoxia from 4750 m on [26].

Other studies or position papers, mainly from high-altitude researcher John West, concerning persons hiking in high altitudes or living or working at high altitudes, have also shown that the negative aspects of hypoxia like acute mountain sickness or reduced levels of cognitive function can be reduced or eliminated with supplemental oxygen [27–29].

If the oxidative stress from supplemental oxygen in flight operations is unlikely the cause of neuronal cell damage and white matter densities occurring in pilots, especially fighter pilots with crews and astronauts, the question remains of what the possible causes could be. Older and newer discussions circle around the effect of high G-forces on fighter airplane crews and in astronauts repeatedly occurring during training and flight operations [30–32]. Another hypothesis is that weightlessness, as it occurs not only in space flight, but also repeatedly in dog fighting in military flight operations with new fighter jets, could be the cause for neuronal damages [33]. However, these discussed causes happen extremely rarely in civilian flight operations with commercial airlines or private or transport flights.

An interesting side aspect of the outcome of our study is the reason for the drop out of one subject. Hyperventilation and the consequences of hypocapnia and alkalosis with dizziness and possible tetany (paw position) due to stressful inflight situations with the necessity of supplemental oxygen might be a bigger problem in sensible passengers or flight crew members than high concentrations of inspired oxygen or subsequently high arterial PO_2 levels. According to several publications, hyperventilation, especially in hypoxic situations, occurs not only in the passengers of commercial flights, but also in fighter pilots and professional commercial airline pilots [34–36]. Therefore, flight crews and pilots should be trained in breathing techniques, which they can use for themselves and also to help passengers in need.

Our study has a few limitations. One is, of course, the relatively low number of subjects, which is due to the large expense in such a hypobaric chamber study including a constant trial-placed radial line. However, although we were not able to confirm our hypothesis that oxidative stress and arterial PO_2 levels correlate, we could show with the means and in the single subjects that there is no significant increase of the oxidative stress biomarkers even with these high levels of arterial PO_2 in each subject in this two-to-three-hour period. Another limitation is that for feasibility and cost reasons, we could only analyze two oxidative stress biomarkers at baseline and two other points of measurement. However, we think that these two oxidative stress biomarkers and the level of 60% inspired oxygen are representative of a real civilian flight situation where supplemental oxygen is asked for.

5. Conclusions

Supplemental oxygen up to 60% FiO_2 via a venturi valve-equipped face mask in a civilian flight simulation with a three-hour duration at a barometric pressure equivalent to altitudes of 2500 m and 4500 m does not lead to an increase in the oxidative stress markers malondialdehyde-modified low-density lipoprotein and glutathione peroxidase 1 in the arterial blood of healthy subjects despite high arterial PO_2 levels. The recommended supplemental oxygen for flight crews in non-pressurized airplane cabins or in situations with pressure loss does not cause a high level of oxidative stress and is safe.

Author Contributions: Conceptualization, N.C.N., R.P., M.D., C.S. and S.P.; Methodology, N.C.N., R.P., M.D., R.T. and S.P.; Formal analysis, N.C.N., H.J., R.P., J.M.G., R.T. and P.N.; Investigation, N.C.N., R.P., F.E., R.T., C.P., C.S., M.C., F.F. and S.P.; Resources, N.C.N., C.S. and M.C.; Writing—original draft, N.C.N. and P.N.; Supervision, N.C.N. and C.S.; Project administration, N.C.N.; Funding acquisition, N.C.N. and C.S. All authors have read and agreed to the published version of the manuscript.

Funding: Expenses for the study were funded from internal sources of the Hermann Buhl Institute, Eurac Terra X Cube, Eurac Institute for Mountain Emergency Medicine, and the Medical University of Innsbruck. Publication funding was provided by the University of Innsbruck Open Access Publication fund.

Institutional Review Board Statement: The review boards of Bozen/South Tyrol (106-2020) and in addition the ethical review board of the University of Innsbruck, Austria (42-2019) approved the study protocol and amendmends.

Informed Consent Statement: Informed consent was obtained from all participants.

Data Availability Statement: RAW data is available on request by the corresponding author.

Acknowledgments: The authors would like to thank Giulia Roveri for medical support during the study.

Conflicts of Interest: The authors declare no conflicts of interest.

References

1. Bouak, F.; Vartanian, O.; Hofer, K.; Cheung, B. Acute mild hypoxic hypoxia effects on cognitive and simulated aircraft pilot performance. *Aerosp. Med. Hum. Perform.* **2018**, *89*, 526–535. [CrossRef] [PubMed]
2. West, J.B. A strategy for in-flight measurements of physiology of pilots of high-performance fighter aircraft. *J. Appl. Physiol.* **1985**, *115*, 145–149. [CrossRef] [PubMed]

3. Tamma, G.; Valenti, G.; Grossini, E.; Donnini, S.; Marino, A.; Marinelli, R.A.; Calamita, G. Aquaporin Membrane Channels in Oxidative Stress, Cell Signaling, and Aging: Recent Advances and Research Trends. *Oxid. Med. Cell Longev.* **2018**, *2018*, 1501847. [CrossRef] [PubMed]
4. Vottier, G.; Pham, H.; Pansiot, J.; Biran, V.; Gressens, P.; Charriaut-Marlangue, C.; Baud, O. Deleterious effect of hyperoxia at birth on white matter damage in the newborn rat. *Dev. Neurosci.* **2011**, *33*, 261–269. [CrossRef] [PubMed]
5. Machado, R.S.; Tenfen, L.; Joaquim, L.; Lanzzarin, E.V.R.; Bernardes, G.C.; Bonfante, S.R.; Mathias, K.; Biehl, E.; Bagio, É.; de Souza Stork, S.; et al. Hyperoxia by short-term promotes oxidative damage and mitochondrial dysfunction in rat brain. *Respir. Physiol. Neurobiol.* **2022**, *306*, 103963. [CrossRef] [PubMed]
6. Lee, J.K.; Koppelmans, V.; Riascos, R.F.; Hasan, K.M.; Pasternak, O.; Mulavara, A.P.; Bloomberg, J.J.; Seidler, R.D. Spaceflight-Associated Brain White Matter Microstructural Changes and Intracranial Fluid Redistribution. *JAMA Neurol.* **2019**, *76*, 412–419. [CrossRef] [PubMed]
7. Damato, E.G.; Fillioe, S.J.; Margevicius, S.P.; Mayes, R.S.; Somogyi, J.E.; Vannix, I.S.; Abdollahifar, A.; Turner, A.M.; Ilcus, L.S.; Decker, M.J. Increased Serum Levels of Proinflammatory Cytokines Are Accompanied by Fatigue in Military T-6A Texan II Instructor Pilots. *Front. Physiol.* **2022**, *13*, 876750. [CrossRef] [PubMed]
8. Damato, E.G.; Fillioe, S.J.; Vannix, I.S.; Norton, L.K.; Margevicius, S.P.; Beebe, J.L.; Decker, M.J. Characterizing the Dose Response of Hyperoxia with Brain Perfusion. *Aerosp. Med. Hum. Perform.* **2022**, *93*, 493–498. [CrossRef]
9. Damato, E.G.; Flak, T.A.; Mayes, R.S.; Strohl, K.P.; Ziganti, A.M.; Abdollahifar, A.; Flask, C.A.; LaManna, J.C.; Decker, M.J. Neurovascular and cortical responses to hyperoxia: Enhanced cognition and electroencephalographic activity despite reduced perfusion. *J. Physiol.* **2020**, *598*, 3941–3956. [CrossRef]
10. FAR 91.211-Supplemental Oxygen. Available online: http://www.faa-aircraft-certification.com (accessed on 20 October 2023).
11. Available online: http://www.ainonline.com/aviation-news/air-transport/2020-05-13/faa-amends-oxygen-mask-rule-not-without-controversy (accessed on 20 October 2020).
12. Liu, J.; Li, S.; Qian, L.; Xu, X.; Zhang, Y.; Cheng, J.; Zhang, W. Effects of acute mild hypoxia on cerebral blood flow in pilots. *Neurol. Sci.* **2021**, *42*, 673–680. [CrossRef]
13. Saito, I.; Fujiwara, O.; Utsuki, N.; Mizumoto, C.; Arimori, T. Hypoxia-induced fatal aircraft accident revealed by voice analysis. *Aviat. Space Environ. Med.* **1980**, *51*, 402–406. [PubMed]
14. Rayman, R.B. Sudden incapacitation in flight, 1 January 1966–30 November 1971. *Aerosp. Med.* **1973**, *44*, 953–955. [PubMed]
15. Rayman, R.B.; McNaughton, G.B. Hypoxia: USAF experience 1970–1980. *Aviat. Space Environ. Med.* **1983**, *54*, 357–359. [PubMed]
16. Martin-Saint-Laurent, A.; Lavernhe, J.; Casano, G.; Simkoff, A. Clinical aspects of inflight incapacitations in commercial aviation. *Aviat. Space Environ. Med.* **1990**, *61*, 256–260. [PubMed]
17. Available online: http://www.longislandpress.com/2023/06/06/loss-oxygen-virginia-plane-crash/ (accessed on 20 October 2023).
18. Aerospace Medical Association; Aviation Safety Committee; Civil Aviation Subcommittee. Cabin cruising altitudes for regular transport aircraft. *Aviat. Space Environ. Med.* **2008**, *79*, 433–439. [CrossRef] [PubMed]
19. Sies, H. Oxidative stress: From basic research to clinical application. *Am. J. Med.* **1991**, *91*, 31S–38S. [CrossRef] [PubMed]
20. Sies, H. Oxidative stress: Oxidants and antioxidants. *Exp. Physiol.* **1997**, *82*, 291–295. [CrossRef] [PubMed]
21. Sies, H. Biological redox systems and oxidative stress. *Cell Mol. Life Sci.* **2007**, *64*, 2181–2188. [CrossRef] [PubMed]
22. Katerji, M.; Filippova, M.; Duerksen-Hughes, P. Approaches and Methods to Measure Oxidative Stress in Clinical Samples: Research Applications in the Cancer Field. *Oxid. Med. Cell Longev.* **2019**, *2019*, 1279250. [CrossRef] [PubMed]
23. Robert Koch-Institut. Oxidativer Stress und Möglichkeiten seiner Messung aus umweltmedizinischer Sicht: Mitteilung der Kommission "Methoden und Qualitätssicherung in der Umweltmedizin". *Bundesgesundheitsblatt Gesundheitsforschung Gesundheitsschutz* **2008**, *51*, 1464–1482. [CrossRef]
24. Hemmerich, W. StatisticGuru: Poweranalyse für Korrelationen 2018. Available online: https://statisticguru.deTrechner/poweranalyse-korrelationen.html (accessed on 15 March 2023).
25. Petraki, K.; Grammatikopoulou, M.G.; Tekos, F.; Skaperda, Z.; Orfanou, M.; Mesnage, R.; Vassilakou, T.; Kouretas, D. Estimation of Redox Status in Military Pilots during Hypoxic Flight-Simulation Conditions—A Pilot Study. *Antioxidants* **2022**, *11*, 1241. [CrossRef] [PubMed]
26. Shaw, D.M.; Cabre, G.; Gant, N. Hypoxic Hypoxia and Brain Function in Military Aviation: Basic Physiology and Applied Perspectives. *Front. Physiol.* **2021**, *12*, 665821. [CrossRef] [PubMed]
27. West, J.B. Cognitive Impairment of School Children at High Altitude: The Case for Oxygen Conditioning in Schools. *High. Alt. Med. Biol.* **2016**, *17*, 203–207. [CrossRef] [PubMed]
28. West, J.B. High-altitude medicine. *Am. J. Respir. Crit. Care Med.* **2012**, *186*, 1229–1237. [CrossRef] [PubMed]
29. West, J.B. A new approach to very-high-altitude land travel: The train to Lhasa, Tibet. *Ann. Intern. Med.* **2008**, *149*, 898–900. [CrossRef] [PubMed]
30. Koppelmans, V.; Mulavara, A.P.; Seidler, R.D.; De Dios, Y.E.; Bloomberg, J.J.; Wood, S.J. Cortical thickness of primary motor and vestibular brain regions predicts recovery from fall and balance directly after spaceflight. *Brain Struct. Funct.* **2022**, *227*, 2073–2086. [CrossRef] [PubMed]
31. Li, J.S.; Sun, X.Q.; Wu, X.Y.; Rao, Z.R.; Liu, H.L.; Xie, X.P. Influences of repeated lower +Gz exposures on high +Gz exposure induced brain injury in rats. *Space Med. Med. Eng.* **2002**, *15*, 339–342.

32. Pashchenko, P.S.; Risman, B.V. Strukturnye preobrazovaniia v serom veshchestve spinnogo mozga posle vozdeĭstviia gravitatsionnykh peregruzok [Structural changes in spinal cord gray matter induced by gravitational overloads]. *Morfologiia* **2002**, *121*, 49–54. [PubMed]
33. Clément, G.R.; Boyle, R.D.; George, K.A.; Nelson, G.A.; Reschke, M.F.; Williams, T.J.; Paloski, W.H. Challenges to the central nervous system during human spaceflight missions to Mars. *J. Neurophysiol.* **2020**, *123*, 2037–2063. [CrossRef]
34. Kramer, K.E.P.; Anderson, E.E. Hyperventilation-Induced Hypocapnia in an Aviator. *Aerosp. Med. Hum. Perform.* **2022**, *93*, 470–471. [CrossRef]
35. Leinonen, A.M.; Varis, N.O.; Kokki, H.J.; Leino, T.K. A New Method for Combined Hyperventilation and Hypoxia Training in a Tactical Fighter Simulator. *Aerosp. Med. Hum. Perform.* **2022**, *93*, 681–687. [CrossRef] [PubMed]
36. Varis, N.; Leinonen, A.; Parkkola, K.; Leino, T.K. Hyperventilation and Hypoxia Hangover during Normobaric Hypoxia Training in Hawk Simulator. *Front. Physiol.* **2022**, *13*, 942249. [CrossRef] [PubMed]

 *biomolecules*

Review

Monogenic Disorders of ROS Production and the Primary Anti-Oxidative Defense

Nana-Maria Grüning [1,*] and Markus Ralser [1,2,3]

1   Department of Biochemistry, Charité Universitätsmedizin Berlin, 10117 Berlin, Germany; markus.ralser@charite.de
2   The Wellcome Centre for Human Genetics, Nuffield Department of Medicine, University of Oxford, Oxford OX3 7BN, UK
3   Max Planck Institute for Molecular Genetics, 14195 Berlin, Germany
*   Correspondence: nanam.gruening@gmail.com

Abstract: Oxidative stress, characterized by an imbalance between the production of reactive oxygen species (ROS) and the cellular anti-oxidant defense mechanisms, plays a critical role in the pathogenesis of various human diseases. Redox metabolism, comprising a network of enzymes and genes, serves as a crucial regulator of ROS levels and maintains cellular homeostasis. This review provides an overview of the most important human genes encoding for proteins involved in ROS generation, ROS detoxification, and production of reduced nicotinamide adenine dinucleotide phosphate (NADPH), and the genetic disorders that lead to dysregulation of these vital processes. Insights gained from studies on inherited monogenic metabolic diseases provide valuable basic understanding of redox metabolism and signaling, and they also help to unravel the underlying pathomechanisms that contribute to prevalent chronic disorders like cardiovascular disease, neurodegeneration, and cancer.

Keywords: oxidative stress; reactive oxygen species (ROS); cellular redox balance; monogenic disorder; inherited disease

Citation: Grüning, N.-M.; Ralser, M. Monogenic Disorders of ROS Production and the Primary Anti-Oxidative Defense. *Biomolecules* **2024**, *14*, 206. https://doi.org/10.3390/biom14020206

Academic Editor: Fabio Di Domenico

Received: 1 January 2024
Revised: 25 January 2024
Accepted: 1 February 2024
Published: 9 February 2024

1. Introduction

Reactive oxygen species (ROS, [1]) are formed by incomplete reduction of oxygen, and comprise a variety of highly reactive, oxygen-containing compounds, with the superoxide anion radical ($O_2^{\cdot-}$), hydrogen peroxide (H_2O_2) and the hydroxyl radical ($OH^{\cdot}$) being the most prominent types in aerobic organisms [2]. ROS can be subdivided into free radicals and nonradicals. The former carry an unpaired electron (e.g., $O_2^{\cdot-}$, $OH^{\cdot}$) and can combine to form nonradicals (e.g., H_2O_2) [3]. Cells can experience ROS exposure either through their production by endogenous systems or through external perturbations. At physiological concentrations, ROS serve important functions. For example, H_2O_2 acts as a second messenger to communicate pro-inflammatory and growth-stimulating signals [4–6]. Furthermore, ROS are also employed in the defense against pathogens when immune cells produce high levels of $O_2^{\cdot-}$ or H_2O_2 to kill bacteria or fungi [7].

However, in order to stabilize themselves, ROS take up electrons and oxidize, and thereby unspecifically damage biomolecules such as nucleic acids, metabolites, lipids, and proteins. In order to keep oxidative damage low, cells have evolved anti-oxidative defense systems. When the level of ROS overflows the capacity of these defense mechanisms, oxidative stress occurs, with the redox balance of the cell, or of its different compartments, shifting away from the physiologically ideal state. If this deviation is strong, disruption of cellular processes may occur [8]. Hence, oxidative stress has implications in various human diseases such as cancer, stroke, or late-onset neurodegenerative disorders caused by environmental stresses or congenital dysfunction of the proteins involved in redox balance [9].

When endogenous ROS production or the anti-oxidative machinery are inherently impaired, the cell has limited capacities to tackle oxidative stress from the start. In this review, we summarize the main monogenic disorders that are caused by germline pathogenic variants in genes that are part of endogenous ROS production and the primary anti-oxidative defense machinery. Our emphasis is on the initial encounters of cells with ROS, encompassing ROS production, scavenging, and the recycling of ROS scavengers. While repair mechanisms for damaged biomolecules associated with prolonged adaptation to oxidative stress are vital components, their abundance demands dedicated exploration beyond the scope of this review.

This overview of monogenic disorders related to ROS production and the primary anti-oxidative defense encompasses a spectrum ranging from prevalent enzymopathies to some of the rarest inherited disorders. Putting the spotlight onto these rare enzymopathies describes the intricacies of redox metabolism and might inspire more research attention to underexplored diseases.

2. Sources of ROS

ROS are either produced through cellular processes or environmental factors. Of note, oxidative stress has been described as a secondary effect within the pathology of several rare monogenic diseases and sometimes been called a common denominator [10,11]. For example, the Friedreich's ataxia protein FXN was shown to play a role in mitochondria biogenesis [12] and iron–sulfur cluster synthesis [13]. Via these functions, mutation of FXN leads to oxidative stress, which contributes to the pathology of Friedreich's ataxia [14]. Such effects in which oxidative stress appears as a molecular symptom of the original genetic disruption have been extensively discussed elsewhere (e.g., [10,15–17]). Herein, we focus on pathogenic genetic variants that affect the main genes directly responsible for internal ROS production and hence have the immediate ability to cause oxidative imbalance in either direction and impair physiological function.

2.1. The Mitochondrial Respiratory Chain

The mitochondrial respiratory chain, or electron transport chain (ETC), is one of the biggest endogenous sources of ROS in eukaryotic cells which use aerobic metabolism (Figure 1) [18,19]. In humans, the enzyme complexes I (NADH:ubiquinone oxidoreductase), II (succinate:ubiquinone oxidoreductase), and III (ubiquinol:cytochrome c oxidoreductase) are somewhat "leaky", leading to direct one-electron transfer to molecular oxygen and the perpetual production of the superoxide anion radical ($O_2{\cdot}^-$) as side reactions [19,20].

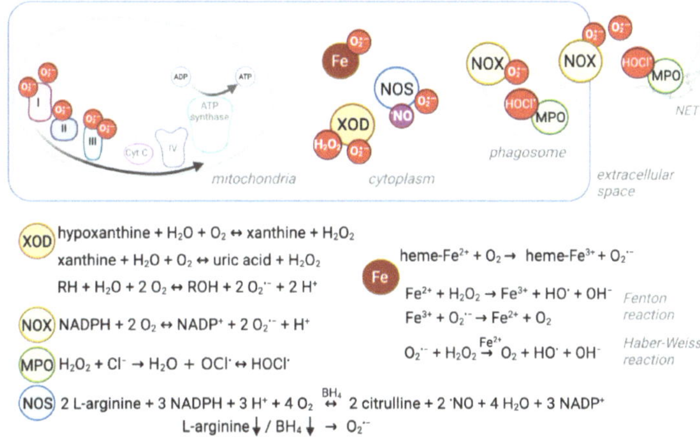

XOD: hypoxanthine + H_2O + O_2 ↔ xanthine + H_2O_2

xanthine + H_2O + O_2 ↔ uric acid + H_2O_2

RH + H_2O + 2 O_2 ↔ ROH + 2 $O_2{\cdot}^-$ + 2 H^+

NOX: NADPH + 2 O_2 ↔ $NADP^+$ + 2 $O_2{\cdot}^-$ + H^+

MPO: H_2O_2 + Cl^- → H_2O + OCl^- ↔ $HOCl^-$

heme-Fe^{2+} + O_2 → heme-Fe^{3+} + $O_2{\cdot}^-$

Fe^{2+} + H_2O_2 → Fe^{3+} + $HO^{\cdot}$ + OH^- *Fenton reaction*

Fe^{3+} + $O_2{\cdot}^-$ → Fe^{2+} + O_2

$O_2{\cdot}^-$ + H_2O_2 $\xrightarrow{Fe^{2+}}$ O_2 + $HO^{\cdot}$ + OH^- *Haber-Weiss reaction*

NOS: 2 L-arginine + 3 NADPH + 3 H^+ + 4 O_2 $\xrightarrow{BH_4}$ 2 citrulline + 2 $^{\cdot}NO$ + 4 H_2O + 3 $NADP^+$

L-arginine↓ / BH_4↓ → $O_2{\cdot}^-$

Figure 1. Schematic overview of ROS production sites and reactions. Several enzymatic reactions as well as interactions of iron with oxygen give rise to reactive oxygen species. Their localization and production level can differ between cell compartments and cell types. For example, red blood cells experience high levels of superoxide due to their heme-bound iron (Fe^2+) and oxygen interaction, whereas the mitochondrial electron transport chain is a major source of ROS in other cells. Under ischemic conditions, xanthine oxidase (XOD) produces superoxide anions ($O_2^{\cdot-}$) and hydrogen peroxide (H_2O_2) in the cytosol. NADPH oxidases (NOX) mainly produce $O_2^{\cdot-}$ to kill pathogens in the phagosome or extracellular space. Similarly, myeloperoxidase (MPO) produces hypochlorous acid (HOCl·) from H_2O_2 which derives from NOX-produced $O_2^{\cdot-}$ in phagosomes and neutrophil extracellular traps (NETs). Instead, nitric oxide synthases (NOS) normally produce nitric oxide (·NO) as a signaling molecule. Depletion of its substrate arginine or cofactor (6R)-5,6,7,8-tetrahydro-L-biopterin (BH_4) can cause the enzyme to uncouple and produce $O_2^{\cdot-}$.

Complex I is encoded by more than 40 genes [21], complex II by four genes [22] and complex III by 11 genes [23]. However, the specific sites and levels of ROS production within the different protein complexes depend on the respiration substrate, whether the cell experiences norm- or hypoxic conditions, and on the inhibitor used in experimental setups [24]. Furthermore, respiratory complexes can assemble into supercomplexes, which decreases ROS production [25]. Astrocytes contain high amounts of free complex I and thus potentially higher levels of ROS compared to neurons, which display complex I and III assembly [26]. Thus, the diversity of conditions as well as of cell type specificities make it often hard to pin down the specific set of genes of respiratory complex subunits which involve pathomechanisms related to ROS production in humans. This is also exemplified by the fact that some pathogenic mutations within subunits do not lead to ROS elevation, as shown for H_2O_2 in a mouse model for Leigh syndrome [27]. By contrast, in other experiments, mutations of genes involved in complex I, II, or III formation were indeed shown to create greater "leakiness", and cause elevated ROS levels and oxidative stress [21,28–30].

Since the ETC generates a strong proton gradient used for oxidative phosphorylation, transport processes, and heat production, it can be difficult to disentangle whether the pathophysiological mechanism triggered by mutation of ETC genes is based on compromised primary functions, or elevated ROS levels. However, at least for complex I deficiency, elevated ROS production and its consequent oxidative damage were shown to induce apoptotic molecular pathways leading to neuron degeneration and hence neurological symptoms, the main features of the disorder [28].

Deficiencies in respiratory complexes have similar but wide-ranging symptoms from neonatal death, lactic acidosis, myopathy, hepatopathy, encephalopathy, Leigh syndrome [31], Leber hereditary optic neuropathy (LHON) [32], to adult-onset neurological symptoms such as some forms of Parkinson disease [33,34]. Isolated complex I deficiency is the most prevalent genetic disorder of oxidative phosphorylation [35].

As outlined above, it is challenging to disentangle the underlying factors, but the numerous potential ROS production sites and levels might be part of the explanation of the broad range of symptoms and disorder severities.

2.2. Heme in Red Blood Cells

Another ROS formation hotspot, despite being devoid of mitochondria, is the cytosol of red blood cells (RBCs). RBCs transport oxygen from the lung to peripheral tissues and superoxide can be formed when O_2 interacts with the iron (Fe^{2+}) of the heme group of hemoglobin (Hb) [36]. Spontaneous autoxidation results in $O^{2\cdot-}$ and methemoglobin (Hb-Fe^{3+}, MetHb, Hb M) (Figure 1) [35,37], and ca. 1% of all Hb is present as Hb M in healthy individuals [38].

This normally occurring rate of superoxide production can be exceeded by pathogenic genetic variants that cause conformational changes to the globin chain proteins that contain heme as a prosthetic group. This phenomenon is well described for, e.g., autosomal recessive sickle cell disease (SCD) (Table 1). SCD is the most common severe hemoglobinopathy worldwide and is caused by a missense pathogenic variant in the globin beta-chain (*HBB*), resulting in Hb S (NM_000518.5(*HBB*):c.20A>T (p.Glu7Val); E6V) [39], which is unstable, prone to polymerization and autoxidation [40]. Besides other ROS sources like iron release, this autoxidation is the primary source for oxidative stress in sickle cells, and it leads to loss of membrane structure and function and consequent multisystem disease [41–44]. When iron is released, it impacts the oxidative balance through transfer of single electrons via the Fenton reaction ($Fe^{2+} + H_2O_2 \rightarrow Fe^{3+} + HO^{\cdot} + OH^{-}$) and the Haber–Weiss reaction ($O_2^{\cdot -} + H_2O_2 \rightarrow O_2 + HO^{\cdot} + OH^{-}$ catalyzed by iron) (Figure 1) [45].

Also, other *HBB* variants have been described to cause methemoglobinemia. For example, the so-called Hb M-Hyde Park (Hb M-Akita) (NM_000518.4(*HBB*):c.277C>T (p.His93Tyr)) variant leads to conformational changes at the heme binding site and hence higher rates of methemoglobin formation. Inheritance is autosomal dominant [38,46].

2.3. NADPH Oxidases and Myeloperoxidase

Besides these cellular processes in which ROS are formed passively as byproducts, the cell can actively produce high levels of ROS through dedicated enzymes. NADPH oxidases (NOX) use reduced nicotinamide adenine dinucleotide phosphate (NADPH) to generate superoxide (Figure 1) [47]. NOX play an important role in the innate host defense. For the so-called "respiratory burst", superoxide is released into the extracellular space or phagosomes to fight off pathogenic bacteria or fungi [48]. Furthermore, NOX enzymes were found to produce intracellular ROS at lower levels, which is believed to serve signaling functions [49] and to control the cellular redox balance by oxidizing NADPH and through ROS formation [50]. The importance of the ability to generate high ROS levels is exemplified by the detrimental consequences of NOX disruption.

Several monogenic disorders are related to NOX subunits. The genes that encode the NOX2 complex in phagocytes are related to chronic granulomatous disease [51,52]: *CYBB* pathogenic variants cause X-linked recessive chronic granulomatous disease (CGDX, [53]) and immunodeficiency 34 [54]), *CYBA* pathogenic variants cause autosomal recessive (AR) CGD4 [55], *NCF1* pathogenic variants cause AR CGD1 [56], *NCF2* pathogenic variants cause AR CGD2 [57], and *NCF4* pathogenic variants cause AR CGD3 [58] (Table 1). These rare primary immunodeficiencies increase susceptibility to life-threatening bacterial and fungal infections and lead to development of granulomas [59].

Pathogenic variants in other NADPH oxidases were shown to increase susceptibility to inflammatory bowel disease (*NOX1* and *DUOX2*, [60]), or cause congenital hypothyroidism (*DUOX2*, Thyroid dyshormonogenesis 6, AR, [61]; *DUOXA2*, Thyroid dyshormonogenesis 5, AR, [62], Table 1). The latter one is the result of disrupted H_2O_2 production through mutated DUOX2, which would be required for organification of iodide for thyroid hormone synthesis catalyzed by thyroid peroxidase [52,63].

Another enzyme of the innate immune response is the heme-containing enzyme myeloperoxidase (MPO, Table 1, Figure 1). It is highly abundant in azurophilic granules of neutrophils [64] and is a crucial component in neutrophil extracellular nets [65]; it is also found—to a lesser extent—in monocytes [66]. To kill pathogens, MPO creates strongly reactive species, such as hypochlorous acid ($HOCl^{\cdot}$) from hydrogen peroxide (H_2O_2) and chloride (Cl^{-}) [67,68]. Autosomal recessive MPO deficiency is the most common phagocyte defect but is asymptomatic in the majority of patients, suggesting compensatory mechanisms [69]. However, higher susceptibility to fungal infections, especially *Candida albicans*, has been described [70]. This vulnerability might become apparent only in combination with comorbidities like diabetes, which itself increases the risk of infections [68]. Similarly, a recent study described a case of partial DiGeorge syndrome together with MPO deficiency. DiGeorge syndrome is characterized by immunodeficiency among other symptoms. However, the patient had more frequent and severe infections than expected for partial DiGeorge alone, which the authors explained by co-occurrence of MPO deficiency [71]. On the other hand, a protective mechanism against cardiovascular disease by absence of the potentially oxidative stress causing MPO enzyme has also been discussed [69].

2.4. Other ROS Sources

In addition to the above-described reactions and enzymes, there are additional sources that can produce significant amounts of reactive species under certain physiological circumstances.

For example, xanthine oxidoreductase (XDH/XOD), encoded by the *XDH* gene (Table 1, Figure 1), exists in two interconvertible isoforms. Both forms utilize hypoxanthine or xanthine to produce uric acid. However, the predominant form has xanthine dehydrogenase (XDH) activity, and uses NAD^+ as cofactor to produce NADH, whereas the xanthine oxidase (XOD) form uses oxygen to produce the superoxide anion and H_2O_2 [72]. XDH can be converted to XOD by irreversible limited proteolysis or reversibly by thiol oxidation (reviewed in [73]), e.g., in hypoxic/ischemic tissue [74]. Homozygous or compound heterozygous pathogenic *XDH* variants cause type I xanthinuria (XAN1) [75]. The symptoms, low serum and urine uric acid and xanthinuria leading to urolithiasis [76], result from XDH's primary function. Mutations that lead to elevated ROS production are not described for this enzyme to the best of our knowledge.

Furthermore, nitric oxide synthases (NOS1-3) can become sources of $O_2{}^{\cdot-}$ (Figure 1). Normally, they use L-arginine to produce nitric oxide ($\cdot$NO), which belongs to the group of reactive nitrogen species (RNS) and serves as an important signaling molecule, especially for the vascular tone [77]. However, persistent oxidative stress, and thus reduced levels of the cofactor (6R)-5,6,7,8-tetrahydro-L-biopterin (BH_4), lead to uncoupling of endothelial NOS (eNOS, NOS3) so that the enzyme produces $O_2{}^{\cdot-}$ instead of $\cdot$NO [78]. Furthermore, $O_2{}^{\cdot-}$ is produced by NOS2 in arginine-depleted macrophages [79]. Although some *NOS3* variants have been described to increase the risk for certain conditions like pregnancy-induced hypertension [80] or ischemic stroke [81], no clear monogenic disorder has been described for *NOS1*, *NOS2*, or *NOS3*.

At lower amounts, ROS can also be byproducts of other enzymes like cytochrome P450 or cyclooxygenase (reviewed in [82]).

Environmental factors that create elevated ROS levels are, e.g., UV radiation, ionizing radiation, smoking and air pollution, chemicals such as drugs, and certain types of food like the fava bean [2,83]. Often, genetic defects in the anti-oxidative machinery remain unnoticed until such environmental stresses hit and overwhelm the residual anti-oxidative capacity of the cell, e.g., in hemolytic anemia due to glucose 6-phosphate dehydrogenase (G6PDH) deficiency (see below).

3. The Anti-Oxidative Defense System

Protective mechanisms have evolved that keep the redox balance in check or clear ROS-induced damage. There are several ways to categorize the different parts of the overall cellular anti-oxidative defense system. Ighodaro and Akinloye [84] propose to categorize the individual parts of the anti-oxidative machinery into first, second, third, and fourth line defense.

First line antioxidants prevent or reduce ROS formation. The most important agents in this category are superoxide dismutase (SOD), catalase (CAT), glutathione peroxidase (GPX), and peroxiredoxins (PRDX). SOD dismutates the superoxide anion $O^{2\cdot-}$ to H_2O_2, and CAT detoxifies H_2O_2 to water and oxygen. GPX and PRDX convert H_2O_2 to water, are oxidized themselves, and need to be reduced again to stay functional (Figure 2, [3]). Second-line anti-oxidants act as so-called "scavengers" and prevent ROS from damaging biomolecules or forming further ROS. These are either small, non-enzyme molecules that are produced by human cells, like glutathione (GSH), uric acid, and ubiquinol, or proteins with thiol groups like thioredoxins (TXN) or peroxiredoxins [85,86]. Also, nitric oxide ($\cdot$NO) falls into this category as it scavenges $O_2{}^{\cdot-}$ to form peroxynitrite ($ONOO^-$), which itself is highly reactive at physiological pH and needs to be further converted, e.g., by GSH [87]. Scavengers can also be small molecules taken up by nutrition, like ascorbic acid (vitamin C) and alpha-tocopherol (vitamin E). Third-line anti-oxidants repair already damaged structures, and fourth-line anti-oxidants build up prevention and adaptation mechanisms, e.g., upregulate expression of relevant genes [84] or, even faster, metabolic switches like the inactivation of glyceraldehyde 3-phosphate dehydrogenase (GAPDH), which leads to metabolic rerouting into the pentose phosphate pathway and increased NADPH production [88,89].

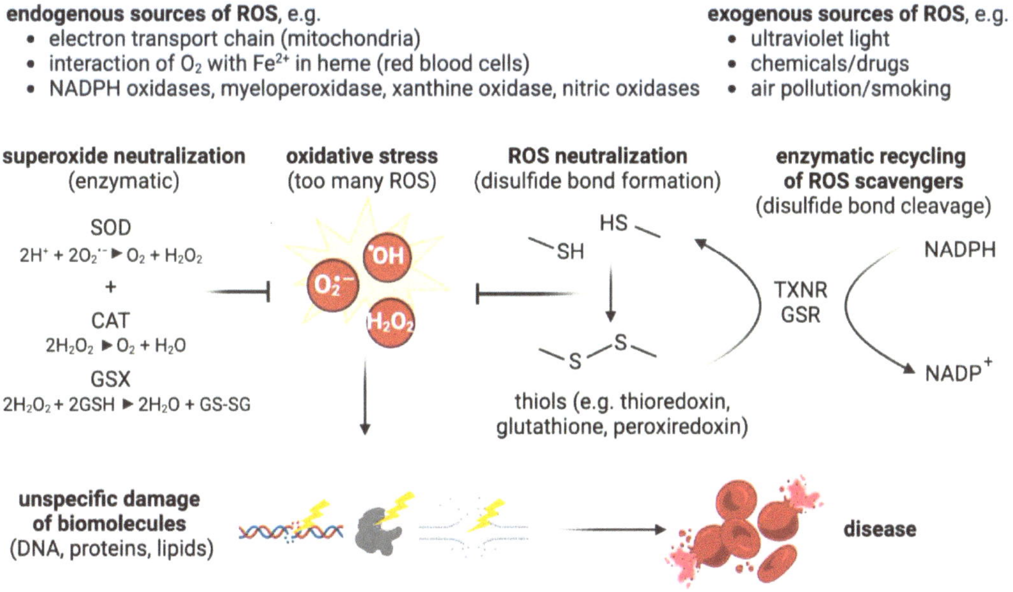

Figure 2. The cellular anti-oxidative machinery. Oxidative stress from exogenous or endogenous sources would damage biomolecules and cause disease if not counteracted by a set of direct enzymatic (SOD, CAT, GSX) and non-enzymatic (GSH, TXN) anti-oxidative molecular systems that are regenerated by different reductases (TXNR, GSR).

First- and second-line defense can be summarized as primary anti-oxidative defense, which serves for immediate protection [90]. Within the primary antioxidative defense, the thiols GSH and TXN play crucial roles. The tripeptide GSH scavenges ROS or, like the polypeptide TXN, serves as an electron donor for peroxidases. The peroxidases either detoxify H_2O_2 (GPX) or reduce oxidized cysteine residues and cleave ROS-caused disulfide bonds in proteins. In order to maintain a constant pool of reduced GSH and TXN, respective reductases recycle GSH and TXN under consumption of NADPH (Figure 2) [91,92].

3.1. Enzymes of the ROS Scavenger System

3.1.1. Superoxide Dismutases

Three genes in humans (*SOD1*, *SOD2*, *SOD3*) encode for superoxide dismutases, of which only *SOD1* is described as disease associated. Amyotrophic lateral sclerosis (ALS) is a heterogenous, late-onset neurodegenerative disorder for which up to 95% of cases are sporadic. However, up to 10% of cases seem to have familial predisposition, and *SOD1* is a top hit among 40 associated genes. It is not unequivocally defined whether *SOD1* variants are causative for ALS, co-causative or modifying [93], but hypothetical pathomechanisms explain the disease with mostly heterozygous gain-of-function variants that lead to the accumulation of aggregated SOD1 protein [94,95]. Elevated oxidative stress was also described in ALS, but it is not clear whether both mechanisms are linked [96].

Interestingly, more recently, homozygous *SOD1* variants were detected in children with progressive severe developmental delay, axial hypotonia, and spastic tetraplegia in at least three families. Biochemical testing confirmed total absence of enzyme activity in the patients and 50% of enzyme activity in the parents who were heterozygous for the respective familial variant [97–99].

3.1.2. Catalase

Once SOD has converted $O^{2\cdot-}$ into H_2O_2, the latter needs to be further detoxified. This task is efficiently accomplished by catalase that is encoded by the *CAT* gene, and which can convert millions of hydrogen peroxide molecules to water and oxygen per second [100]. Homozygous pathogenic *CAT* variants are associated with the rare condition acatalasemia, also known as acatalasia, which leads to total or near total loss of CAT activity in erythrocytes [101]. Probands with heterozygous variants are called hypocatalasemic and have blood catalase activity of about 50%. The disorder was coincidentally discovered by a Japanese doctor named Takahara in 1946, who performed oral surgery on a girl and used H_2O_2 to disinfect the wound. He noticed the absence of oxygen bubbles that would normally be formed due to CAT activity. Additionally, the blood turned black [102]. A dark chocolate brown color of arterial blood refers to methemoglobinemia [103]—the aforementioned phenomenon when some of the ferrous heme iron [Fe^{2+}] is converted to ferric [Fe^{3+}] iron in hemoglobin and hence oxygen transport is impaired. Takahara also described gangrenes, oral ulceration, and gingival necrosis in CAT-deficient patients. In the 1940s and 50s, oral hygiene was still poor in Japan and the symptoms were attributed to hydrogen peroxide produced by phagocytic cells and bacterial infections. When oral hygiene improved, these symptoms were seen less often. But besides these, the disorder used to be considered asymptomatic [104]. However, animal models displayed a lower hydrogen peroxide removal rate and elevated oxidative stress. In fact, CAT-deficient patients develop diabetes more often and earlier, and it is known that insulin-producing pancreatic beta cells are especially vulnerable to oxidative damage. Indeed, acatalasemia and hypocatalasemic patients are prone to age- and oxidative stress-related diseases such as type II diabetes, tumors, Parkinson's disease, vitiligo, schizophrenia, or loss of soft tissues (reviewed in [105,106]).

3.1.3. Glutathione Peroxidases

Glutathione peroxidases (GSX) are a family of oxidoreductases that reduce H_2O_2, organic hydroperoxides and lipid peroxides by using GSH as a reductant. In humans, eight GPX enzymes are known to be encoded by the *GPX1, GPX2, GPX3, GPX4, GPX5, GPX6, GPX7,* and *GPX8* genes [107]. Of these, *GPX1* and *GPX4* were mentioned in association with rare monogenic disorders.

Biallelic pathogenic variants of *GPX1* have been described as associated with hemolytic anemia [108]. Hemolysis due to lipid peroxidation is a well-described consequence of oxidative stress [109]. However, a clear genotype–phenotype relationship could not be unequivocally established to date.

Biallelic pathogenic *GPX4* variants cause the rare and fatal disorder Sedaghatian type of spondylometaphyseal dysplasia (SSMD, [110]), with only 24 cases described to date. SSMD is a severe metaphyseal chondrodysplasia with, e.g., delayed epiphyseal ossification, mild limb shortening, platyspondyly, pulmonary hemorrhage, and brain anomalies resulting in death due to respiratory failure or cardiac abnormalities in the perinatal period [111]. Disruption of GPX4 causes buildup of lipid peroxidation and ferroptosis, which is an iron-dependent, non-apoptotic form of cell death [112,113].

3.1.4. Peroxiredoxins

Peroxiredoxins are ubiquitous cysteine-containing proteins that are highly abundant, especially in RBCs, and play crucial roles in regulating intracellular peroxide levels [114]. In humans, there are six genes: *PRDX1, PRDX2, PRDX3, PRDX4, PRDX5,* and *PRDX6.* Of these, *PRDX1, PRDX2* and *PRDX4-6* have not been described as related to a monogenic disorder.

To the best of our knowledge, thus far, only *PRDX3*, which is localized to mitochondria [115], is associated with monogenic disorders. In only eight cases, and rather recently, homozygous pathogenic *PRDX3* variants have been described to be causative for spinocerebellar ataxia-32 (SCAR32) [116–118]. Patients presented with gait ataxia, cerebellar ataxia, upper limb ataxia, dysarthria, dysphagia, oculomotor signs, sometimes hyper- or hypokinetic movement abnormalities, and cerebellar atrophy in one patient. The disease mechanism was shown to be oxidative stress and mitochondrial damage due to absent or nearly absent expression of PRDX3 protein [116,118].

Heterozygous pathogenic *PRDX3* variants have been described in segregation studies as related to punctiform and polychromatic pre-Descemet corneal dystrophy (PPPCD). The phenotype, which was not reported in patients with homozygous pathogenic *PRDX3* variants, is characterized by hyperreflective deposits in the posterior stroma of the cornea of the eye. Affected individuals are asymptomatic with no visual impairment [119–121].

PRDX1 together with the *MMACHC* gene have been described as associated with the digenic disorder autosomal recessive combined methylmalonic aciduria and homocystinuria type cblC. *PRDX1* participates in the disease pathology by affecting *MMACHC* expression [122].

3.2. Electron Donors

3.2.1. Glutathione

GSH is a small tripeptide composed of cysteine, glutamate, and glycine. Due to the cysteine's thiol group, two GSH molecules can form a disulfide bond and thus serve as electron donors. GSH synthesis is catalyzed by the enzymes glutamate–cysteine ligase (*GCLC*, catalytic subunit, γ-glutamylcysteine synthetase) and glutathione synthetase (*GSS*), which are part of the γ-glutamyl cycle [123]. Pathogenic variants in both genes are described to be related to ultra-rare metabolic disorders.

Glutamate–cysteine ligase deficiency is an autosomal recessive disorder that is characterized by hemolytic anemia and variable other features like myopathy or late-onset spinocerebellar degeneration [124]. Only very few patients have been described to date: five until 1996 and none more recently to our knowledge. The GSH content was very low in the RBCs of all five patients (between 2 and 13.1% of normal) [125].

For glutathione synthetase deficiency, the OMIM database distinguishes between two forms: isolated hemolytic anemia [126] and a severe, multi-tissue form characterized by hemolytic anemia, central nervous system symptoms (e.g., mental retardation, ataxia), metabolic acidosis and severe urinary excretion of 5-oxoproline [127]. In the latter case, different cell types have severely reduced levels of enzyme activity and GSH content, whereas nucleated cells are able to maintain substantial levels of GSH in patients with isolated hemolytic anemia [128]. It was proposed that accumulation of the GSS substrate γ-glutamylcysteine substitutes GSH at least partially as an antioxidant [129].

No null alleles have been described for both disorders; and complete absence of GSH due to *Gclc* or *Gss* knockout is lethal in mice [130–132].

3.2.2. Thioredoxins

Thioredoxins are small proteins that carry neighboring thiol groups in their active site that can form a disulfide bond and thereby serve as electron donors. Two thioredoxin genes have been described in humans: *TXN* (cytosolic) and *TXN2* (mitochondrial). *Txn* as well as *Txn2* knockout is embryonically lethal in mice [133,134].

While no genetic disorder has been described for *TXN*, the OMIM database mentions a phenotype named "Combined oxidative phosphorylation deficiency 29" (Table 1) for *TXN2*. So far, only one patient with absent TXN2 protein in fibroblasts, disrupted ROS homeostasis, and impaired mitochondrial ATP generation due to a homozygous *TXN2* loss-of-function variant was identified by whole exome sequencing [135]. These cellular phenotypes could be rescued by lentiviral expression of the wild-type TXN2 protein, and supplementation of antioxidants had beneficial effects on the patient's condition in short-term follow-up. The

16-year-old patient's main symptoms were microcephaly, optic atrophy, abnormal muscle tone, global developmental delay, seizures, cerebellar atrophy, spasticity, and increased serum lactate at the time [135]. Nevertheless, the case remains interesting as no other pathogenic or likely pathogenic variant has been described for *TXN2*, and loss of TXN has been described as lethal in mice, which points to compensatory mechanisms in the patient described by Holzerova et al. [135].

3.2.3. Reductases for Recycling of Electron Donors

Under constant ROS production, the pools of anti-oxidant electron donors would quickly deplete. Hence, reductases recycle the reduced forms of glutathione and thioredoxin, utilizing NADPH.

Glutathione Reductase, GSR

Glutathione reductase is encoded by only one gene in humans (*GSR*), or mice (*Gsr*). Knockout of *Gsr* in mice results in a healthy phenotype [136,137]. However, homozygous pathogenic *GSR* variants in humans result in hemolytic anemia and cataract or hyperbilirubinemia and jaundice (Table 1) [138]. GSR activity was very low or absent in tested cells in patients from two families [139,140]. Normal glucose 6-phosphate dehydrogenase (G6PDH) activity, whose disruption could result in similar symptoms, was shown in the infant with jaundice described by Kamerbeek et al. [140]. However, in the few studies available, no other possible genetic causes for the described symptoms were genetically excluded, e.g., by whole exome analysis. That humans can be rather robust toward reduced GSR activity is also exemplified by the high frequency of moderate GSR deficiency due to malnutrition, specifically due to the lack of riboflavin, which is a precursor of the GSR cofactor flavin adenine dinucleotide (FAD) in some populations [141,142].

Thioredoxin Reductase, TXNR

Humans express three thioredoxin reductase isoforms: TXNRD1, TXNRD2, TXNRD3. *Txnrd1* and *Txnrd2* knockout were shown to be embryonically lethal in mice [143]. *Txnrd3* knockout mice are viable and without discernible phenotype. However, male mice have impaired fertility, which points to a rather specific function of Txnrd3 in spermatogenesis [144]. In humans, to our knowledge, no disease phenotype has been described for all three genes.

3.2.4. NADPH—The Key Coenzyme of the Anti-Oxidant Response

Regeneration of reduced thiols from disulfides by reductases requires a constant supply of the universal redox coenzyme NADPH. Several cytosolic and mitochondrial enzymes convert $NADP^+$ into NADPH.

The Oxidative Pentose Phosphate Pathway, oxPPP

The most prominent enzyme in the category of NADPH production is the pentose phosphate pathway enzyme glucose 6-phosphate dehydrogenase (G6PD). G6PD is generally considered to be the main source of cytoplasmic NADPH [145]. Complete G6PD knockout was shown to be embryonically lethal in mice [146]. In humans, more than 200 pathogenic *G6PD* variants have been described. The most severe manifestations have residual enzyme activity of around 5% [147]. With over 400 million cases worldwide, X-linked recessive G6PD deficiency (Table 1) is the most common enzymopathy in humans, and the most common symptoms are neonatal jaundice and acute hemolytic anemia [148]. Hemolysis is mainly triggered by external factors such as drugs, certain foods (e.g., fava beans), or infection [149]. Otherwise, individuals with G6PD deficiency are mostly symptom-free [148]. The high frequencies of G6PD alleles in populations that geographically overlap with the distribution of malaria-causing *Plasmodium falciparum* indicate at least partial protective effects of these alleles against the infectious disease [147]. G6PD deficiency is an example of a redox disorder that only displays symptoms when the anti-oxidative capacity of the cell is overwhelmed due to acute oxidative stress and insufficient NADPH supplies. Otherwise, the residual activity of G6PD is sufficient to maintain health. Additionally, it was shown that, although G6PD is a major source of NADPH, it is dispensable for pentose production [150].

The second NADPH-producing enzyme of the oxPPP is 6-phosphogluconate dehydrogenase (PGD). PGD deficiency (Table 1) was found to cause transient hemolysis and well-compensated chronic nonspherocytic hemolytic anemia. Inheritance was described as autosomal dominant [151]. However, the genotype-phenotype relationship is not unequivocally established and individuals who were found in population surveys who had PGD activity of less than 5% in RBC were completely healthy [152].

Malic Enzyme and Glutamate Dehydrogenase

There are two NADPH-producing isoforms of malic enzyme (ME) in humans: cytosolic ME1 and mitochondrial ME3. ME1 and 3 catalyze the reversible decarboxylation of malate to pyruvate. In mice, knockout of *Me1* causes decreased body weight [153,154], and no specific phenotype was published for knockout of *Me3* to our knowledge. Neither of the two genes has been described as associated with monogenic disorders in humans.

Glutamate dehydrogenase (GLUD) catalyzes the conversion of glutamate to α-ketoglutarate and ammonia while producing NADPH [155]. There are two mitochondrial isoforms, GLUD1 and GLUD2. Only *GLUD1* has been described in association with a genetic phenotype—autosomal dominant hyperinsulinism-hyperammonemia syndrome (Table 1) [156].

Isocitrate Dehydrogenase

Humans possess two IDH isoforms that produce NADPH by converting isocitrate to α-ketoglutarate (α-KG): cytosolic IDH1 and mitochondrial IDH2. *Idh1* knockout is viable in mice, but the animals display greater susceptibility to oxidative stress [157]. No monogenic disorder has been described for *IDH1* in humans, although pathogenic IDH1 variants are common in human neoplastic disorders [158]. *Idh2* knockout mice show susceptibility to mitochondrial dysfunction, cardiac hypertrophy, and accelerated age-related hearing loss [159,160]. However, in humans, *IDH2* has been described as related to D-2-hydroxyglutaric aciduria 2 (D2HGA2, Table 1). Kranendijk et al. identified 15 unrelated patients with specific de novo heterozygous variants affecting codon R140 (NM_002168.2). D2HGA2 has a wide phenotypic spectrum from asymptomatic to severe neurological disorder with developmental delay, epilepsy, dysmorphic features, and cardiomyopathy. The phenotype was not described to be caused by elevated oxidative stress but by an increase in a promiscuous newly introduced enzymatic function that is of low activity in the wild type enzyme: the described variants reduce the enzyme's ability to convert isocitrate to α-KG but increase the conversion of α-KG into D-2-hydroxyglutarate [158].

Folate Metabolism

The importance and amount of NADPH derived from folate metabolism have been described in several publications in 2014 [145,161,162]. In proliferating cells—hence not in RBCs—the NADPH production from folate metabolism almost equals that from the oxidative pentose phosphate pathway [161].

The folate cycle gene methylenetetrahydrofolate dehydrogenase (MTHFD) converts 5,10-methylenetetrahydrofolate and $NADP^+$ into 5,10-methenyltetrahydrofolate and NADPH [163]. Knockout of the genes *Mthfd1* or *Mthfd2* is embryonically lethal in mice [164,165]. No genetic disease has been described for the human mitochondrial isoform MTHFD2. However, autosomal recessive pathogenic *MTHFD1* variants cause the ultra-rare disorder "combined immunodeficiency and megaloblastic anemia with or without hyperhomocysteinemia" (Table 1) [166,167]. Functional studies on patient fibroblasts revealed severely impaired methionine production but did not analyze for potential oxidative stress in these cells [168].

The folate cycle enzyme formyltetrahydrofolate dehydrogenase, alternatively named aldehyde dehydrogenase 1 family member L (ALDH1L), produces tetrahydrofolate and CO_2 from 10-formyltetrahydrofolate under simultaneous NADPH production [169]. There are two isoforms in humans: cytosolic ALDH1L1 and mitochondrial ALDH1L2 [170]. Knockout of *Aldh1l1* in mice resulted in no obvious phenotype [171], and no inherited monogenic disorder has been described in humans. Interestingly, for *ALDH1L2*, one patient was described with a neuro-ichthyotic syndrome characterized by ichthyosis, thick fingers and toes, dysmorphic features, and developmental delay. Two *ALDH1L2* variants were detected in the index patient in compound heterozygosity, which indicates autosomal recessive inheritance. Functional analyses in the patient's fibroblasts confirmed impaired mitochondrial metabolism with impact on β-oxidation of fatty acids [172]. Knockout of *Aldh1l2* resulted in normally appearing mice. However, in-depth physiological analysis confirmed, e.g., reduced NADPH, abnormal lipid metabolism, showed drastically reduced GSH levels, and increased oxidative stress which impacted β-oxidation of fatty acids [173]. Nevertheless, further patients from different families would support an unequivocal genotype-phenotype association.

Table 1. Summary of discussed genes and phenotypes. AR, autosomal recessive, AD, autosomal dominant, XLR, X-linked recessive, OMIM®, Online Mendelian Inheritance in Man database.

Protein	Gene (OMIM® no.)	Monogenic Disease (OMIM® no·)	Inheritance	Described by
		ROS production		
Hemoglobin beta-locus	*HBB* (141900)	Sickle cell disease (603903)	AR	[39]
		Methemoglobinemia, beta type (617971)	AD	[38]
Cytochrome b(-245), beta subunit, p91-phox	*CYBB* (300481)	Chronic granulomatous disease (306400)	XLR	[53]
		Immunodeficiency 34, mycobacteriosis (300645)	XLR	[54]
Cytochrome b(-245), alpha subunit, p22-phox	*CYBA* (608508)	Chronic granulomatous disease 4 (233690)	AR	[55]
Neutrophil cytosolic factor-1, p47-phox	*NCF1* (608512)	Chronic granulomatous disease 1 (233700)	AR	[56]
Neutrophil cytosolic factor-2, p67-phox	*NCF2* (608515)	Chronic granulomatous disease 2 (233710)	AR	[57]
Neutrophil cytosolic factor-4, p40-phox	*NCF4* (601488)	Chronic granulomatous disease 3 (613960)	AR	[58]
Myeloperoxidase	*MPO* (606989)	Myeloperoxidase deficiency (254600)	AR	[70]
Xanthine dehydrogenase	*XDH* (607633)	Xanthinuria, type I (278300)	AR	[75]
Nitric oxide synthase 1	*NOS1* (163731)	-		
Nitric oxide synthase 2	*NOS2* (163730)	-		
Nitric oxide synthase 3	*NOS3* (163729)	-		
		enzymatic ROS clearance		
superoxide dismutase 1	*SOD1* (147450)	Amyotrophic lateral sclerosis 1 (105400)	AD, AR	[93]
		Spastic tetraplegia and axial hypotonia, progressive (618598)	AR	[99]
superoxide dismutase 2	*SOD2* (147460)	-		
superoxide dismutase 3	*SOD3* (185490)	-		
catalase	*CAT* (115500)	Acatalasemia (614097)	AR	[102]
glutathione peroxidase 1	*GPX1* (138320)	Hemolytic anemia due to glutathione peroxidase deficiency (614164)	AR	[108]
glutathione peroxidase 4	*GPX4* (138322)	Spondylometaphyseal dysplasia, Sedaghatian type (250220)	AR	[110]
glutathione peroxidase 2, 3, 5-8	*GPX2* (138319) *GPX3* (138321) *GPX5* (603435) *GPX6* (607913) *GPX7* (615784) *GPX8* (617172)	-		
peroxiredoxin 3	*PRDX3* (604769)	Corneal dystrophy, punctiform, and polychromatic pre-Descemet (619871)	AD	[120]
		Spinocerebellar ataxia, autosomal recessive 32 (619862)	AR	[116]

Table 1. *Cont.*

Protein	Gene (OMIM® no.)	Monogenic Disease (OMIM® no·)	Inheritance	Described by
		enzymatic ROS clearance		
peroxiredoxin 1, 2, 4, 5, 6	*PRDX1* (176763) *PRDX2* (600538) *PRDX4* (300927) *PRDX5* (606583) *PRDX6* (602316)	-		
		thiols and thiol production		
gamma-glutamylcysteine synthetase, catalytic subunit	*GCLC* (606857)	Hemolytic anemia due to gamma-glutamylcysteine synthetase deficiency (230450)	AR	[124]
glutathione synthetase	*GSS* (601002)	Glutathione synthetase deficiency (266130)	AR	[127]
		Hemolytic anemia due to glutathione synthetase deficiency (231900)	AR	[126]
thioredoxin 2	*TXN2* (609063)	Combined oxidative phosphorylation deficiency 29 (616811)	AR	[135]
thioredoxin	*TXN* (187700)	-		
		thiol recycling		
glutathione reductase	*GSR* (138300)	Hemolytic anemia due to glutathione reductase deficiency (618660)	AR	[138]
thioredoxin reductase 1, 2, 3	*TXNRD1* (601112) *TXNRD2* (606448) *TXNRD3* (606235)	-		
		NADPH production		
glucose 6P-dehydrogenase	*G6PD* (305900)	Hemolytic anemia, G6PD deficient (favism) (611162)	XL	[147]
6-phosphogluconate dehydrogenase	*PGD* (172200)	Phosphogluconate dehydrogenase deficiency (619199)	AD	[151]
malic enzyme 1	*ME1* (154250)	-		
malic enzyme 3	*ME3* (604626)	-		
glutamate dehydrogenase 1	*GLUD1* (138130)	Hyperinsulinism-hyperammonemia syndrome (606762)	AD	[156]
glutamate dehydrogenase 2	*GLUD2* (300144)	-		
isocitrate dehydrogenase 2	*IDH2* (147650)	D-2-hydroxyglutaric aciduria 2 (613657)	AD	[158]
isocitrate dehydrogenase 1	*IDH1* (147700)	-		
methylenetetrahydrofolate dehydrogenase 1	*MTHFD1* (172460)	Combined immunodeficiency and megaloblastic anemia with or without hyperhomocysteinemia (617780)	AR	[166]
methylenetetrahydrofolate dehydrogenase 2	*MTHFD2* (604887)	-		
10-formyltetrahydrofolate dehydrogenase (mitochondrial)	*ALDH1L2* (613584)	neuro-ichthyotic syndrome?	AR	[172]
10-formyltetrahydrofolate dehydrogenase (cytosolic)	*ALDH1L1* (600249)	-		

4. Conclusions

In this review, we explored monogenic diseases related to genes associated with endogenous ROS production and anti-oxidative defense systems, and we described the most common as well as some of the rarest genetic disorders. Of note, this article aimed at providing an overview of the most significant genes involved in redox metabolism and its regulation—while there are numerous further metabolic pathway enzymes located upstream of the described genes that could also have a potential impact on the discussed functions. Herein, attention was drawn to extremely rare genetic conditions. For some disorders, only few patients have been described despite the recent surge in sample throughput by more and more affordable sequencing technologies, as well as fast progress in other analytical technologies, including metabolomics and proteomics. Since a vast number of samples are processed by diagnostic companies and not research active clinics or institutes, comprehensive re-analysis of proprietary databases might identify further patients that would back up weak genotype-phenotype links. This would not only improve diagnostics of rare diseases but also provide valuable information for basic biochemical research.

Funding: M.R. acknowledges support by the Deutsche Forschungsgemeinschaft (DFG, German Research Foundation) through the CRC-TRR186, 'Molecular Switches'.

Acknowledgments: We would like to thank Michael Breitenbach for his invaluable support, extensive knowledge, and mentorship throughout the years. Breitenbach's guidance has played a pivotal role in shaping the direction of our work, and his unwavering commitment to advancing scientific understanding has been a constant source of inspiration. We would also like to thank Julia Muenzner and Johannes Hartl for critical reading of this manuscript.

Conflicts of Interest: The authors declare no conflict of interest.

References

1. Fenton, H.J.H. LXXIII.—Oxidation of Tartaric Acid in Presence of Iron. *J. Chem. Soc. Trans.* **1894**, *65*, 899–910. [CrossRef]
2. Tavassolifar, M.J.; Vodjgani, M.; Salehi, Z.; Izad, M. The Influence of Reactive Oxygen Species in the Immune System and Pathogenesis of Multiple Sclerosis. *Autoimmune Dis.* **2020**, *2020*, 5793817. [CrossRef]
3. Birben, E.; Sahiner, U.M.; Sackesen, C.; Erzurum, S.; Kalayci, O. Oxidative Stress and Antioxidant Defense. *World Allergy Organ. J.* **2012**, *5*, 9–19. [CrossRef]
4. Devasagayam, T.P.A.; Tilak, J.C.; Boloor, K.K.; Sane, K.S.; Ghaskadbi, S.S.; Lele, R.D. Free Radicals and Antioxidants in Human Health: Current Status and Future Prospects. *J. Assoc. Physicians India* **2004**, *52*, 794–804.
5. Hensley, K.; Robinson, K.A.; Gabbita, S.P.; Salsman, S.; Floyd, R.A. Reactive Oxygen Species, Cell Signaling, and Cell Injury. *Free Radic. Biol. Med.* **2000**, *28*, 1456–1462. [CrossRef]
6. D'Autréaux, B.; Toledano, M.B. ROS as Signalling Molecules: Mechanisms That Generate Specificity in ROS Homeostasis. *Nat. Rev. Mol. Cell Biol.* **2007**, *8*, 813–824. [CrossRef]
7. Yang, Y.; Bazhin, A.V.; Werner, J.; Karakhanova, S. Reactive Oxygen Species in the Immune System. *Int. Rev. Immunol.* **2013**, *32*, 249–270. [CrossRef]
8. Sies, H.; Berndt, C.; Jones, D.P. Oxidative Stress. *Annu. Rev. Biochem.* **2017**, *86*, 715–748. [CrossRef]
9. Favier, A. Oxidative stress in human diseases. *Ann. Pharm. Fr.* **2006**, *64*, 390–396. [CrossRef]
10. Espinós, C.; Galindo, M.I.; García-Gimeno, M.A.; Ibáñez-Cabellos, J.S.; Martínez-Rubio, D.; Millán, J.M.; Rodrigo, R.; Sanz, P.; Seco-Cervera, M.; Sevilla, T.; et al. Oxidative Stress, a Crossroad between Rare Diseases and Neurodegeneration. *Antioxidants* **2020**, *9*, 313. [CrossRef]
11. Matschke, V.; Theiss, C.; Matschke, J. Oxidative Stress: The Lowest Common Denominator of Multiple Diseases. *Neural Regen. Res.* **2019**, *14*, 238–241. [CrossRef] [PubMed]
12. Jasoliya, M.J.; McMackin, M.Z.; Henderson, C.K.; Perlman, S.L.; Cortopassi, G.A. Frataxin Deficiency Impairs Mitochondrial Biogenesis in Cells, Mice and Humans. *Hum. Mol. Genet.* **2017**, *26*, 2627–2633. [CrossRef] [PubMed]
13. Vaubel, R.A.; Isaya, G. Iron-Sulfur Cluster Synthesis, Iron Homeostasis and Oxidative Stress in Friedreich Ataxia. *Mol. Cell. Neurosci.* **2013**, *55*, 50–61. [CrossRef] [PubMed]
14. Chantrel-Groussard, K.; Geromel, V.; Puccio, H.; Koenig, M.; Munnich, A.; Rötig, A.; Rustin, P. Disabled Early Recruitment of Antioxidant Defenses in Friedreich's Ataxia. *Hum. Mol. Genet.* **2001**, *10*, 2061–2067. [CrossRef]
15. Romá-Mateo, C.; García-Giménez, J.L. Oxidative Stress and Rare Diseases: From Molecular Crossroads to Therapeutic Avenues. *Antioxidants* **2021**, *10*, 617. [CrossRef]
16. Domènech, B.E.; Marfany, G. The Relevance of Oxidative Stress in the Pathogenesis and Therapy of Retinal Dystrophies. *Antioxid. Redox Signal.* **2020**, *9*, 347. [CrossRef]
17. Maciejczyk, M.; Mikoluc, B.; Pietrucha, B.; Heropolitanska-Pliszka, E.; Pac, M.; Motkowski, R.; Car, H. Oxidative Stress, Mitochondrial Abnormalities and Antioxidant Defense in Ataxia-Telangiectasia, Bloom Syndrome and Nijmegen Breakage Syndrome. *Redox Biol.* **2017**, *11*, 375–383. [CrossRef]
18. Muller, F. The Nature and Mechanism of Superoxide Production by the Electron Transport Chain: Its Relevance to Aging. *J. Am. Aging Assoc.* **2000**, *23*, 227–253. [CrossRef]
19. Turrens, J.F. Mitochondrial Formation of Reactive Oxygen Species. *J. Physiol.* **2003**, *552*, 335–344. [CrossRef]
20. Hadrava Vanova, K.; Kraus, M.; Neuzil, J.; Rohlena, J. Mitochondrial Complex II and Reactive Oxygen Species in Disease and Therapy. *Redox Rep.* **2020**, *25*, 26–32. [CrossRef]
21. Sharma, L.K.; Fang, H.; Liu, J.; Vartak, R.; Deng, J.; Bai, Y. Mitochondrial Respiratory Complex I Dysfunction Promotes Tumorigenesis through ROS Alteration and AKT Activation. *Hum. Mol. Genet.* **2011**, *20*, 4605–4616. [CrossRef]
22. Fullerton, M.; McFarland, R.; Taylor, R.W.; Alston, C.L. The Genetic Basis of Isolated Mitochondrial Complex II Deficiency. *Mol. Genet. Metab.* **2020**, *131*, 53–65. [CrossRef]
23. Iwata, S.; Lee, J.W.; Okada, K.; Lee, J.K.; Iwata, M.; Rasmussen, B.; Link, T.A.; Ramaswamy, S.; Jap, B.K. Complete Structure of the 11-Subunit Bovine Mitochondrial Cytochrome bc1 Complex. *Science* **1998**, *281*, 64–71. [CrossRef]
24. Hernansanz-Agustín, P.; Enríquez, J.A. Generation of Reactive Oxygen Species by Mitochondria. *Antioxidants* **2021**, *10*, 415. [CrossRef]
25. Maranzana, E.; Barbero, G.; Falasca, A.I.; Lenaz, G.; Genova, M.L. Mitochondrial Respiratory Supercomplex Association Limits Production of Reactive Oxygen Species from Complex I. *Antioxid. Redox Signal.* **2013**, *19*, 1469–1480. [CrossRef]

26. Lopez-Fabuel, I.; Le Douce, J.; Logan, A.; James, A.M.; Bonvento, G.; Murphy, M.P.; Almeida, A.; Bolaños, J.P. Complex I Assembly into Supercomplexes Determines Differential Mitochondrial ROS Production in Neurons and Astrocytes. *Proc. Natl. Acad. Sci. USA* **2016**, *113*, 13063–13068. [CrossRef]

27. Jain, I.H.; Zazzeron, L.; Goldberger, O.; Marutani, E.; Wojtkiewicz, G.R.; Ast, T.; Wang, H.; Schleifer, G.; Stepanova, A.; Brepoels, K.; et al. Leigh Syndrome Mouse Model Can Be Rescued by Interventions That Normalize Brain Hyperoxia, but Not HIF Activation. *Cell Metab.* **2019**, *30*, 824–832.e3. [CrossRef]

28. Perier, C.; Tieu, K.; Guégan, C.; Caspersen, C.; Jackson-Lewis, V.; Carelli, V.; Martinuzzi, A.; Hirano, M.; Przedborski, S.; Vila, M. Complex I Deficiency Primes Bax-Dependent Neuronal Apoptosis through Mitochondrial Oxidative Damage. *Proc. Natl. Acad. Sci. USA* **2005**, *102*, 19126–19131. [CrossRef] [PubMed]

29. Tretter, L.; Sipos, I.; Adam-Vizi, V. Initiation of Neuronal Damage by Complex I Deficiency and Oxidative Stress in Parkinson's Disease. *Neurochem. Res.* **2004**, *29*, 569–577. [CrossRef] [PubMed]

30. Liu, Y.; Fiskum, G.; Schubert, D. Generation of Reactive Oxygen Species by the Mitochondrial Electron Transport Chain. *J. Neurochem.* **2002**, *80*, 780–787. [CrossRef] [PubMed]

31. Lake, N.J.; Compton, A.G.; Rahman, S.; Thorburn, D.R. Leigh Syndrome: One Disorder, More than 75 Monogenic Causes. *Ann. Neurol.* **2016**, *79*, 190–203. [CrossRef] [PubMed]

32. Riordan-Eva, P.; Harding, A.E. Leber's Hereditary Optic Neuropathy: The Clinical Relevance of Different Mitochondrial DNA Mutations. *J. Med. Genet.* **1995**, *32*, 81–87. [CrossRef] [PubMed]

33. Di Monte, D.A. Mitochondrial DNA and Parkinson's Disease. *Neurology* **1991**, *41*, 38–42; discussion 42–43. [CrossRef]

34. Ikebe, S.; Tanaka, M.; Ozawa, T. Point Mutations of Mitochondrial Genome in Parkinson's Disease. *Brain Res. Mol. Brain Res.* **1995**, *28*, 281–295. [CrossRef]

35. Tucker, E.J.; Compton, A.G.; Calvo, S.E.; Thorburn, D.R. The Molecular Basis of Human Complex I Deficiency. *IUBMB Life* **2011**, *63*, 669–677. [CrossRef]

36. Misra, H.P.; Fridovich, I. The Generation of Superoxide Radical during the Autoxidation of Hemoglobin. *J. Biol. Chem.* **1972**, *247*, 6960–6962. [CrossRef] [PubMed]

37. Winterbourn, C.C. Oxidative Reactions of Hemoglobin. In *Methods in Enzymology*; Academic Press: Cambridge, MA, USA, 1990; Volume 186, pp. 265–272.

38. Schnedl, W.J.; Queissner, R.; Schenk, M.; Enko, D.; Mangge, H. Hereditary Methemoglobinemia Caused by Hb M-Hyde Park (Hb M-Akita) (*HBB*:c.277C > T; p.His93Tyr). *Wien. Klin. Wochenschr.* **2019**, *131*, 381–384. [CrossRef]

39. Mears, J.G.; Lachman, H.M.; Cabannes, R.; Amegnizin, K.P.; Labie, D.; Nagel, R.L. Sickle Gene. Its Origin and Diffusion from West Africa. *J. Clin. Investig.* **1981**, *68*, 606–610. [CrossRef]

40. Rees, D.C.; Williams, T.N.; Gladwin, M.T. Sickle-Cell Disease. *Lancet* **2010**, *376*, 2018–2031. [CrossRef]

41. Hebbel, R.P.; Morgan, W.T.; Eaton, J.W.; Hedlund, B.E. Accelerated Autoxidation and Heme Loss due to Instability of Sickle Hemoglobin. *Proc. Natl. Acad. Sci. USA* **1988**, *85*, 237–241. [CrossRef]

42. Hebbel, R.P.; Ney, P.A.; Foker, W. Autoxidation, Dehydration, and Adhesivity May Be Related Abnormalities of Sickle Erythrocytes. *Am. J. Physiol.* **1989**, *256*, C579–C583. [CrossRef] [PubMed]

43. Wang, Q.; Zennadi, R. The Role of RBC Oxidative Stress in Sickle Cell Disease: From the Molecular Basis to Pathologic Implications. *Antioxidants* **2021**, *10*, 1608. [CrossRef] [PubMed]

44. Orrico, F.; Laurance, S.; Lopez, A.C.; Lefevre, S.D.; Thomson, L.; Möller, M.N.; Ostuni, M.A. Oxidative Stress in Healthy and Pathological Red Blood Cells. *Biomolecules* **2023**, *13*, 1262. [CrossRef] [PubMed]

45. Papanikolaou, G.; Pantopoulos, K. Iron Metabolism and Toxicity. *Toxicol. Appl. Pharmacol.* **2005**, *202*, 199–211. [CrossRef] [PubMed]

46. Thom, C.S.; Dickson, C.F.; Gell, D.A.; Weiss, M.J. Hemoglobin Variants: Biochemical Properties and Clinical Correlates. *Cold Spring Harb. Perspect. Med.* **2013**, *3*, a011858. [CrossRef] [PubMed]

47. Panday, A.; Sahoo, M.K.; Osorio, D.; Batra, S. NADPH Oxidases: An Overview from Structure to Innate Immunity-Associated Pathologies. *Cell. Mol. Immunol.* **2014**, *12*, 5–23. [CrossRef]

48. Iles, K.E.; Forman, H.J. Macrophage Signaling and Respiratory Burst. *Immunol. Res.* **2002**, *26*, 95–105. [CrossRef]

49. Nauseef, W.M. Biological Roles for the NOX Family NADPH Oxidases. *J. Biol. Chem.* **2008**, *283*, 16961–16965. [CrossRef]

50. Schröder, K. NADPH Oxidases: Current Aspects and Tools. *Redox Biol* **2020**, *34*, 101512. [CrossRef]

51. Matute, J.D.; Arias, A.A.; Wright, N.A.M.; Wrobel, I.; Waterhouse, C.C.M.; Li, X.J.; Marchal, C.C.; Stull, N.D.; Lewis, D.B.; Steele, M.; et al. A New Genetic Subgroup of Chronic Granulomatous Disease with Autosomal Recessive Mutations in p40phox and Selective Defects in Neutrophil NADPH Oxidase Activity. *Blood* **2009**, *114*, 3309–3315. [CrossRef]

52. O'Neill, S.; Brault, J.; Stasia, M.-J.; Knaus, U.G. Genetic Disorders Coupled to ROS Deficiency. *Redox. Biol.* **2015**, *6*, 135–156. [CrossRef] [PubMed]

53. Zurro, N.B.; Tavares de Albuquerque, J.A.; França, T.T.; Vendramini, P.; Arslanian, C.; Tavares-Scancetti, F.; Condino-Neto, A. A Novel Mutation in *CYBB* Gene in a Patient with Chronic Colitis and Recurrent Pneumonia due to X-Linked Chronic Granulomatous Disease. *Pediatr. Blood Cancer* **2018**, *65*, e27382. [CrossRef]

54. Bustamante, J.; Picard, C.; Fieschi, C.; Filipe-Santos, O.; Feinberg, J.; Perronne, C.; Chapgier, A.; de Beaucoudrey, L.; Vogt, G.; Sanlaville, D.; et al. A Novel X-Linked Recessive Form of Mendelian Susceptibility to Mycobaterial Disease. *J. Med. Genet.* **2007**, *44*, e65. [CrossRef] [PubMed]

55. Teimourian, S.; Zomorodian, E.; Badalzadeh, M.; Pouya, A.; Kannengiesser, C.; Mansouri, D.; Cheraghi, T.; Parvaneh, N. Characterization of Six Novel Mutations in *CYBA*: The Gene Causing Autosomal Recessive Chronic Granulomatous Disease. *Br. J. Haematol.* **2008**, *141*, 848–851. [CrossRef] [PubMed]

56. Curnutte, J.T.; Berkow, R.L.; Roberts, R.L.; Shurin, S.B.; Scott, P.J. Chronic Granulomatous Disease due to a Defect in the Cytosolic Factor Required for Nicotinamide Adenine Dinucleotide Phosphate Oxidase Activation. *J. Clin. Investig.* **1988**, *81*, 606–610. [CrossRef] [PubMed]

57. Patiño, P.J.; Rae, J.; Noack, D.; Erickson, R.; Ding, J.; de Olarte, D.G.; Curnutte, J.T. Molecular Characterization of Autosomal Recessive Chronic Granulomatous Disease Caused by a Defect of the Nicotinamide Adenine Dinucleotide Phosphate (reduced Form) Oxidase Component p67-Phox. *Blood* **1999**, *94*, 2505–2514. [CrossRef] [PubMed]

58. van de Geer, A.; Nieto-Patlán, A.; Kuhns, D.B.; Tool, A.T.; Arias, A.A.; Bouaziz, M.; de Boer, M.; Franco, J.L.; Gazendam, R.P.; van Hamme, J.L.; et al. Inherited p40phox Deficiency Differs from Classic Chronic Granulomatous Disease. *J. Clin. Investig.* **2018**, *128*, 3957–3975. [CrossRef] [PubMed]

59. de Oliveira-Junior, E.B.; Bustamante, J.; Newburger, P.E.; Condino-Neto, A. The Human NADPH Oxidase: Primary and Secondary Defects Impairing the Respiratory Burst Function and the Microbicidal Ability of Phagocytes. *Scand. J. Immunol.* **2011**, *73*, 420–427. [CrossRef]

60. Hayes, P.; Dhillon, S.; O'Neill, K.; Thoeni, C.; Hui, K.Y.; Elkadri, A.; Guo, C.H.; Kovacic, L.; Aviello, G.; Alvarez, L.A.; et al. Defects in NADPH Oxidase Genes *NOX1* and *DUOX2* in Very Early Onset Inflammatory Bowel Disease. *Cell Mol. Gastroenterol. Hepatol.* **2015**, *1*, 489–502. [CrossRef]

61. Moreno, J.C.; Bikker, H.; Kempers, M.J.E.; van Trotsenburg, A.S.P.; Baas, F.; de Vijlder, J.J.M.; Vulsma, T.; Ris-Stalpers, C. Inactivating Mutations in the Gene for Thyroid Oxidase 2 (*THOX2*) and Congenital Hypothyroidism. *N. Engl. J. Med.* **2002**, *347*, 95–102. [CrossRef]

62. Zamproni, I.; Grasberger, H.; Cortinovis, F.; Vigone, M.C.; Chiumello, G.; Mora, S.; Onigata, K.; Fugazzola, L.; Refetoff, S.; Persani, L.; et al. Biallelic Inactivation of the Dual Oxidase Maturation Factor 2 (*DUOXA2*) Gene as a Novel Cause of Congenital Hypothyroidism. *J. Clin. Endocrinol. Metab.* **2008**, *93*, 605–610. [CrossRef] [PubMed]

63. De Deken, X.; Wang, D.; Many, M.C.; Costagliola, S.; Libert, F.; Vassart, G.; Dumont, J.E.; Miot, F. Cloning of Two Human Thyroid cDNAs Encoding New Members of the NADPH Oxidase Family. *J. Biol. Chem.* **2000**, *275*, 23227–23233. [CrossRef] [PubMed]

64. Schultz, J.; Kaminker, K. Myeloperoxidase of the Leucocyte of Normal Human Blood. I. Content and Localization. *Arch. Biochem. Biophys.* **1962**, *96*, 465–467. [CrossRef]

65. Metzler, K.D.; Fuchs, T.A.; Nauseef, W.M.; Reumaux, D.; Roesler, J.; Schulze, I.; Wahn, V.; Papayannopoulos, V.; Zychlinsky, A. Myeloperoxidase Is Required for Neutrophil Extracellular Trap Formation: Implications for Innate Immunity. *Blood* **2011**, *117*, 953–959. [CrossRef]

66. Bos, A.; Wever, R.; Roos, D. Characterization and Quantification of the Peroxidase in Human Monocytes. *Biochim. Biophys. Acta* **1978**, *525*, 37–44. [CrossRef]

67. Odobasic, D.; Kitching, A.R.; Holdsworth, S.R. Neutrophil-Mediated Regulation of Innate and Adaptive Immunity: The Role of Myeloperoxidase. *J. Immunol. Res.* **2016**, *2016*, 2349817. [CrossRef]

68. van der Veen, B.S.; de Winther, M.P.J.; Heeringa, P. Myeloperoxidase: Molecular Mechanisms of Action and Their Relevance to Human Health and Disease. *Antioxid. Redox Signal.* **2009**, *11*, 2899–2937. [CrossRef]

69. Kutter, D.; Devaquet, P.; Vanderstocken, G.; Paulus, J.M.; Marchal, V.; Gothot, A. Consequences of Total and Subtotal Myeloperoxidase Deficiency: Risk or Benefit? *Acta Haematol.* **2000**, *104*, 10–15. [CrossRef]

70. Lehrer, R.I.; Cline, M.J. Leukocyte Myeloperoxidase Deficiency and Disseminated Candidiasis: The Role of Myeloperoxidase in Resistance to Candida Infection. *J. Clin. Investig.* **1969**, *48*, 1478–1488. [CrossRef]

71. Abraitytė, S.; Kotsi, E.; Devlin, L.A.; Edgar, J.D.M. Unexpected Combination: DiGeorge Syndrome and Myeloperoxidase Deficiency. *BMJ Case Rep.* **2020**, *13*, e232741. [CrossRef]

72. Chung, H.Y.; Baek, B.S.; Song, S.H.; Kim, M.S.; Huh, J.I.; Shim, K.H.; Kim, K.W.; Lee, K.H. Xanthine Dehydrogenase/xanthine Oxidase and Oxidative Stress. *Age* **1997**, *20*, 127–140. [CrossRef]

73. Nishino, T.; Okamoto, K.; Eger, B.T.; Pai, E.F.; Nishino, T. Mammalian Xanthine Oxidoreductase—Mechanism of Transition from Xanthine Dehydrogenase to Xanthine Oxidase. *FEBS J.* **2008**, *275*, 3278–3289. [CrossRef]

74. McCord, J.M. Oxygen-Derived Free Radicals in Postischemic Tissue Injury. *N. Engl. J. Med.* **1985**, *312*, 159–163.

75. Tsukasa, K.; Toshihiro, N.; Motoshi, K.; Tatsuo, H.; Kusuki, N. Biochemical Studies on the Purine Metabolism of Four Cases with Hereditary Xanthinuria. *Clin. Chim. Acta* **1984**, *137*, 189–198. [CrossRef]

76. Sebesta, I.; Stiburkova, B.; Krijt, J. Hereditary Xanthinuria Is Not so Rare Disorder of Purine Metabolism. *Nucleosides Nucleotides Nucleic Acids* **2018**, *37*, 324–328. [CrossRef] [PubMed]

77. Förstermann, U. Nitric Oxide and Oxidative Stress in Vascular Disease. *Pflug. Arch.* **2010**, *459*, 923–939. [CrossRef]

78. Förstermann, U.; Münzel, T. Endothelial Nitric Oxide Synthase in Vascular Disease: From Marvel to Menace. *Circulation* **2006**, *113*, 1708–1714. [CrossRef]

79. Xia, Y.; Zweier, J.L. Superoxide and Peroxynitrite Generation from Inducible Nitric Oxide Synthase in Macrophages. *Proc. Natl. Acad. Sci. USA* **1997**, *94*, 6954–6958. [CrossRef]

80. Arngrímsson, R.; Hayward, C.; Nadaud, S.; Baldursdóttir, A.; Walker, J.J.; Liston, W.A.; Bjarnadóttir, R.I.; Brock, D.J.; Geirsson, R.T.; Connor, J.M.; et al. Evidence for a Familial Pregnancy-Induced Hypertension Locus in the *eNOS*-Gene Region. *Am. J. Hum. Genet.* **1997**, *61*, 354–362. [CrossRef] [PubMed]
81. Berger, K.; Stögbauer, F.; Stoll, M.; Wellmann, J.; Huge, A.; Cheng, S.; Kessler, C.; John, U.; Assmann, G.; Ringelstein, E.B.; et al. The glu298asp Polymorphism in the Nitric Oxide Synthase 3 Gene Is Associated with the Risk of Ischemic Stroke in Two Large Independent Case-Control Studies. *Hum. Genet.* **2007**, *121*, 169–178. [CrossRef] [PubMed]
82. Gwozdzinski, K.; Pieniazek, A.; Gwozdzinski, L. Reactive Oxygen Species and Their Involvement in Red Blood Cell Damage in Chronic Kidney Disease. *Oxid. Med. Cell. Longev.* **2021**, *2021*, 6639199. [CrossRef]
83. Kattamis, C.A.; Kyriazakou, M.; Chaidas, S. Favism: Clinical and Biochemical Data. *J. Med. Genet.* **1969**, *6*, 34–41. [CrossRef]
84. Ighodaro, O.M.; Akinloye, O.A. First Line Defence Antioxidants-Superoxide Dismutase (SOD), Catalase (CAT) and Glutathione Peroxidase (GPX): Their Fundamental Role in the Entire Antioxidant Defence Grid. *Alex. J. Med.* **2018**, *54*, 287–293. [CrossRef]
85. Graves, J.A.; Metukuri, M.; Scott, D.; Rothermund, K.; Prochownik, E.V. Regulation of Reactive Oxygen Species Homeostasis by Peroxiredoxins and c-Myc. *J. Biol. Chem.* **2009**, *284*, 6520–6529. [CrossRef]
86. Talwar, D.; Miller, C.G.; Grossmann, J.; Szyrwiel, L.; Schwecke, T.; Demichev, V.; Mikecin Drazic, A.-M.; Mayakonda, A.; Lutsik, P.; Veith, C.; et al. The GAPDH Redox Switch Safeguards Reductive Capacity and Enables Survival of Stressed Tumour Cells. *Nat. Metab.* **2023**, *5*, 660–676. [CrossRef]
87. Modun, D.; Giustarini, D.; Tsikas, D. Nitric Oxide-Related Oxidative Stress and Redox Status in Health and Disease. *Oxid. Med. Cell. Longev.* **2014**, *2014*, 129651. [CrossRef]
88. Ralser, M.; Wamelink, M.M.; Kowald, A.; Gerisch, B.; Heeren, G.; Struys, E.A.; Klipp, E.; Jakobs, C.; Breitenbach, M.; Lehrach, H.; et al. Dynamic Rerouting of the Carbohydrate Flux Is Key to Counteracting Oxidative Stress. *J. Biol.* **2007**, *6*, 10. [CrossRef] [PubMed]
89. Ralser, M.; Wamelink, M.M.C.; Latkolik, S.; Jansen, E.E.W.; Lehrach, H.; Jakobs, C. Metabolic Reconfiguration Precedes Transcriptional Regulation in the Antioxidant Response. *Nat. Biotechnol.* **2009**, *27*, 604–605. [CrossRef]
90. Grune, T. Oxidants and Antioxidative Defense. *Hum. Exp. Toxicol.* **2002**, *21*, 61–62. [CrossRef] [PubMed]
91. Lu, J.; Holmgren, A. The Thioredoxin Antioxidant System. *Free Radic. Biol. Med.* **2014**, *66*, 75–87. [CrossRef] [PubMed]
92. Lubos, E.; Loscalzo, J.; Handy, D.E. Glutathione Peroxidase-1 in Health and Disease: From Molecular Mechanisms to Therapeutic Opportunities. *Antioxid. Redox Signal.* **2011**, *15*, 1957–1997. [CrossRef]
93. Berdyński, M.; Miszta, P.; Safranow, K.; Andersen, P.M.; Morita, M.; Filipek, S.; Żekanowski, C.; Kuźma-Kozakiewicz, M. *SOD1* Mutations Associated with Amyotrophic Lateral Sclerosis Analysis of Variant Severity. *Sci. Rep.* **2022**, *12*, 103. [CrossRef] [PubMed]
94. Culik, R.M.; Sekhar, A.; Nagesh, J.; Deol, H.; Rumfeldt, J.A.O.; Meiering, E.M.; Kay, L.E. Effects of Maturation on the Conformational Free-Energy Landscape of SOD1. *Proc. Natl. Acad. Sci. USA* **2018**, *115*, E2546–E2555. [CrossRef]
95. Sau, D.; De Biasi, S.; Vitellaro-Zuccarello, L.; Riso, P.; Guarnieri, S.; Porrini, M.; Simeoni, S.; Crippa, V.; Onesto, E.; Palazzolo, I.; et al. Mutation of SOD1 in ALS: A Gain of a Loss of Function. *Hum. Mol. Genet.* **2007**, *16*, 1604–1618. [CrossRef] [PubMed]
96. Valentine, J.S. Do Oxidatively Modified Proteins Cause ALS? *Free Radic. Biol. Med.* **2002**, *33*, 1314–1320. [CrossRef] [PubMed]
97. Ezer, S.; Daana, M.; Park, J.H.; Yanovsky-Dagan, S.; Nordström, U.; Basal, A.; Edvardson, S.; Saada, A.; Otto, M.; Meiner, V.; et al. Infantile SOD1 Deficiency Syndrome Caused by a Homozygous *SOD1* Variant with Absence of Enzyme Activity. *Brain* **2022**, *145*, 872–878. [CrossRef] [PubMed]
98. Andersen, P.M.; Nordström, U.; Tsiakas, K.; Johannsen, J.; Volk, A.E.; Bierhals, T.; Zetterström, P.; Marklund, S.L.; Hempel, M.; Santer, R. Phenotype in an Infant with *SOD1* Homozygous Truncating Mutation. *N. Engl. J. Med.* **2019**, *381*, 486–488. [CrossRef] [PubMed]
99. Park, J.H.; Elpers, C.; Reunert, J.; McCormick, M.L.; Mohr, J.; Biskup, S.; Schwartz, O.; Rust, S.; Grüneberg, M.; Seelhöfer, A.; et al. SOD1 Deficiency: A Novel Syndrome Distinct from Amyotrophic Lateral Sclerosis. *Brain* **2019**, *142*, 2230–2237. [CrossRef]
100. Smejkal, G.B.; Kakumanu, S. Enzymes and Their Turnover Numbers. *Expert Rev. Proteom.* **2019**, *16*, 543–544. [CrossRef]
101. Ogata, M. Acatalasemia. *Hum. Genet.* **1991**, *86*, 331–340. [CrossRef]
102. Takahara, S. Progressive Oral Gangrene Probably due to Lack of Catalase in the Blood (acatalasaemia); Report of Nine Cases. *Lancet* **1952**, *2*, 1101–1104. [CrossRef]
103. Ludlow, J.T.; Wilkerson, R.G.; Nappe, T.M. *Methemoglobinemia*; StatPearls Publishing: Treasure Island, FL, USA, 2023.
104. Takahara, S.; Ogata, M. Metabolism in Japanese Acatalasemia with Special Reference to Superoxide Dismutase and Glutathione Peroxidase. *Biochem. Med. Asp. Act. Oxyg.* **1978**, *3*, 279–292.
105. Góth, L.; Nagy, T. Inherited Catalase Deficiency: Is It Benign or a Factor in Various Age Related Disorders? *Mutat. Res.* **2013**, *753*, 147–154. [CrossRef]
106. Nandi, A.; Yan, L.-J.; Jana, C.K.; Das, N. Role of Catalase in Oxidative Stress- and Age-Associated Degenerative Diseases. *Oxid. Med. Cell. Longev.* **2019**, *2019*, 9613090. [CrossRef] [PubMed]
107. Brigelius-Flohé, R.; Maiorino, M. Glutathione Peroxidases. *Biochim. Biophys. Acta* **2013**, *1830*, 3289–3303. [CrossRef] [PubMed]
108. Necheles, T.F.; Maldonado, N.; Barquet-Chediak, A.; Allen, D.M. Homozygous Erythrocyte Glutathione-Peroxidase Deficiency: Clinical and Biochemical Studies. *Blood* **1969**, *33*, 164–169. [CrossRef] [PubMed]
109. Fibach, E.; Rachmilewitz, E. The Role of Oxidative Stress in Hemolytic Anemia. *Curr. Mol. Med.* **2008**, *8*, 609–619. [CrossRef]

110. Sedaghatian, M.R. Congenital Lethal Metaphyseal Chondrodysplasia: A Newly Recognized Complex Autosomal Recessive Disorder. *Am. J. Med. Genet.* **1980**, *6*, 269–274. [CrossRef] [PubMed]
111. Peshimam, N.; Farah, H.; Caswell, R.; Ellard, S.; Jan, W.; Calder, A.D.; Cobben, J.; Kariholu, U.; Leitch, H.G. Sedaghatian Spondylometaphyseal Dysplasia in Two Siblings. *Eur. J. Med. Genet.* **2022**, *65*, 104541. [CrossRef]
112. Friedmann Angeli, J.P.; Schneider, M.; Proneth, B.; Tyurina, Y.Y.; Tyurin, V.A.; Hammond, V.J.; Herbach, N.; Aichler, M.; Walch, A.; Eggenhofer, E.; et al. Inactivation of the Ferroptosis Regulator Gpx4 Triggers Acute Renal Failure in Mice. *Nat. Cell Biol.* **2014**, *16*, 1180–1191. [CrossRef]
113. Fedida, A.; Ben Harouch, S.; Kalfon, L.; Abunassar, Z.; Omari, H.; Mandel, H.; Falik-Zaccai, T.C. Sedaghatian-Type Spondylometaphyseal Dysplasia: Whole Exome Sequencing in Neonatal Dry Blood Spots Enabled Identification of a Novel Variant in *GPX4*. *Eur. J. Med. Genet.* **2020**, *63*, 104020. [CrossRef] [PubMed]
114. Perkins, A.; Nelson, K.J.; Parsonage, D.; Poole, L.B.; Karplus, P.A. Peroxiredoxins: Guardians against Oxidative Stress and Modulators of Peroxide Signaling. *Trends Biochem. Sci.* **2015**, *40*, 435–445. [CrossRef] [PubMed]
115. Chiribau, C.B.; Cheng, L.; Cucoranu, I.C.; Yu, Y.-S.; Clempus, R.E.; Sorescu, D. FOXO3A Regulates Peroxiredoxin III Expression in Human Cardiac Fibroblasts. *J. Biol. Chem.* **2008**, *283*, 8211–8217. [CrossRef] [PubMed]
116. Rebelo, A.P.; Eidhof, I.; Cintra, V.P.; Guillot-Noel, L.; Pereira, C.V.; Timmann, D.; Traschütz, A.; Schöls, L.; Coarelli, G.; Durr, A.; et al. Biallelic Loss-of-Function Variations in *PRDX3* Cause Cerebellar Ataxia. *Brain* **2021**, *144*, 1467–1481. [CrossRef] [PubMed]
117. Rebelo, A.P.; Bender, B.; Haack, T.B.; Zuchner, S.; PREPARE consortium; Basak, A.N.; Synofzik, M. Expanding PRDX3 Disease: Broad Range of Onset Age and Infratentorial MRI Signal Changes. *Brain* **2022**, *145*, e95–e98. [CrossRef] [PubMed]
118. Martínez-Rubio, D.; Rodríguez-Prieto, Á.; Sancho, P.; Navarro-González, C.; Gorría-Redondo, N.; Miquel-Leal, J.; Marco-Marín, C.; Jenkins, A.; Soriano-Navarro, M.; Hernández, A.; et al. Protein Misfolding and Clearance in the Pathogenesis of a New Infantile Onset Ataxia Caused by Mutations in *PRDX3*. *Hum. Mol. Genet.* **2022**, *31*, 3897–3913. [CrossRef] [PubMed]
119. Alió Del Barrio, J.L.; Chung, D.D.; Al-Shymali, O.; Barrington, A.; Jatavallabhula, K.; Swamy, V.S.; Yébana, P.; Angélica Henríquez-Recine, M.; Boto-de-Los-Bueis, A.; Alió, J.L.; et al. Punctiform and Polychromatic Pre-Descemet Corneal Dystrophy: Clinical Evaluation and Identification of the Genetic Basis. *Am. J. Ophthalmol.* **2020**, *212*, 88–97. [CrossRef]
120. Henríquez-Recine, M.A.; Marquina-Lima, K.S.; Vallespín-García, E.; García-Miñaur, S.; Benitez Del Castillo, J.M.; Boto de Los Bueis, A. Heredity and in Vivo Confocal Microscopy of Punctiform and Polychromatic Pre-Descemet Dystrophy. *Graefes Arch. Clin. Exp. Ophthalmol.* **2018**, *256*, 1661–1667. [CrossRef]
121. Choo, C.H.; Boto de Los Bueis, A.; Chung, D.D.; Aldave, A.J. Confirmation of PRDX3 c.568G>C as the Genetic Basis of Punctiform and Polychromatic Pre-Descemet Corneal Dystrophy. *Cornea* **2022**, *41*, 779–781. [CrossRef]
122. Guéant, J.-L.; Chéry, C.; Oussalah, A.; Nadaf, J.; Coelho, D.; Josse, T.; Flayac, J.; Robert, A.; Koscinski, I.; Gastin, I.; et al. *APRDX1* Mutant Allele Causes a MMACHC Secondary Epimutation in cblC Patients. *Nat. Commun.* **2018**, *9*, 67. [CrossRef]
123. Bachhawat, A.K.; Yadav, S. The Glutathione Cycle: Glutathione Metabolism beyond the γ-Glutamyl Cycle. *IUBMB Life* **2018**, *70*, 585–592. [CrossRef] [PubMed]
124. Konrad, P.N.; Richards, F., 2nd; Valentine, W.N.; Paglia, D.E. -Glutamyl-Cysteine Synthetase Deficiency. A Cause of Hereditary Hemolytic Anemia. *N. Engl. J. Med.* **1972**, *286*, 557–561. [CrossRef] [PubMed]
125. Hirono, A.; Iyori, H.; Sekine, I.; Ueyama, J.; Chiba, H.; Kanno, H.; Fujii, H.; Miwa, S. Three Cases of Hereditary Nonspherocytic Hemolytic Anemia Associated with Red Blood Cell Glutathione Deficiency. *Blood* **1996**, *87*, 2071–2074. [CrossRef] [PubMed]
126. Mohler, D.N.; Majerus, P.W.; Minnich, V.; Hess, C.E.; Garrick, M.D. Glutathione Synthetase Deficiency as a Cause of Hereditary Hemolytic Disease. *N. Engl. J. Med.* **1970**, *283*, 1253–1257. [CrossRef] [PubMed]
127. Jellum, E.; Kluge, T.; Börresen, H.C.; Stokke, O.; Eldjarn, L. Pyroglutamic Aciduria—A New Inborn Error of Metabolism. *Scand. J. Clin. Lab. Investig.* **1970**, *26*, 327–335. [CrossRef] [PubMed]
128. Spielberg, S.P.; Garrick, M.D.; Corash, L.M.; Butler, J.D.; Tietze, F.; Rogers, L.; Schulman, J.D. Biochemical Heterogeneity in Glutathione Synthetase Deficiency. *J. Clin. Investig.* **1978**, *61*, 1417–1420. [CrossRef] [PubMed]
129. Ristoff, E.; Hebert, C.; Njålsson, R.; Norgren, S.; Rooyackers, O.; Larsson, A. Glutathione Synthetase Deficiency: Is Gamma-Glutamylcysteine Accumulation a Way to Cope with Oxidative Stress in Cells with Insufficient Levels of Glutathione? *J. Inherit. Metab. Dis.* **2002**, *25*, 577–584. [CrossRef]
130. Shi, Z.Z.; Osei-Frimpong, J.; Kala, G.; Kala, S.V.; Barrios, R.J.; Habib, G.M.; Lukin, D.J.; Danney, C.M.; Matzuk, M.M.; Lieberman, M.W. Glutathione Synthesis Is Essential for Mouse Development but Not for Cell Growth in Culture. *Proc. Natl. Acad. Sci. USA* **2000**, *97*, 5101–5106. [CrossRef]
131. Dalton, T.P.; Dieter, M.Z.; Yang, Y.; Shertzer, H.G.; Nebert, D.W. Knockout of the Mouse Glutamate Cysteine Ligase Catalytic Subunit (*Gclc*) Gene: Embryonic Lethal When Homozygous, and Proposed Model for Moderate Glutathione Deficiency When Heterozygous. *Biochem. Biophys. Res. Commun.* **2000**, *279*, 324–329. [CrossRef]
132. Winkler, A.; Njålsson, R.; Carlsson, K.; Elgadi, A.; Rozell, B.; Abraham, L.; Ercal, N.; Shi, Z.-Z.; Lieberman, M.W.; Larsson, A.; et al. Glutathione Is Essential for Early Embryogenesis--Analysis of a Glutathione Synthetase Knockout Mouse. *Biochem. Biophys. Res. Commun.* **2011**, *412*, 121–126. [CrossRef]
133. Matsui, M.; Oshima, M.; Oshima, H.; Takaku, K.; Maruyama, T.; Yodoi, J.; Taketo, M.M. Early Embryonic Lethality Caused by Targeted Disruption of the Mouse Thioredoxin Gene. *Dev. Biol.* **1996**, *178*, 179–185. [CrossRef] [PubMed]
134. Nonn, L.; Williams, R.R.; Erickson, R.P.; Powis, G. The Absence of Mitochondrial Thioredoxin 2 Causes Massive Apoptosis, Exencephaly, and Early Embryonic Lethality in Homozygous Mice. *Mol. Cell. Biol.* **2003**, *23*, 916–922. [CrossRef] [PubMed]

135. Holzerova, E.; Danhauser, K.; Haack, T.B.; Kremer, L.S.; Melcher, M.; Ingold, I.; Kobayashi, S.; Terrile, C.; Wolf, P.; Schaper, J.; et al. Human Thioredoxin 2 Deficiency Impairs Mitochondrial Redox Homeostasis and Causes Early-Onset Neurodegeneration. *Brain* **2016**, *139*, 346–354. [CrossRef] [PubMed]

136. Pretsch, W. Glutathione Reductase Activity Deficiency in Homozygous Gr1a1Neu Mice Does Not Cause Haemolytic Anaemia. *Genet. Res.* **1999**, *73*, 1–5. [CrossRef]

137. Rogers, L.K.; Tamura, T.; Rogers, B.J.; Welty, S.E.; Hansen, T.N.; Smith, C.V. Analyses of Glutathione Reductase Hypomorphic Mice Indicate a Genetic Knockout. *Toxicol. Sci.* **2004**, *82*, 367–373. [CrossRef]

138. Loehr, G.W.; Waller, H.D. A new enzymopenic hemolytic anemia with glutathione reductase deficiency. *Med. Klin.* **1962**, *57*, 1521–1525.

139. Loos, H.; Roos, D.; Weening, R.; Houwerzijl, J. Familial Deficiency of Glutathione Reductase in Human Blood Cells. *Blood* **1976**, *48*, 53–62. [CrossRef] [PubMed]

140. Kamerbeek, N.M.; van Zwieten, R.; de Boer, M.; Morren, G.; Vuil, H.; Bannink, N.; Lincke, C.; Dolman, K.M.; Becker, K.; Schirmer, R.H.; et al. Molecular Basis of Glutathione Reductase Deficiency in Human Blood Cells. *Blood* **2007**, *109*, 3560–3566. [CrossRef]

141. el-Hazmi, M.A.; Warsy, A.S. Glutathione Reductase in the South-Western Province of Saudi Arabia--Genetic Variation vs. Acquired Deficiency. *Haematologia* **1989**, *22*, 37–42.

142. Warsy, A.S.; el-Hazmi, M.A. Glutathione Reductase Deficiency in Saudi Arabia. *East. Mediterr. Health J.* **1999**, *5*, 1208–1212. [CrossRef]

143. Conrad, M.; Bornkamm, G.W.; Brielmeier, M. Mitochondrial and Cytosolic Thioredoxin Reductase Knockout Mice. In *Selenium: Its Molecular Biology and Role in Human Health*; Hatfield, D.L., Berry, M.J., Gladyshev, V.N., Eds.; Springer: Boston, MA, USA, 2006; pp. 195–206. ISBN 9780387338279.

144. Dou, Q.; Turanov, A.A.; Mariotti, M.; Hwang, J.Y.; Wang, H.; Lee, S.-G.; Paulo, J.A.; Yim, S.H.; Gygi, S.P.; Chung, J.-J.; et al. Selenoprotein TXNRD3 Supports Male Fertility via the Redox Regulation of Spermatogenesis. *J. Biol. Chem.* **2022**, *298*, 102183. [CrossRef] [PubMed]

145. Maddocks, O.D.K.; Labuschagne, C.F.; Vousden, K.H. Localization of NADPH Production: A Wheel within a Wheel. *Mol. Cell* **2014**, *55*, 158–160. [CrossRef] [PubMed]

146. Longo, L.; Vanegas, O.C.; Patel, M.; Rosti, V.; Li, H.; Waka, J.; Merghoub, T.; Pandolfi, P.P.; Notaro, R.; Manova, K.; et al. Maternally Transmitted Severe Glucose 6-Phosphate Dehydrogenase Deficiency Is an Embryonic Lethal. *EMBO J.* **2002**, *21*, 4229–4239. [CrossRef] [PubMed]

147. Gómez-Manzo, S.; Marcial-Quino, J.; Vanoye-Carlo, A.; Serrano-Posada, H.; Ortega-Cuellar, D.; González-Valdez, A.; Castillo-Rodríguez, R.A.; Hernández-Ochoa, B.; Sierra-Palacios, E.; Rodríguez-Bustamante, E.; et al. Glucose-6-Phosphate Dehydrogenase: Update and Analysis of New Mutations around the World. *Int. J. Mol. Sci.* **2016**, *17*, 2069. [CrossRef] [PubMed]

148. Cappellini, M.D.; Fiorelli, G. Glucose-6-Phosphate Dehydrogenase Deficiency. *Lancet* **2008**, *371*, 64–74. [CrossRef] [PubMed]

149. Luzzatto, L.; Arese, P. Favism and Glucose-6-Phosphate Dehydrogenase Deficiency. *N. Engl. J. Med.* **2018**, *378*, 60–71. [CrossRef] [PubMed]

150. Pandolfi, P.P.; Sonati, F.; Rivi, R.; Mason, P.; Grosveld, F.; Luzzatto, L. Targeted Disruption of the Housekeeping Gene Encoding Glucose 6-Phosphate Dehydrogenase (G6PD): G6PD Is Dispensable for Pentose Synthesis but Essential for Defense against Oxidative Stress. *EMBO J.* **1995**, *14*, 5209–5215. [CrossRef]

151. Brewer, G.J.; Dern, R.J. A new inherited enzymatic deficiency of human erythrocytes: 6-phosphogluconate dehydrogenase deficiency. *Am. J. Hum. Genet.* **1964**, *16*, 472–476.

152. Parr, C.W.; Fitch, L.I. Inherited Quantitative Variations of Human Phosphogluconate Dehydrogenase. *Ann. Hum. Genet.* **1967**, *30*, 339–353. [CrossRef]

153. Johnson, F.M.; Chasalow, F.; Lewis, S.E.; Barnett, L.; Lee, C.-Y. A Null Allele at the Mod-1 Locus of the Mouse. *J. Hered.* **1981**, *72*, 134–136. [CrossRef]

154. Yang, X.; Deignan, J.L.; Qi, H.; Zhu, J.; Qian, S.; Zhong, J.; Torosyan, G.; Majid, S.; Falkard, B.; Kleinhanz, R.R.; et al. Validation of Candidate Causal Genes for Obesity That Affect Shared Metabolic Pathways and Networks. *Nat. Genet.* **2009**, *41*, 415–423. [CrossRef] [PubMed]

155. Smith, T.J.; Peterson, P.E.; Schmidt, T.; Fang, J.; Stanley, C.A. Structures of Bovine Glutamate Dehydrogenase Complexes Elucidate the Mechanism of Purine Regulation. *J. Mol. Biol.* **2001**, *307*, 707–720. [CrossRef]

156. Brandt, A.; Agarwal, N.; Giri, D.; Yung, Z.; Didi, M.; Senniappan, S. Hyperinsulinism Hyperammonaemia (HI/HA) Syndrome due to *GLUD1* Mutation: Phenotypic Variations Ranging from Late Presentation to Spontaneous Resolution. *J. Pediatr. Endocrinol. Metab.* **2020**, *33*, 675–679. [CrossRef] [PubMed]

157. Itsumi, M.; Inoue, S.; Elia, A.J.; Murakami, K.; Sasaki, M.; Lind, E.F.; Brenner, D.; Harris, I.S.; Chio, I.I.C.; Afzal, S.; et al. Idh1 Protects Murine Hepatocytes from Endotoxin-Induced Oxidative Stress by Regulating the Intracellular NADP+/NADPH Ratio. *Cell Death Differ.* **2015**, *22*, 1837–1845. [CrossRef]

158. Kranendijk, M.; Struys, E.A.; van Schaftingen, E.; Gibson, K.M.; Kanhai, W.A.; van der Knaap, M.S.; Amiel, J.; Buist, N.R.; Das, A.M.; de Klerk, J.B.; et al. IDH2 Mutations in Patients with D-2-Hydroxyglutaric Aciduria. *Science* **2010**, *330*, 336. [CrossRef] [PubMed]

159. Ku, H.J.; Ahn, Y.; Lee, J.H.; Park, K.M.; Park, J.-W. IDH2 Deficiency Promotes Mitochondrial Dysfunction and Cardiac Hypertrophy in Mice. *Free Radic. Biol. Med.* **2015**, *80*, 84–92. [CrossRef] [PubMed]

160. White, K.; Kim, M.-J.; Han, C.; Park, H.-J.; Ding, D.; Boyd, K.; Walker, L.; Linser, P.; Meneses, Z.; Slade, C.; et al. Loss of IDH2 Accelerates Age-Related Hearing Loss in Male Mice. *Sci. Rep.* **2018**, *8*, 5039. [CrossRef] [PubMed]
161. Fan, J.; Ye, J.; Kamphorst, J.J.; Shlomi, T.; Thompson, C.B.; Rabinowitz, J.D. Quantitative Flux Analysis Reveals Folate-Dependent NADPH Production. *Nature* **2014**, *510*, 298–302. [CrossRef]
162. Lewis, C.A.; Parker, S.J.; Fiske, B.P.; McCloskey, D.; Gui, D.Y.; Green, C.R.; Vokes, N.I.; Feist, A.M.; Vander Heiden, M.G.; Metallo, C.M. Tracing Compartmentalized NADPH Metabolism in the Cytosol and Mitochondria of Mammalian Cells. *Mol. Cell* **2014**, *55*, 253–263. [CrossRef]
163. Hum, D.W.; Bell, A.W.; Rozen, R.; MacKenzie, R.E. Primary Structure of a Human Trifunctional Enzyme. Isolation of a cDNA Encoding Methylenetetrahydrofolate Dehydrogenase-Methenyltetrahydrofolate Cyclohydrolase-Formyltetrahydrofolate Synthetase. *J. Biol. Chem.* **1988**, *263*, 15946–15950. [CrossRef]
164. Di Pietro, E.; Sirois, J.; Tremblay, M.L.; MacKenzie, R.E. Mitochondrial NAD-Dependent Methylenetetrahydrofolate Dehydrogenase-Methenyltetrahydrofolate Cyclohydrolase Is Essential for Embryonic Development. *Mol. Cell. Biol.* **2002**, *22*, 4158–4166. [CrossRef]
165. MacFarlane, A.J.; Perry, C.A.; Girnary, H.H.; Gao, D.; Allen, R.H.; Stabler, S.P.; Shane, B.; Stover, P.J. *Mthfd1* Is an Essential Gene in Mice and Alters Biomarkers of Impaired One-Carbon Metabolism. *J. Biol. Chem.* **2009**, *284*, 1533–1539. [CrossRef]
166. Watkins, D.; Schwartzentruber, J.A.; Ganesh, J.; Orange, J.S.; Kaplan, B.S.; Nunez, L.D.; Majewski, J.; Rosenblatt, D.S. Novel Inborn Error of Folate Metabolism: Identification by Exome Capture and Sequencing of Mutations in the *MTHFD1* Gene in a Single Proband. *J. Med. Genet.* **2011**, *48*, 590–592. [CrossRef]
167. Keller, M.D.; Ganesh, J.; Heltzer, M.; Paessler, M.; Bergqvist, A.G.C.; Baluarte, H.J.; Watkins, D.; Rosenblatt, D.S.; Orange, J.S. Severe Combined Immunodeficiency Resulting from Mutations in *MTHFD1*. *Pediatrics* **2013**, *131*, e629–e634. [CrossRef]
168. Burda, P.; Kuster, A.; Hjalmarson, O.; Suormala, T.; Bürer, C.; Lutz, S.; Roussey, G.; Christa, L.; Asin-Cayuela, J.; Kollberg, G.; et al. Characterization and Review of MTHFD1 Deficiency: Four New Patients, Cellular Delineation and Response to Folic and Folinic Acid Treatment. *J. Inherit. Metab. Dis.* **2015**, *38*, 863–872. [CrossRef]
169. Anguera, M.C.; Field, M.S.; Perry, C.; Ghandour, H.; Chiang, E.-P.; Selhub, J.; Shane, B.; Stover, P.J. Regulation of Folate-Mediated One-Carbon Metabolism by 10-Formyltetrahydrofolate Dehydrogenase. *J. Biol. Chem.* **2006**, *281*, 18335–18342. [CrossRef]
170. Strickland, K.C.; Krupenko, N.I.; Dubard, M.E.; Hu, C.J.; Tsybovsky, Y.; Krupenko, S.A. Enzymatic Properties of ALDH1L2, a Mitochondrial 10-Formyltetrahydrofolate Dehydrogenase. *Chem. Biol. Interact.* **2011**, *191*, 129–136. [CrossRef] [PubMed]
171. Sharma, J.; Krupenko, N.I.; Krupenko, S.A. Impact of Aldh1l1 Knockout on Metabolic Phenotype in Mouse Liver. *FASEB J.* **2019**, *33*, lb249. [CrossRef]
172. Sarret, C.; Ashkavand, Z.; Paules, E.; Dorboz, I.; Pediaditakis, P.; Sumner, S.; Eymard-Pierre, E.; Francannet, C.; Krupenko, N.I.; Boespflug-Tanguy, O.; et al. Deleterious Mutations in *ALDH1L2* Suggest a Novel Cause for Neuro-Ichthyotic Syndrome. *NPJ Genom. Med.* **2019**, *4*, 17. [CrossRef] [PubMed]
173. Krupenko, N.I.; Sharma, J.; Pediaditakis, P.; Helke, K.L.; Hall, M.S.; Du, X.; Sumner, S.; Krupenko, S.A. Aldh1l2 Knockout Mouse Metabolomics Links the Loss of the Mitochondrial Folate Enzyme to Deregulation of a Lipid Metabolism Observed in Rare Human Disorder. *Hum. Genom.* **2020**, *14*, 41. [CrossRef] [PubMed]

biomolecules

Article

Elevated Levels of Mislocalised, Constitutive Ras Signalling Can Drive Quiescence by Uncoupling Cell-Cycle Regulation from Metabolic Homeostasis

Elliot Piper-Brown [†], Fiona Dresel [†], Eman Badr and Campbell W. Gourlay *

Kent Fungal Group, School of Biosciences, University of Kent, Canterbury CT2 7NZ, UK
* Correspondence: c.w.gourlay@kent.ac.uk
[†] These authors contributed equally to this work.

Abstract: The small GTPase Ras plays an important role in connecting external and internal signalling cues to cell fate in eukaryotic cells. As such, the loss of RAS regulation, localisation, or expression level can drive changes in cell behaviour and fate. Post-translational modifications and expression levels are crucial to ensure Ras localisation, regulation, function, and cell fate, exemplified by RAS mutations and gene duplications that are common in many cancers. Here, we reveal that excessive production of yeast Ras2, in which the phosphorylation-regulated serine at position 225 is replaced with alanine or glutamate, leads to its mislocalisation and constitutive activation. Rather than inducing cell death, as has been widely reported to be a consequence of constitutive Ras2 signalling in yeast, the overexpression of $RAS2^{S225A}$ or $RAS2^{S225E}$ alleles leads to slow growth, a loss of respiration, reduced stress response, and a state of quiescence. These effects are mediated via cAMP/PKA signalling and transcriptional changes, suggesting that quiescence is promoted by an uncoupling of cell-cycle regulation from metabolic homeostasis. The quiescent cell fate induced by the overexpression of $RAS2^{S225A}$ or $RAS2^{S225E}$ could be rescued by the deletion of *CUP9*, a suppressor of the dipeptide transporter Ptr2, or the addition of peptone, implying that a loss of metabolic control, or a failure to pass a metabolic checkpoint, is central to this altered cell fate. Our data suggest that the combination of an increased *RAS2* copy number and a dominant active mutation that leads to its mislocalisation can result in growth arrest and add weight to the possibility that approaches to retarget RAS signalling could be employed to develop new therapies.

Keywords: yeast signalling; Ras; metabolism; yquiescence; cell fate

Citation: Piper-Brown, E.; Dresel, F.; Badr, E.; Gourlay, C.W. Elevated Levels of Mislocalised, Constitutive Ras Signalling Can Drive Quiescence by Uncoupling Cell-Cycle Regulation from Metabolic Homeostasis. *Biomolecules* **2023**, *13*, 1619. https://doi.org/10.3390/biom13111619

Academic Editor: Jürg Bähler

Received: 13 October 2023
Revised: 28 October 2023
Accepted: 30 October 2023
Published: 6 November 2023

1. Introduction

The small guanosine triphosphate hydrolases (GTPases) of the Ras superfamily function as molecular switches within cells to convey extracellular stimuli to intracellular effectors. *S. cerevisiae* carries two Ras genes, *Ras1* and *Ras2*, which share significant homology with the Ras proteins found in mammals [1]. Indeed, yeast cells lacking endogenous Ras function can be rescued by the expression of mammalian H- or N-Ras [2]. This simple yeast has therefore been used to explore fundamental principles of Ras protein function in eukaryotes. In addition to its well-known role as an oncogene within mammalian systems [3], Ras proteins are important in regulating the virulence of several human and plant fungal pathogens [4]. Although no strategies to manipulate Ras proteins directly have emerged as therapeutic or antifungal options to date, this possibility remains enticing and an active area of research.

Post-translational modifications (PTMs) can regulate Ras's subcellular localisation, affecting the interaction of Ras with specific classes of effectors and regulators [5]. Processing of the CAAX motif located at the C-terminal domain and the subsequent palmitoylation of Ras are vital for endomembrane and plasma membrane targeting [6,7]. Phosphorylation has also been shown to be important in the regulation of Ras's localisation and may offer a new approach to targeting oncogenic Ras within cancer therapy [8]. In *S. cerevisiae*, the two Ras proteins, Ras1 and Ras2, have also been shown to rely on phosphorylation to direct localisation and activity [9], and several putative serine phosphorylation sites have been identified that may play important roles. For example in studies where Ras2 serine 214 was replaced with alanine, cells showed reduced glycogen accumulation and increased activity of the canonical downstream cAMP/PKA pathway that connects Ras signalling to cellular responses in yeast [9,10].

Ras proteins have been shown to localise to several compartments of the cell. For example, in *S. cerevisiae*, Ras2's association with the ER has been shown to be modulated by the Ras inhibitor 1 (Eri1) [11,12]. Furthermore, Ras2 has been shown to accumulate at mitochondrial membranes [13]. Ras accumulation was observed at the mitochondrial compartment in strains lacking either *WHI2*, a phosphatase activator that is important in stress responses, [14] or *COX4*, which is essential for cytochrome c oxidase assembly and oxidative phosphorylation [15], where constitutive signalling promotes ROS production and cell death. Ras2 has also been reported to localise to the nucleus when grown in the presence of glucose [16]. In each case, changes in Ras2 localisation led to dramatic changes in downstream signalling and cell fate, suggesting that spatial and temporal control are crucial regulatory factors in yeast cells.

Here, we explore the outcomes of increasing the level of expression of Ras2 within the context of both its mislocalisation and inappropriate activation. We report that the introduction of mutations at serine position 225 that interfere with its phosphorylation status lead to its mislocalisation and the cAMP/PKA-dependent uncoupling of cell-cycle regulation from metabolic control when overexpressed. These effects could be rescued by the deletion of *CUP9*, which represses the dipeptide transporter *PTR2* and elevates the levels of a number of metabolite transporters, suggesting that Ras^{ser225}-induced signalling effects are rooted in a loss of metabolic homeostasis. Our findings illustrate the importance of both appropriate expression levels and phosphorylation control of Ras2 within the coordination of cellular homeostasis and cell fate in *S. cerevisiae*. The finding that the mislocalisation of a constitutively active form of Ras can drive a quiescent state in yeast has implications for possible antifungal approaches that could target Ras activity outside of its "un-druggable" [17] GTPase regulatory domain.

2. Materials and Methods

2.1. Yeast Growth and Manipulation

All experiments were carried out using the wild-type BY4741 (*MATa hisΔ1 leu2Δ0 met15Δ0 ura3Δ0*) grown at 30 °C in either YPD or in defined minimal media (Formedium) lacking uracil or leucine to allow for plasmid selection and maintenance. The strain lacking *CUP9* (MATa his3Δ1 leu2Δ0 met15Δ0 ura3Δ0 Δcup9::KANMX) was obtained from the yeast deletion collection (Open Biosystems). Cell growth was measured using a BMG LABTECH SPECTROstar Nano automatic plate reader. The plasmid used to overexpress *PDE2* has been described previously [18].

2.2. Site-Directed Mutagenesis and Ras Expression Plasmids

The QuikChange Lightning Site-Directed Mutagenesis Kit (Agilent, Cheshire, UK) was used to perform mutagenesis of the RAS2 gene cloned within the Gateway entry plasmid PDONR221 RAS2 KAN (Addgene, Watertown, MA, USA). Primers were constructed using the QuikChange Primer Design software (http://www.genomics.agilent.com/primerDesignProgram.jsp) to introduce mutations encoding a change from serine position 225 of Ras2 to either alanine (Ras2S225A forward primer 5′-ggcattcacgacagtt gtggcactgttcacattttaccg-3′ and reverse primer 5′-cggtaaaaatgtgaacagtgccacaactgtcgtgaatgcc-3′) or glutamate (Ras2S225E forward primer 5′-cattcacgacagttgtctcactgttcacattttaccgttggcag cattg-3′ and reverse primer 5′-caatgctgccaacggtaaaaatgtgaacagtgagacaactgtcgtgaatg-3′). The mutations were verified by Sanger sequencing. Low- and high-copy-expression plasmids expressing Ras2, Ras2S225A, or Ras2S225E were generated using a gateway

LR cloning reaction (Thermo Fisher, Leicestershire, UK) using the destination vectors pAG416GPD-EGFP-ccdB and pAG426GPD-EGFP-ccdB.

2.3. GFP-Atg8 Autophagy Assay

An overnight culture was grown in defined minimal medium lacking uracil and containing 2% glucose at 30 °C with shaking at 200 rpm. The overnight culture was diluted to an OD_{600} of 0.1 in 5 mL of fresh minimal medium lacking uracil and containing 2% glucose. Cells were incubated at 30 °C with shaking at 200 rpm for 24 h. To induce autophagy via nitrogen starvation, cells were incubated at 30 °C with 200 rpm shaking for 6 h in nitrogen starvation medium (2% glucose, 0.675% yeast nitrogen base without amino acids and ammonium sulphate, 0.193% synthetic dropout media supplement lacking uracil, Formedium). Total protein was extracted from cells before and after the induction of autophagy and subjected to Western blotting using an anti-GFP antibody.

2.4. Western Blotting

The primary antibody used for Ras2 detection, goat anti-Ras2 polyclonal IgG, was purchased from Santa Cruz Biotechnology (Santa Cruz Biotechnology, Inc. Bergheimer Str. 89-2, 69115 Heidelberg, Germany) and was used at a dilution of 1/1000. The secondary antibody for Ras2 detection was anti-sheep IgG HRP (Sigma, Hertfordshire, UK, catalogue number A3415) and was used at a dilution of 1/5000. The primary antibody used for the detection of Pgk1p, rabbit anti-Pgk1p, was a kind gift from Professor Mick Tuite of the University of Kent and was used at a 1/10,000 dilution. For the detection of GFP-Atg8, a mouse anti-GFP antibody (Sigma, 1181446000) at a dilution of 1/1000 was used, with secondary antibody detection with anti-mouse IgG HRP (Sigma, A4416) at a dilution of 1/5000. Where quantified for qualitative comparison, protein band intensities were assessed using the densitometry function of the open-source ImageJ platform (https://imagej.net/ij/), and the bands were normalised to the appropriate PGK band within each lane to take account of loading differences.

2.5. Chronological Ageing Assay

Cells were grown overnight in 5 mL of minimal medium lacking uracil and containing 2% glucose, and then they were sub-cultured to an OD_{600} of 0.1 in 10 mL of fresh minimal medium lacking uracil and containing 2% glucose in 100 mL conical flasks. The samples were incubated at 30 °C, with shaking at 200 rpm.

After 24 h of incubation, an aliquot of the culture was taken to calculate the colony-forming units and for flow cytometry analysis to measure the levels of propidium iodide (necrotic cells) and dihydroethidium fluorescence (ROS). Aliquots of cells were taken from the same culture sequentially at 24 h intervals for 6 days, and colony-forming unit counts and flow cytometry analysis were performed. The final measurement was taken after 12 days of sample incubation. To check for the presence of necrotic cells, propidium iodide was added at a concentration of 0.4 μM for 30 s before being placed into a BD Accuri flow cytometer and assessed for excitation at 535 nm and emission at 617 nm. To assess ROS levels, the dihydroethidium (DHE) dye (Invitrogen, Waltham, MA, USA, D23107) was used for the detection of superoxide radicals; 2×10^6 cells were resuspended in PBS containing 10 μM DHE and incubated for 15 min before detection with a BD Accuri flow cytometer (500 nm and an emission peak at 582 nm). For the assessment of PI and DHE staining, a baseline was set using an unlabelled control, and all experiments were repeated in biological triplicate.

2.6. CFU and Viability Assays

Cells from an overnight culture were diluted to an OD_{600} of 0.1 and incubated at 30 °C for 24 h. The cells were then diluted to 2×10^3 cells/mL, and 300 cells were plated onto the appropriate medium in technical and biological triplicates. The percentage viability was determined by the number of arising colonies.

2.7. Fluorescence Microscopy

For the visualisation of GFP fluorescent cells (excitation/emission 488/512 nm), an Olympus 1X81 inverted microscope with a Cool LED pE4000 illumination system and an Andor Zyla 4.2 PLUS sCMOS camera was used. The acquisition software used to obtain the images was Micro-Manger, version 1.4.22. The experiments were conducted in biological triplicate, and representative images are presented.

2.8. High-Resolution Respirometry

The Oroboros O2K Oxygraph High-Resolution Respirometer was used to determine the consumption of O_2 in the cell suspensions. Datlab 4 software was used for the analysis and acquisition of data generated by the respirometer. Cell suspensions were diluted to a cell density of 10^6 cells/mL and a routine respiration value (routine) was obtained, followed by the administration of the following respiratory chain inhibitors, in order: 0.2 mM triethiltyn bromide (TET, Sigma-Aldrich, UK), a complex V inhibitor used to assess respiration levels not linked to proton movement across the ATP synthase, referred to as LEAK respiration; 12 µM carbonylcyanidep-trifluoromethoxyphenylhydrazone (FCCP, Fluka, UK), a proton ionophore that reveals maximum (or uncoupled) respiration, referred to as ETS respiration; 2 µM antimycin A (AntA, Sigma-Aldrich), a complex III inhibitor that shows NMT or non-mitochondrial respiration. All experiments were performed in biological triplicate, as indicated.

2.9. RNA Sequencing and Gene Set Enrichment Analysis

Total RNA was extracted from 5×10^7 yeast cells grown in biological triplicate at 30 °C to log phase in SD-LEU medium containing 2% glucose using the E.Z.N.A. Yeast RNA Kit Spin protocol (Omega Bio-tek, London, UK). An Illumina NextSeq 500 platform was utilised, producing 75 bp single-end reads (Novogene, Cambridge, UK). For each library, 20–30 M reads were generated. The analysis of RNA-Seq samples was performed using the Galaxy web platform (www.usegalaxy.org). The quality of the RNA sequencing reads was checked using FastQC v0.11.5 with default settings. Low-quality ends (Phred score < 20) and any adaptor sequences were trimmed using TrimGalore! v0.4.3. Reads shorter than 40 bp after trimming were not progressed for further analysis. After trimming, the quality was checked again using FastQC v0.11.5 to ensure correct trimming. Processed reads were aligned with the reference *S. cerevisiae* genome S288C version R64-2-1_20150113 (SacCer3) using HISAT2 v2.1.0 with single-end reads and reverse-strand settings. After alignment, the number of mapped reads was determined using the featureCounts plugin v1.6.3 with default settings. Reads aligning to multiple positions or overlapping more than one gene were discarded, counting only reads that mapped unambiguously to a single gene. Differential gene expression analysis between conditions was performed using DESeq2 v1.18.1 with default settings. To identify the cellular pathways containing differentially expressed genes, we conducted gene set enrichment analysis (GSEA; Broad Institute, Cambridge, MA, USA) [19]. To conduct GSEA, a list of significantly ($q \leq 0.05$) differentially expressed genes, ranked from the most upregulated to the most downregulated, was uploaded into the GSEA database and compared to a gene set database (S288C version R64-2-1_20150113_features gtf annotation file).

3. Results

We wished to determine novel phosphorylation events that may influence Ras2's localisation in *S. cerevisiae*. As a starting point, we made use of the PhosphoPep version 2.0 project database (www.phosphopep.org/index.php) [20]. This database allows users to search mass spectrometry data that compare the presence and levels of phosphorylated peptides between the proteomes of wild-type and gene-knockout yeast strains. The use of this database also allows for the prediction of regulatory kinases and phosphatases based on known consensus sequences. Using Ras2 as the search term and default parameters, the database's output suggested that the loss of several enzymes led to changes in Ras2-derived phosphorylated peptides. The loss of kinases Mck1, Cka1, Ypk1, Pkp1, Bud32, Ssn33, Ctk1, and Pho85 or phosphatases Psr1, Psr2, and Sit4 led to changes in phosphorylated Ras2 peptide levels (Table S1). Upon examination of Ras2's localisation in strains deleted for each enzyme predicted to change Ras's phosphorylation status, only cells lacking *SSN3*, *SIT4*, or *PHO85* led to changes in GFP-Ras2 localisation (Supplementary Figure S1). Interestingly, the loss of either *SSN3* or *SIT4* led to an increase in the number of Ras2 peptides phosphorylated at the same residue: serine 225 (Table S1). To investigate whether the phosphorylation of Ras2 at serine 225 is important for its localisation, we replaced the amino acid with either glutamate (*RAS2^{S225E}*, phospho-mimic) or alanine (*RAS2^{S225A}*, non-phosphorylatable) via site-directed mutagenesis. We also cloned *RAS2*, *RAS2^{S225A}*, and *RAS2^{S225E}* into both low- and high-copy-expression plasmids driven by a constitutive GPD promoter, to investigate the effects of gene expression levels, and introduced these into wild-type cells.

To validate the expression levels of *RAS2*, *RAS2^{S225A}*, and *RAS2^{S225E}*, we conducted Western blots (Figure 1A, Supplementary Figures S2 and S3). Low-copy plasmids led to a tenfold increase in the expression level of *RAS2* and a fifteenfold increase in the levels of *RAS2^{S225A}* and *RAS2^{S225E}* (Figure 1A). The expression level was further increased to twenty-fivefold in strains expressing *RAS2*, *RAS2^{S225A}*, and *RAS2^{S225E}* from a multi-copy plasmid (Figure 1A). Despite the high levels of expression of *RAS2*, *RAS2^{S225A}*, or *RAS2^{S225E}* from a low-copy plasmid, we did not observe any changes in growth (Figure 1B). In contrast, the further increase in expression observed from the introduction of a high-copy plasmid led to a significant reduction in growth, but only in cells expressing *RAS2^{S225A}* or *RAS2^{S225E}* (Figure 1C). The viability of wild-type cells containing an empty high-copy plasmid grown to the stationary phase in the minimal selective medium used was around 50% when assayed at 24 h (Figure 1D), which represents the stationary phase of growth (Figure 1C). The overexpression of *RAS2* led to a significant increase in viability when compared to the wild-type control; however, viability was significantly reduced by the overexpression of *RAS2^{S225A}* or *RAS2^{S225E}* (Figure 1D). These data suggest that the modification of Ras2 at S225 can have a dominant negative impact on viability, but that the expression levels need to exceed a regulatory threshold to achieve this.

To investigate whether the expression of *RAS2^{S225A}* or *RAS2^{S225E}* influenced Ras2's localisation and activity, we made use of a construct that expresses the Ras-binding domain of human Raf1 fused to GFP [14]. As the RBD domain binds to Ras in its active GTP-bound state, this probe can give an account of the localisation of active Ras in living cells. The localisation of active Ras was analysed during the logarithmic and stationary phases of growth in cells overexpressing *RAS2*, *RAS2^{S225A}*, and *RAS2^{S225E}* from a high-copy plasmid (Figure 2A). During logarithmic growth, the wild-type controls and those overexpressing Ras2 showed localisation of GFP-RBD and, thus, active Ras at the plasma membrane and within the nucleus, as has been previously reported [14] (Figure 2A). In contrast, cells expressing *RAS2^{S225A}* or *RAS2^{S225E}* showed a strong GFP-RBD signal at the nuclear envelope (Figure 2A). During the stationary phase the RBD-GFP probe signal was seen as a diffuse cytoplasmic signal in wild-type and Ras2-overexpressing cells, as has been previously reported, reflecting the reduction in Ras activity during this phase of growth [14] (Figure 2A). In contrast, wild-type cells overexpressing either *RAS2^{S225A}* or *RAS2^{S225E}* maintained the nuclear envelope GFP-RBD signal, indicating the sustained localisation of active Ras at this location (Figure 2A). Intracellular foci were also observed in *RAS2^{S225A}*- and *RAS2^{S225E}*-expressing strains within the stationary phase (Figure 2A). These data suggest that the overexpression of *RAS2^{S225A}* or *RAS2^{S225E}* leads to constitutive activation, most notably at the nuclear envelope, throughout growth.

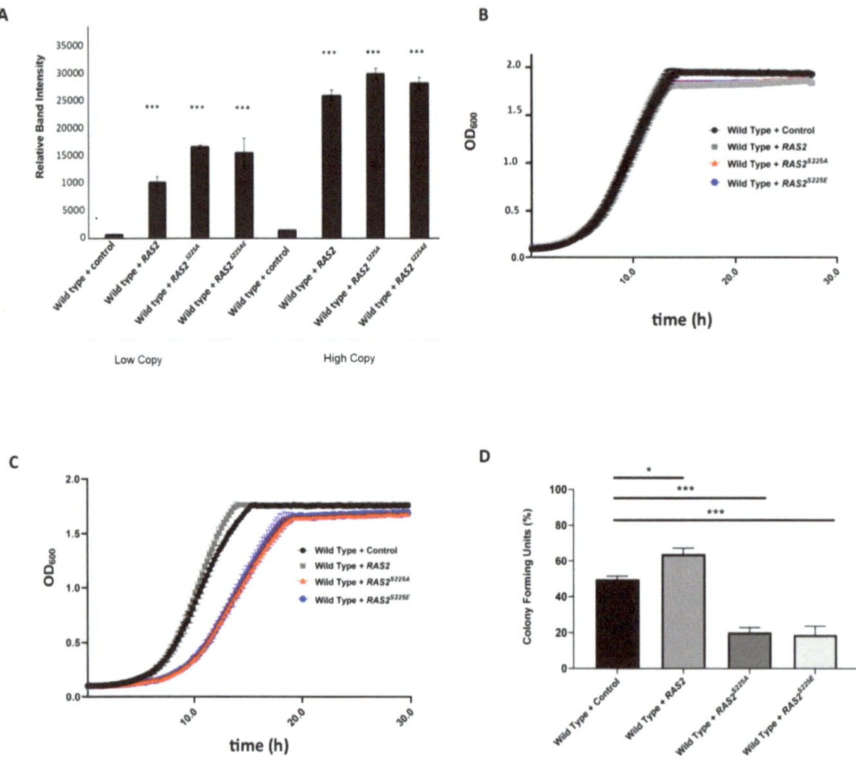

Figure 1. Cells expressing *RAS2*, *RAS2^{S225A}*, and *RAS2^{S225E}* from either a low-copy (CEN) or high-copy (2 μ) plasmid were grown in selective SD-URA minimal medium for 24 h at 30 °C before the total protein was extracted and probed by Western blotting with an anti-Ras2 and anti-Pgk1 antibody (loading control). Graphs represent the normalised relative band intensity from three biological replicates (**A**). Growth analysis of wild-type *S. cerevisiae* cells overexpressing *RAS2*, *RAS2^{S225A}*, *RAS2^{S225E}*, or an empty plasmid control (EV) from either a low-copy (CEN) (**B**) or high-copy (2 μ) (**C**) plasmid, representing an average of three biological replicates. Colony-forming unit assay of cells overexpressing *RAS2*, *RAS2^{S225A}*, *RAS2^{S225E}*, or an empty plasmid control grown in SD-URA medium (**D**). A one-way ANOVA using Dunnett's multiple comparison test was used to determine statistical significance; * $p \leq 0.05$, *** $p \leq 0.001$. Error bars represent standard deviations.

The constitutive activation of Ras2 is often associated with a reduced ability to respond to oxidative stress. To test this, we grew strains in the presence of either hydrogen peroxide or copper sulphate, both of which led to a significant inhibition of growth in *RAS2^{S225A}*- and *RAS2^{S225E}*-expressing strains when compared to wild-type or *RAS2*-overexpressing strains (Figure 2B,C).

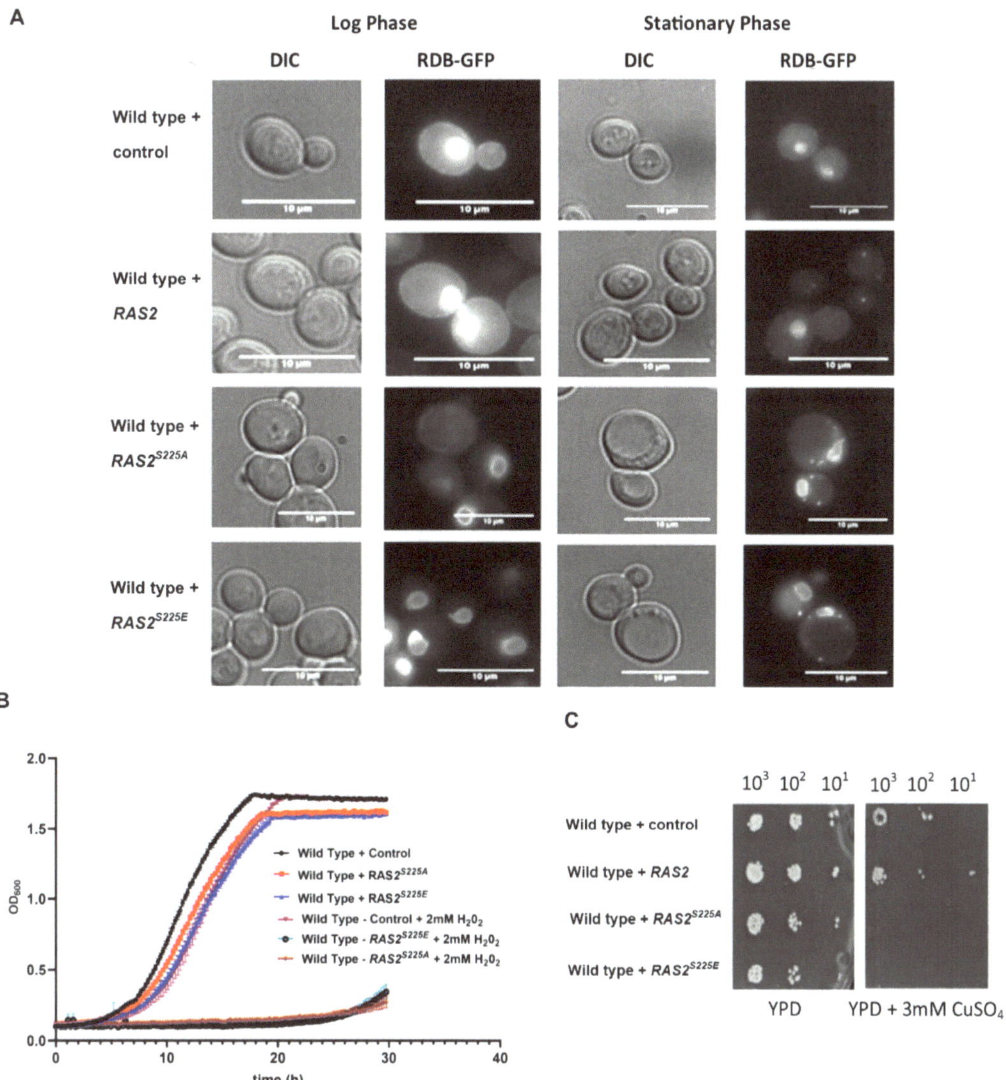

Figure 2. Active Ras was visualised in cells overexpressing *RAS2*, *RAS2*^{S225A}, *RAS2*^{S225E}, or a control containing an empty plasmid using a 3xGFP-RBD probe during the logarithmic and stationary phases of growth. Cells were cultured in SD-URA/-LEU growth media. The experiment was repeated three times, and a representative dataset is shown. Scale bar—10 μm (**A**). Growth of wild-type cells overexpressing *RAS2*, *RAS2*^{S225A}, *RAS2*^{S225E}, or containing an empty plasmid control was carried out in SD-URA or SD-URA + 2 mM H_2O_2 media; n = 3, error bars represent the standard deviation (**B**). Wild-type cells overexpressing *RAS2*, *RAS2*^{S225A}, *RAS2*^{S225E}, or containing an empty plasmid control were serially diluted from 2×10^6/mL to 2×10^3/mL and plated onto SD-URA plates supplemented with increasing concentrations of copper sulphate. This experiment was completed three times, and a representative result is shown (**C**).

3.1. Overexpression of RAS2$^{S225A/E}$ Promotes a Quiescent Cell Fate

Changes in Ras's localisation and activity often lead to an alternate cell fate. We therefore sought to determine the cell fate associated with the loss of viability observed upon overexpression of $RAS2^{S225A}$ or $RAS2^{S225E}$. Loss of viability within a culture can occur as a result of cell death, which can be determined by the presence of cell-death-associated markers such as high levels of reactive oxygen species (ROS) or a loss of plasma membrane integrity (necrosis), as assessed by the uptake of the impermeant dye propidium iodide. The presence of high ROS containing and necrotic cells increases in a culture that is left to grow and deplete nutrients from the medium over time; this is commonly referred to as chronological ageing. Despite the overexpression of $RAS2^{S225A}$ or $RAS2^{S225E}$ leading to an increase in sensitivity to oxidative stress and exhibiting a loss of viability, we did not observe an increase in ROS production or propidium iodide uptake during chronological ageing when compared to the wild-type (Figure 3A,B). In addition, autophagy, a key process that is activated upon nutrient depletion and necessary for survival during chronological ageing, remained fully responsive, as assessed by the hydrolysis of GTP-ATG8, when $RAS2$, $RAS2^{S225A}$, or $RAS2^{S225E}$ was overexpressed (Supplementary Figure S4). However, the loss of viability associated with $RAS2^{S225A}$ or $RAS2^{S225E}$ overexpression was linked to a dramatic reduction in mitochondrial respiration when assayed at 24 h of growth (Figure 3C).

A lack of cell death markers and reduced respiration profile led us to investigate whether the loss of viability associated with $RAS2^{S225A}$ or $RAS2^{S225E}$ overexpression was in fact a consequence of quiescence. This assumption was supported by the fact that $RAS2^{S225A}$- or $RAS2^{S225E}$-overexpressing cells grown to 24 h in selective minimal medium could be revived by plating them onto an alternative rich medium (YPD) (Figure 3D). The cell viability was also increased when wild-type control cells were plated onto rich YPD medium (Figure 3D). From this, we hypothesised that nutritional limitation, bestowed by the selective media used in our experiments, led to a quiescent state in a proportion of wild-type cells, which was further exacerbated by the overexpression of $RAS2^{S225A}$ or $RAS2^{S225E}$. Our findings also suggest that Ras2's localisation and signalling play important roles in mediating the metabolic plasticity required for growth under minimal medium conditions. Ras2 mediates its signalling by controlling the activity of adenylate cyclase (Cyr1), which generates cAMP, which, in turn, activates protein kinase A (PKA). Activated PKA then phosphorylates a number of downstream targets that control a range of cellular responses. The increase in quiescence observed with the overexpression of $RAS2^{S225A}$ or $RAS2^{S225E}$ could be restored to wild-type levels by overexpressing the high affinity cAMP phosphodiesterase, *PDE2*, which reduces cAMP levels and PKA signalling (Figure 3E). In addition, the deletion of *PDE2*, which elevates cAMP levels, led to a further reduction in viability in cells expressing $RAS2^{S225A}$ or $RAS2^{S225E}$ (Figure 3F). These findings suggest that the overexpression of $RAS2^{S225A}$ or $RAS2^{S225E}$ promotes quiescence via cAMP/PKA signalling.

3.2. RAS2$^{S225A/E}$ Overexpression Uncouples Cell-Cycle Regulation from Metabolic Homeostasis

We conducted RNA sequencing to help determine the cellular state underpinning $RAS2^{S225A}$- and $RAS2^{S225E}$-induced quiescence. To identify the cellular pathways with which differentially expressed genes were associated, we conducted a gene set enrichment analysis (GSEA; Broad Institute) [19]. Significantly ($q \leq 0.05$) differentially expressed genes, ranked from the most upregulated to the most downregulated, from $RAS2^{S225A}$- or $RAS2^{S225E}$-expressing cells were compared to wild-type controls and to the GSEA database. GSEA functions to identify whether the members of each gene set are randomly distributed or significantly overrepresented (enriched) within a Gene Ontology term. using a normalised enrichment score (NES). Our data showed that overexpression of either $RAS2^{S225A}$ or $RAS2^{S225E}$ led to changes in precisely the same gene sets. We observed significant upregulation in pathways involved in growth, including ribosomal biogenesis, transcription, translation, glucose transport, and central metabolism—hallmarks of actively dividing cells

(Figure 4A,B). However, we also observed a significant downregulation in GSEA gene sets in $RAS2^{S225A}$- and $RAS2^{S225E}$-expressing cells linked to cell growth, including cytoskeleton, microtubule, and kinetochore dynamics, nuclear division, chromosome condensation and segregation, cell-cycle phase transition, mitosis, and transcription factors associated with cell-cycle phase progression. (Figure 4A,C). These data suggest that the overexpression of $RAS2^{S225A}$ or $RAS2^{S225E}$ leads to a loss of coordination between processes that are essential for growth and offers a reasonable explanation as to the observed increase in the loss of viability associated with quiescence.

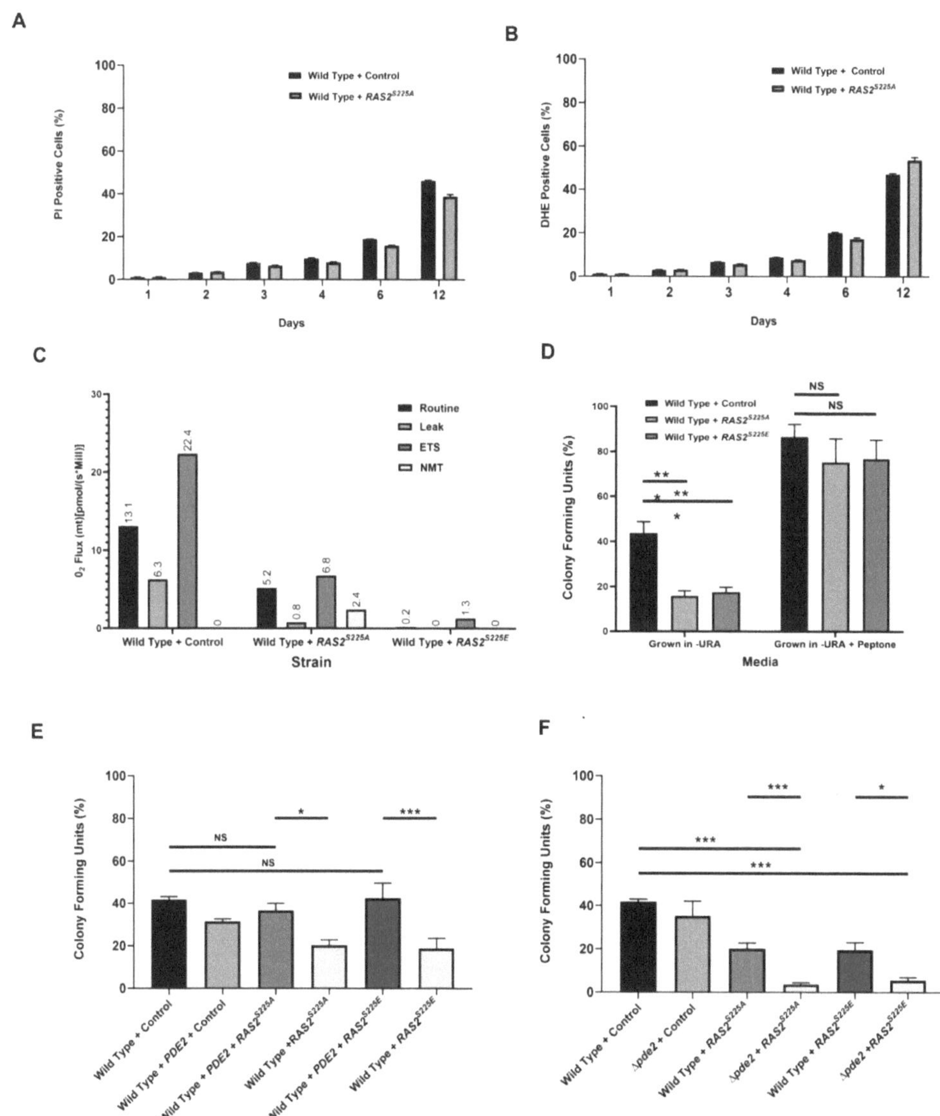

Figure 3. Wild-type cells overexpressing $RAS2^{S225A}$ or an empty plasmid control were grown in SD-URA, and necrosis (PI uptake) (**A**) and ROS (DHE) (**B**) measurements were taken over a 12-day period of continuous incubation. The data displayed are the average of three technical repeats, and the error

bars represent the standard deviation. A bar chart showing the routine, leak, ETS, and NMT O_2 flux values for wild-type cells overexpressing $RAS2^{S225A}$ or $RAS2^{S225E}$, or containing an empty plasmid backbone control. The experiment was conducted in triplicate, and a representative dataset from one experiment is shown (**C**). A colony-forming unit assay of wild-type cells overexpressing $RAS2^{S225A}$ or $RAS2^{S225E}$, or containing an empty plasmid control, grown in SD-URA medium for 24 h at 30 °C, was conducted and plated on either SD-URA or SD-URA + peptone (**D**). A colony-forming efficiency assay of wild-type cells overexpressing $RAS2^{S225A}$, $RAS2^{S225E}$, or an empty plasmid backbone co-expressed with *PDE2* (**E**), or in a strain lacking *PDE2* (**F**). The data presented are the average of three biological replicates, and the error bars represent the standard deviation. A one-way ANOVA using Tukey's multiple comparison test was used to determine statistical significance. Nonsignificant = NS, * = adjusted *p*-value ≤ 0.01, ** adjusted *p*-value of 0.05 and *** = adjusted *p* value ≤ 0.001.

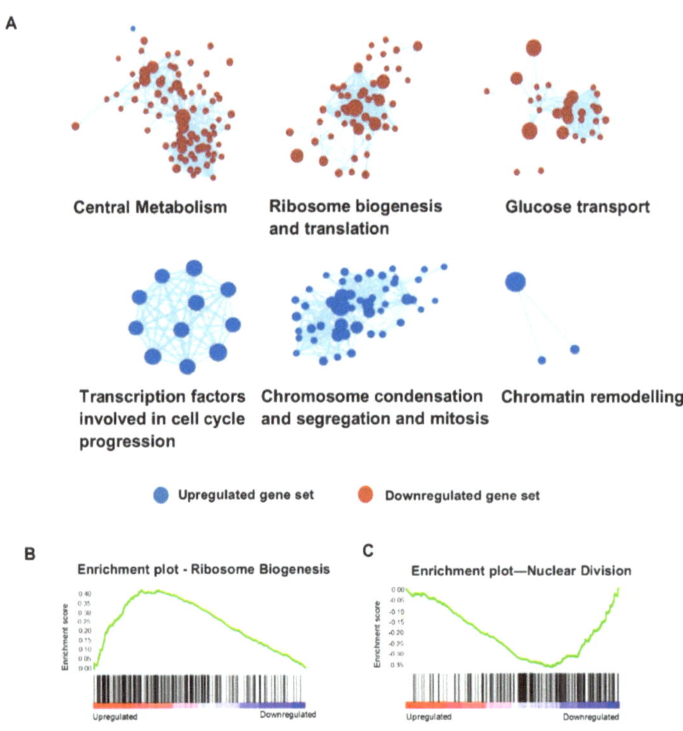

Figure 4. (**A**) Global gene expression changes in wild-type cells overexpressing $RAS2^{S225A}$ when compared to a wild-type control. DESeq2 was used to compare gene expression in wild-type cells overexpressing RAS2S225A to a wild-type control; in total, there were 4133 significantly differentially expressed genes. Gene set cluster maps were created using the Cytoscape plugin, showing the most upregulated and downregulated gene sets, as determined by GSEA analysis, along with their cellular functions; circle size within a cluster represents the change in the expression level of a single gene. Representative gene sets shown to be upregulated (**B**) or downregulated (**C**) upon the overexpression of $RAS2^{S225A}$ when compared to the wild-type control by GSEA. Vertical black lines represent individual genes in the significantly differentially expressed ranked gene list, from upregulated (**left**) to downregulated (**right**). An increase in the enrichment score is seen if there are many genes towards the beginning of the ranked list (upregulated) in the gene set.

3.3. Loss of the Repressor Cup9 Rescues the Effects of Ras2$^{ser225A/E}$ Overexpression

Given that the addition of a rich nutrient source was able to reactivate quiescent cells overexpressing $RAS2^{S225A}$ or $RAS2^{S225E}$, we wished to determine the nutritional queue(s) responsible. The addition of additional glucose did not result in the relief of quiescence when $RAS2^{S225A}$ or $RAS2^{S225E}$ cells grown overnight were re-inoculated into fresh media (Supplementary Figure S5). As di/tripeptides are among the principal components of rich yeast growth media, we tested whether deletion of the repressor Cup9, which increases expression of the di/tripeptide transporter Ptr2, would allow $RAS2^{S225A}$ or $RAS2^{S225E}$ cells to re-enter cell growth when placed onto fresh minimal medium. The deletion of *CUP9* restored growth in cells overexpressing $RAS2^{S225A}$ or $RAS2^{S225E}$ to wild-type levels (Figure 5A). Viability was also restored in cells overexpressing $RAS2^{S225A}$, with a slight increase in cells overexpressing $RAS2^{S225E}$ when *CUP9* was deleted when compared to the wild-type (Figure 5B). As the overexpression of $RAS2^{S225A}$ or $RAS2^{S225E}$ led to essentially the same phenotypes, we proceeded with further detailed analysis of metabolism using only $RAS2^{S225A}$-overexpressing cells. The loss of *CUP9* alone did not affect respiration (Figure 5C). However, in line with the reversal of quiescence, the loss of respiration observed in $RAS2^{S225A}$-overexpressing cells was reversed upon *CUP9*'s deletion (Figure 5C). To determine the nature of the metabolic changes that underpin this rescue, we compared the transcriptomes of $RAS2^{S225A}$ and $\Delta cup9\ RAS2^{S225A}$ cells by GSEA. The significantly ($q \leq 0.05$) differentially expressed genes were ranked from the most upregulated to the most downregulated and analysed by gene set enrichment analysis. Of the 202 genes that were significantly upregulated comparing $RAS2^{S225A}$ and $\Delta cup9\ RAS2^{S225A}$ cells, we observed that in addition to the well-known target of Cup9 repression, *PTR2*, 40 genes were clustered within the term of transmembrane transporters. These included a number of genes involved in amino acid transport (*BAP2*, *BAP3*, *TAT1*, *ATG22*, *MMP1*, *DIP5*, and *SAM3*) and metal ion uptake (*FRE1*, *FRE7*, *CCC1*, *YOR1*, *PIC2*, and *ARN1*) (Figure 5D, Supplementary Table S2). A broader but uncharacterised role for *CUP9* in regulating metabolism was observed when RNA sequencing data from wild-type and $\Delta cup9$ cells were compared using Gene Ontology within the yeastmine database. This highlighted that genes involved in the processes of pyruvate metabolism, glycolysis, carbohydrate catabolism, organic acid metabolism, and cytoplasmic translation were significantly enriched upon the deletion of *CUP9* (Supplementary Table S3). The deletion of *CUP9* also led to a significant downregulation of genes involved in rRNA modification, with a focus on methylation and sulphur metabolism (Supplementary Table S4).

Overall, our data suggest that the overexpression of $RAS2^{S225A}$ or $RAS2^{S225E}$ alleles results in a quiescent state that is driven by changes in metabolism that do not support entry into or maintenance of an active cell cycle within a minimal nutrition environment. The deletion of *CUP9* appears to de-repress genes involved in the transport of metabolites, facilitating escape from this quiescent state (Figure 6). We also highlight a previously uncharacterised role for Cup9 in the control of genes involved in a range of metabolic processes required for normal growth.

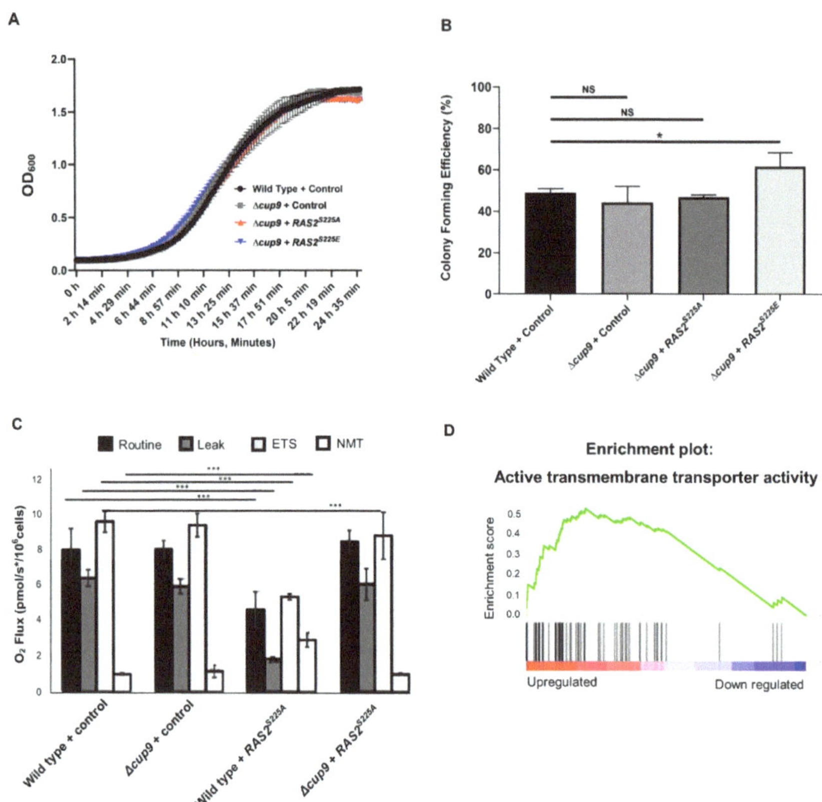

Figure 5. Growth analysis (**A**) and CFU assay (**B**) of wild-type and Δ*cup9* cells overexpressing *RAS2^{S225A}* or *RAS2^{S225E}*, or containing an empty plasmid control. Routine, leak, ETS, and NMT O2 flux values for wild-type and Δ*cup9* cells overexpressing *RAS2^{S225A}* or an empty plasmid backbone control. In each case, the data shown represent an average of three biological repeats, and the error bars indicate the standard deviation. A one-way ANOVA using Tukey's multiple comparison test was used to determine statistical significance; Nonsignificant = NS, * $p \leq 0.01$, *** $p \leq 0.001$ (**C**). A representative GSEA gene set shown to be upregulated upon the overexpression of *RAS2^{S225A}* in a Δ*cup9* background when compared to wild-type cells overexpressing *RAS2^{S225A}* (**D**).

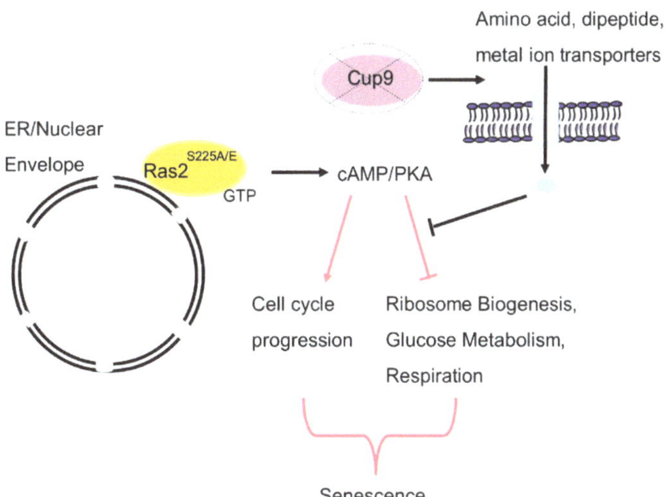

Figure 6. Model depicting a mechanism by which the overexpression of $RAS2^{S225A}$ or $RAS2^{S225E}$ promotes quiescence. The substitution of Ras2 at serine 225 for alanine or glutamate leads to constitutive activation at the nuclear envelope/ER when overexpressed. $RAS2^{S225A/E}$-driven cAMP/PKA signalling from the nuclear envelope/ER, in turn, promotes senescence under conditions of nutritional challenge by uncoupling the control of the expression of cell-cycle control for core metabolic processes. The addition of peptone or deletion of *CUP9*, which leads to the upregulation of a battery of metabolite transporters, can counteract quiescence driven by $RAS2^{S225A/E}$ signalling, potentially by overcoming an essential metabolic checkpoint that is required to re-enter the cell cycle.

4. Discussion

The correct regulation of Ras/cAMP/PKA activity is crucial for the integration of cell growth, cell-cycle progression, and metabolic activity [21–24]. As an example, aberrant activation of the Ras/cAMP/PKA pathway can lead to an incorrect diauxic reprogramming of the cell during nutrient depletion [25]. An increase in cAMP levels resulting from constitutive Ras signalling has also been shown to lead to the inability of cells to regulate carbohydrate storage, oxidative phosphorylation, and stress responses during the stationary phase of growth [25,26]. Decreased cAMP/PKA activity can also lead to reduced growth and entry into G_0 [24,27]. Ras2/cAMP/PKA signalling control in yeast is therefore crucial for the coordination of cell growth and an appropriate response to available nutrition.

We sought to further our knowledge of Ras2 regulation by identifying novel phosphorylation events that are important for its localisation. Our initial findings suggested that serine 225, which may be regulated by Ssn3 and Sit4, represents a putative candidate. However, as we observed the same phenotypes when we overexpressed either $RAS2^{S225A}$ (non-phosphorylatable) or $RAS2^{S225E}$ (phospho-mimetic), our findings do not strictly confirm that phosphorylation/dephosphorylation at serine 225 is responsible for directing localisation. The overexpression of either $RAS2^{S225A}$ or $RAS2^{S225E}$ did lead to mislocalisation to the nuclear envelope, which may suggest perturbation of processing the CAAX motif that is required for the transit of Ras2 to the plasma membrane [6]. Alternatively, this aberrant localisation may involve its interaction with the Ras inhibitor 1 (Eri1), which has been reported to enable contact of GTP-bound Ras2 with the ER via a role within the ER-localised GPI-GnT complex [11,12]. Mislocalised $RAS2^{S225A}$ or $RAS2^{S225E}$ was also found to be constitutively active, suggesting that GAP proteins were either not present or insufficient to regulate GDP/GTP activity under the conditions and expression levels used. Similar observations of constitutive activation have been reported under conditions or genetic backgrounds where Ras2 has been found to be localised to the mitochondrial compartment [14]; however, in these studies, this led to cell death consistent with yeast apoptosis. Overall, this adds to the evidence that the location and activity state of Ras2 are crucial in determining cell fate in yeast. The loss of viability and phenotypes attributed to overexpression and mislocalisation of either $RAS2^{S225A}$ or $RAS2^{S225E}$ could be attributed to enhanced cAMP/PKA activity. However, these effects were only observed above a threshold level of expression, suggesting that *S. cerevisiae* possess significant capacity for Ras regulation. It is tempting to speculate that a similarly high threshold exists in human cells, as RAS gene duplication is common in the manifestation of a number of cancers [28].

The loss in viability observed upon the overexpression of $RAS2^{S225A}$ or $RAS2^{S225E}$ was accompanied by an inability to re-engage in cell division upon the addition of fresh medium. One possibility is that the overexpression of $RAS2^{S225A}$ or $RAS2^{S225E}$ led to the dysregulation of essential processes required for growth. In line with this, we observed upregulation of some pathways required to support growth, such as mRNA translation, alongside simultaneous downregulation of others, such as DNA replication and glucose metabolism. The effects of $RAS2^{S225A}$ or $RAS2^{S225E}$ in coordinating cell-cycle entry could be overcome by the addition of peptone or the deletion of *CUP9*, which, in addition to leading to an increase in the expression of the di/tripeptide transporter *PTR2* [29], also led to the upregulation of a range of metabolite transporters. This suggests that Cup9 may be important in the suppression of a range of metabolic process and requires further investigation. One explanation for this finding is that the deletion of *CUP9* allows $RAS2^{S225A}$- or $RAS2^{S225E}$-overexpressing cells trapped within G_0 following nutrient depletion to overcome an essential metabolic checkpoint that is required for cell-cycle re-entry. In yeast, a range of metabolites are known to influence progression through each stage of the cell cycle, in what has been termed the yeast metabolic cycle [30]. However, to the best of our knowledge, this has not been linked to Cup9's function to date.

Clearly, the quiescence observed in $RAS2^{S225A}$- or $RAS2^{S225E}$-overexpressing cells arises as a combination of nutrient limitation and aberrant Ras signalling. These findings highlight the crucial interplay among environmental adaption, signalling, and cell fate determination that is coordinated by Ras2. This is an emerging area of study that holds promise in the field of cancer research, where new approaches to manipulate metabolic checkpoints are being pursued as options for developing new therapies [31], and also in antifungal development. Fungal-specific domains that exist outside of the highly conserved GTPase domain have been reported and may be targeted to develop new antifungal therapies to combat human fungal pathogens such as *Aspergillus fumigatus* [32] and *Candida albicans* [33]. A recent study identified a synthetic peptide that can prevent Ras from interacting with its GEF, Cdc25, at a *C. albicans*-specific domain in the Ras1 protein [33]. As this interaction was shown to be important in regulating the yeast-to-hyphal switch, a key determinant of virulence in *C. albicans*, this represents an exciting new potential route to Ras-targeted infection control. Overall, the recognition that Ras signalling can be manipulated to reliably alter cell fate outside of the "undrugable" GTPase domain [17] is an attractive approach to developing new methods to control its activity.

Supplementary Materials: The following supporting information can be downloaded at: https://www.mdpi.com/article/10.3390/biom13111619/s1. Figure S1: Active Ras localisation in kinase a phosphatase deletions that lead to changes in phosphoryaltion status of Ras2; Figure S2: Western blot showing the detection of Ras2 in cells overexpressing $RAS2$, $RAS2^{S225A}$, $RAS2^{S225E}$ from a low copy plasmid; Figure S3: Western blot showing the detection of Ras2 in cells overexpressing $RAS2$, $RAS2^{S225A}$, $RAS2^{S225E}$ from a high copy plasmid; Figure S4: Western blot showing the detection of GFP in GFP-ATG8 cells overexpressing $RAS2$, $RAS2^{S225A}$, $RAS2^{S225E}$; Figure S5: Colony forming unit assay in cells overexpressing $RAS2$, $RAS2^{S225A}$, $RAS2^{S225E}$ with additional glucose supplement; Table S1: Phospho pep 2.0 output for kinases and phosphatsases whose deletion led to a significant change in Ras2 peptide phosphorylation levels as detected by mass spectrometry; Table S2: Gene Ontology analysis of upregulated genes clustered to the transmembrane transporters function when transcriptomes of $RAS2^{S225A}$ and $\Delta cup9$ $RAS2^{S225A}$ cells were analysed; Table S3: Upregulated genes clustered to the GO Process term by Yeastmine when transcriptomes of wild type and $\Delta cup9$ cells were analysed; Table S4: Downregulated genes clustered to the GO Process term by Yeastmine when transcriptomes of wild type and Δcup9 cells were analysed.

Author Contributions: E.P.-B. and F.D. contributed equally to the experimental procedures, data analysis and manuscript preparation. E.B. contributed to experimental procedures. C.W.G. conceived and supervised the experiments conducted within the study and contributed to the data analysis and manuscript preparation. All authors have read and agreed to the published version of the manuscript.

Funding: The present work was supported by an iCASE PhD studentship to E.P.-B. (BBSRC iCASE PhD Studentship BB/M015785/1).

Institutional Review Board Statement: Not applicable.

Informed Consent Statement: Not applicable.

Data Availability Statement: Data is available at Mendeley Data, V1, doi: 10.17632/pswx38b2zb.1 or upon request.

Conflicts of Interest: The authors declare that we have no conflicts of interest regarding this submission.

References

1. Kataoka, T.; Powers, S.; McGill, C.; Fasano, O.; Strathern, J.; Broach, J.; Wigler, M. Genetic analysis of yeast *RAS1* and *RAS2* genes. *Cell* **1984**, *37*, 437–445. [CrossRef] [PubMed]
2. Boguski, M.S.; McCormick, F. Proteins regulating Ras and its relatives. *Nature* **1993**, *366*, 643–654. [CrossRef] [PubMed]
3. Fernández-Medarde, A.; De Las Rivas, J.; Santos, E. 40 years of RAS—A historic overview. *Genes* **2021**, *12*, 681. [CrossRef] [PubMed]
4. Pentland, D.R.; Piper-Brown, E.; Mühlschlegel, F.A.; Gourlay, C.W. Ras signalling in pathogenic yeasts. *Microb. Cell* **2018**, *5*, 63–73. [CrossRef]
5. Wennerberg, K.; Rossman, K.L.; Der, C.J. The Ras superfamily at a glance. *J. Cell Sci.* **2005**, *118*, 843–846. [CrossRef]
6. Hancock, J.F.; Cadwallader, K.; Paterson, H.; Marshall, C.J. A CAAX or a CAAL motif and a second signal are sufficient for plasma membrane targeting of ras proteins. *EMBO J.* **1991**, *10*, 4033–4039. [CrossRef]
7. Rocks, O.; Peyker, A.; Kahms, M.; Verveer, P.J.; Koerner, C.; Lumbierres, M.; Kuhlmann, J.; Waldmann, H.; Wittinghoferr, A.; Bastiaens, P.I.H. An acylation cycle regulates localization and activity of palmitoylated ras isoforms. *Science* **2005**, *307*, 1746–1752. [CrossRef]
8. Qiu, Y.; Wang, Y.; Chai, Z.; Ni, D.; Li, X.; Pu, J.; Chen, J.; Zhang, J.; Lu, S.; Lv, C.; et al. Targeting RAS phosphorylation in cancer therapy: Mechanisms and modulators. *Acta Pharm. Sin. B* **2021**, *11*, 3433–3446. [CrossRef]
9. Whistler, J.L.; Rine, J. *RAS2* and *RAS1* protein phosphorylation is *Saccharomyces cerevisiae*. *J. Biol. Chem.* **1997**, *272*, 18790–18800. [CrossRef]
10. Cobitz, A.R.; Yim, E.H.; Brown, W.R.; Perou, C.M.; Tamanoi, F. Phosphorylation of *RAS1* and *RAS2* proteins in Saccharomyces cerevisiae. *Proc. Natl. Acad. Sci. USA* **1989**, *86*, 858–862. [CrossRef]
11. Sobering, A.K.; Watanabe, R.; Romeo, M.J.; Yan, B.C.; Specht, C.A.; Orlean, P.; Riezman, H.; Levin, D.E. Yeast Ras Regulates the Complex that Catalyzes the First Step in GPI-Anchor Biosynthesis at the ER. *Cell* **2004**, *117*, 637–648. [CrossRef]
12. Sobering, A.K.; Romeo, M.J.; Vay, H.A.; Levin, D.E. A Novel Ras Inhibitor, Eri1, Engages Yeast Ras at the Endoplasmic Reticulum. *Mol. Cell Biol.* **2003**, *23*, 4983–4990. [CrossRef] [PubMed]
13. Belotti, F.; Tisi, R.; Paiardi, C.; Rigamonti, M.; Groppi, S.; Martegani, E. Localization of Ras signaling complex in budding yeast. *Biochim. Biophys. Acta-Mol. Cell Res.* **2012**, *1823*, 1208–1216. [CrossRef] [PubMed]
14. Leadsham, J.E.; Miller, K.; Ayscough, K.R.; Colombo, S.; Martegani, E.; Sudbery, P.; Gourlay, C.W. Whi2p links nutritional sensing to actin-dependent Ras-cAMP-PKA regulation and apoptosis in yeast. *J. Cell Sci.* **2009**, *122 Pt 5*, 706–715. [CrossRef]
15. Leadsham, J.E.; Sanders, G.; Giannaki, S.; Bastow, E.L.; Hutton, R.; Naeimi, W.R.; Breitenbach, M.; Gourlay, C.W. Loss of Cytochrome c Oxidase Promotes RAS-Dependent ROS Production from the ER Resident NADPH Oxidase, Yno1p, in Yeast. *Cell Metab.* **2013**, *18*, 279–286. [CrossRef] [PubMed]
16. Amigoni, L.; Martegani, E.; Colombo, S. Lack of *HXK2* induces localization of active Ras in mitochondria and triggers apoptosis in the yeast saccharomyces cerevisiae. *Oxidative Med. Cell. Longev.* **2013**, *2013*, 678473. [CrossRef]
17. Whaby, M.; Khan, I.; O'Bryan, J.P.; Johnson, R.H. Targeting the "undruggable" RAS with biologics. *Adv. Cancer Res.* **2022**, *153*, 237–266.
18. Gourlay, C.W.; Ayscough, K.R. Actin-Induced Hyperactivation of the Ras Signaling Pathway Leads to Apoptosis in Saccharomyces cerevisiae. *Mol. Cell Biol.* **2006**, *26*, 6487–6501. [CrossRef]
19. Subramanian, A.; Tamayo, P.; Mootha, V.K.; Mukherjee, S.; Ebert, B.L.; Gillette, M.A.; Paulovich, A.; Pomeroy, S.L.; Golub, T.R.; Lander, E.S.; et al. Gene set enrichment analysis: A knowledge-based approach for interpreting genome-wide expression profiles. *Proc. Natl. Acad. Sci. USA* **2005**, *102*, 15545–15550. [CrossRef]

20. Bodenmiller, B.; Campbell, D.; Gerrits, B.; Lam, H.; Jovanovic, M.; Picotti, P.; Schlapbach, R.; Aebersold, R. PhosphoPep—A Database of Protein Phosphorylation Sites for Systems Level Research in Model Organisms NIH Public Access Author Manuscript. *Nat. Biotechnol.* **2008**, *26*, 1339–1340. [CrossRef]
21. Steyfkens, F.; Zhang, Z.; Van Zeebroeck, G.; Thevelein, J.M. Multiple Transceptors for Macro-and Micro-Nutrients Control Diverse Cellular Properties through the PKA Pathway in Yeast: A Paradigm for the Rapidly Expanding World of Eukaryotic Nutrient Transceptors Up to Those in Human Cells. *Front. Pharmacol.* **2018**, *9*, 191. [CrossRef]
22. Conrad, M.; Schothorst, J.; Kankipati, H.N.; Van Zeebroeck, G.; Rubio-Texeira, M.; Thevelein, J.M. Nutrient sensing and signaling in the yeast Saccharomyces cerevisiae. *FEMS Microbiol. Rev.* **2014**, *38*, 254–299. [CrossRef]
23. Reinders, A.; Bü, N.; Boller, T.; Wiemken, A.; De Virgilio, C. Saccharomyces cerevisiae cAMP-dependent protein kinase controls entry into stationary phase through the Rim15p protein kinase. *Genes Dev.* **1998**, *12*, 2943–2955. [CrossRef] [PubMed]
24. Thevelein, J.M.; de Winde, J.H. Novel sensing mechanisms and targets for the cAMP-protein kinase A pathway in the yeast Saccharomyces cerevisiae. *Mol. Microbiol.* **1999**, *33*, 904–918. [CrossRef] [PubMed]
25. Colombo, S.; Ma, P.; Cauwenberg, L.; Winderickx, J.; Crauwels, M.; Teunissen, A.; Nauwelaers, D.; de Winde, J.H.; Gorwa, M.; Colavizza, D.; et al. Involvement of distinct G-proteins, Gpa2 and Ras, in glucose- and intracellular acidification-induced cAMP signalling in the yeast Saccharomyces cerevisiae. *EMBO J.* **1998**, *17*, 3326–3341. [CrossRef] [PubMed]
26. Leadsham, J.E.; Gourlay, C.W. CAMP/PKA signaling balances respiratory activity with mitochondria dependent apoptosis via transcriptional regulation. *BMC Cell Biol.* **2010**, *11*, 92. [CrossRef]
27. Smets, B.; Ghillebert, R.; De Snijder, P.; Binda, M.; Swinnen, E.; De Virgilio, C.; Winderickx, J. Life in the midst of scarcity: Adaptations to nutrient availability in Saccharomyces cerevisiae. *Curr. Genet.* **2010**, *56*, 1–32. [CrossRef]
28. Burgess, M.R.; Hwang, E.; Mroue, R.; Bielski, C.M.; Wandler, A.M.; Huang, B.J.; Firestone, A.J.; Young, A.; Lacap, J.A.; Crocker, L.; et al. KRAS Allelic Imbalance Enhances Fitness and Modulates MAP Kinase Dependence in Cancer. *Cell* **2017**, *168*, 817–829.e15. [CrossRef]
29. Perry, J.R.; Basrai, M.A.; Steiner, H.-Y.; Naider, F.; Becker, J.M. Isolation and Characterization of a Saccharomyces cerevisiae Peptide Transport Gene. *Mol. Cell. Biol.* **1994**, *14*, 104–115.
30. Burnetti, A.J.; Aydin, M.; Buchler, N.E. Cell cycle Start is coupled to entry into the yeast metabolic cycle across diverse strains and growth rates. *Mol. Biol. Cell* **2016**, *27*, 64–74. [CrossRef]
31. Li, Y.; Tang, J.; Jiang, J.; Chen, Z. Metabolic checkpoints and novel approaches for immunotherapy against cancer. *Int. J. Cancer* **2022**, *150*, 195–207. [CrossRef] [PubMed]
32. Al Abdallah, Q.; Norton, T.S.; Hill, A.M.; LeClaire, L.L.; Fortwendel, J.R. A Fungus-Specific Protein Domain Is Essential for RasA-Mediated Morphogenetic Signaling in *Aspergillus fumigatus*. *mSphere* **2016**, *1*, e00234-16. [CrossRef] [PubMed]
33. Manso, J.A.; Carabias, A.; Sárkány, Z.; de Pereda, J.M.; Pereira, P.J.B.; Macedo-Ribeiro, S. Pathogen-specific structural features of Candida albicans Ras1 activation complex: Uncovering new antifungal drug targets. *MBio* **2023**, *14*, e0063823. [CrossRef] [PubMed]

Review

"Molecular Biology"—Pleonasm or Denotation for a Discipline of Its Own? Reflections on the Origins of Molecular Biology and Its Situation Today

Gregor P. Greslehner 

Department of Philosophy, University of Vienna, Universitätsstraße 7, 1010 Vienna, Austria; gregor.greslehner@univie.ac.at

Abstract: The disciplinary identity of molecular biology has frequently been called into question. Although the debates might sometimes have been more about creating or debunking myths, defending intellectual territory and the distribution of resources, there are interesting underlying questions about this area of biology and how it is conceptually organized. By looking at the history of molecular biology, its origins and development, I examine the possible criteria for its status as a scientific discipline. Doing so allows us to answer the title question in such a way that offers a reasonable middle ground, where molecular biology can be properly viewed as a viable interdisciplinary program that can very well be called a discipline in its own right, even if no strict boundaries can be established. In addition to this historical analysis, a couple of systematic issues from a philosophy of science perspective allow for some assessment of the current situation and the future of molecular biology.

Keywords: molecular biology; history; discipline; central dogma; philosophy of science

Citation: Greslehner, G.P. "Molecular Biology"—Pleonasm or Denotation for a Discipline of Its Own? Reflections on the Origins of Molecular Biology and Its Situation Today. *Biomolecules* **2023**, *13*, 1511. https://doi.org/10.3390/biom13101511

Academic Editor: Laszlo Patthy

Received: 24 June 2023
Revised: 5 October 2023
Accepted: 7 October 2023
Published: 12 October 2023

1. Introduction

The main question of this paper is whether and why molecular biology can be considered a discipline of its own. (This paper is in parts a translation and adaptation of my 2012 bachelor's thesis at the University of Salzburg under Michael Breitenbach's supervision. When discussing my thesis with Michael, he asked me whether I had already published anything, which I had not at the time. Being occupied with other academic pursuits (including a master's thesis in Michael's team on the yeast NADPH oxidase gene *YNO1*), I never actually got to publish on the historical and philosophical trajectory of molecular biology. This special issue "in honor of Prof. Michael Breitenbach" is now a welcome opportunity to finally do so. Thank you, Michael, for your broad and thorough scientific and philosophical interests!) The answer that I develop and defend here is as follows. If by "discipline" one means a stringent and uniform theory or a closed and uniform framework of institutions, methods, theories, etc., then molecular biology would *not* be a discipline of its own. However, the same negative assertion would then also apply to numerous other domains of research that are usually considered disciplines without any controversy. Hardly any of the many fields, such as medicine, meteorology, geology, philosophy, etc., that we ordinarily do refer to as "disciplines", would fall under such a narrow conception of what constitutes a discipline. We may thus rightfully ask what the criteria are to justifiably calling something a discipline. Ideally, such a notion would not be so narrow as to exclude its intended extension. On the other hand, when taking into account that in practically all areas of biological research molecular aspects play an essential role, one might ask whether the boundaries are totally blurred and whether any biology nowadays would be *molecular*. That is to say, "molecular biology" might be considered a pleonasm like "round circle", "unmarried bachelor" or "dead corpse". Putting "molecular" in front of "biology" would thus be verbally redundant.

Between those two extreme positions, pleonasm or clear-cut discipline of its own, one needs to find a balanced and substantially adequate position. This is what I attempt in this paper. By doing so, we will also have to ask the question whether thinking in disciplines is at all appropriate for today's scientific network-landscape. Not only is this important for providing some orientation within the ever-growing landscape of the life sciences but it also provides lessons from the conceptual history and research tradition that molecular biologists are still part of today. In order to address this main question about the disciplinary identity of molecular biology, we will inevitably also touch upon some general questions and problems from the history and philosophy of science. We will do so by investigating the emergence of the field called "molecular biology" (whether or not we ultimately decide that it can be justifiably considered to be a discipline). Addressing this question is not merely a fight about semantics and which terms can be applied—these questions are crucial for navigating the present, past and future of molecular biology, upon which a number of scientific, historical and philosophical questions depend.

This paper will proceed as follows. First, I will argue in Section 2 why the history and philosophy of science are not only relevant to specialists from these fields in the humanities but also to those from the subject at hand, i.e., molecular biologists. Section 3 does a lot of heavy lifting by tracing the historical development of molecular biology in order to eventually answer the title question about the disciplinary status of molecular biology. The answer will be to regard molecular biology neither as a pleonasm nor as a discipline in a narrow sense. Instead, I suggest a wider notion of discipline, in which the central aspect of molecular biology's interdisciplinarity can be embraced as a defining feature. This will be achieved by comparing a number of different positions on the disciplinary status of molecular biology and its development, by addressing the peculiar formulation of the "central dogma of molecular biology", and by developing a nuanced position based on these historical details. Finally, Section 4 closes with some reflections on the current situation and future developments of molecular biology.

2. Why History of Science Matters

Whereas it would be difficult to find anyone openly denying the worth of history of science in general, there are mixed opinions about its value for the everyday practice of science. Why should someone working at the lab bench have to deal with old chestnuts or dig up old, dusty journals and lab notebooks? One might say: "Shouldn't we push the frontiers of science instead of stirring up old stories and outdated ideas that didn't hold up to today's scientific standards? There are better things to do, researchers are busy surviving in their precarious publish-or-perish environment anyway. Let the historians and philosophers have their fun—at least they won't be in the way and prevent us from doing *real* science".

Such an attitude would certainly be detrimental to science and its progress. Just imagine that no one ever cared about the results from a certain Augustinian monk who bred peas in his garden. However, apart from such spectacular examples such as the rediscovery of Mendel's observations by de Vries, Correns and Tschermak in 1900 (but see [1])—which perhaps should have solidified a different attitude towards history in biology—the history of science and the practice of science rarely ever go hand in hand. As Erwin Chargaff remarks, "most papers cite almost nothing older than five years" [2] (p. 193, my translation). However, there is hope that historical roots do get some attention from specialists in the field. Unfortunately, however, most of the time it remains at the level of a few interspersed names, anecdotes, and curiosities: "This is where many scientists begin with history—with an interest in anecdotes and a sketch of the background to current work. Unfortunately, most scientists end there" [3] (pp. 344–345).

This is unfortunate, as historical analysis and rational, logical reconstruction of the process of knowledge generation is a core issue in the philosophy of biology. The Austrian biologist F. M. Wuketits even opined: "Contrary to common opinion, I am convinced that no consideration of modern biology can do without resorting to history" [4] (p. VI, my translation). W. Nachtigall critically remarks that "biology is not history of biology, whereas philosophy is at least four out of five parts history of philosophy" [5] (p. 130). Despite this piece of polemic criticism, Nachtigall's writing is a prime example of how philosophical considerations can be very fruitful for biological reasoning. Perhaps he had in mind certain forms of "philosophy" that indeed have often not been very helpful or interesting for scientists.

S. G. Brush raises the question whether the distortions from history of science might actually be harmful, especially for students entering a discipline: "Should the history of science be rated X?" [6]: "I will examine arguments that young and impressionable students at the start of a scientific career should be shielded from the writings of contemporary science historians for reasons [...] that these writings do violence to the professional ideal and public image of scientists as rational, open-minded investigators, proceeding methodically, grounded incontrovertibly in the outcome of controlled experiments, and seeking objectively for the truth, let the chips fall where they may" [6] (p. 1164).

The criticism might be somewhat anachronistic and aimed at the wrong target. However, perhaps such considerations are at least part of the reason why history of science usually has no place in the curricula of individual disciplines. This is unfortunate, as it would allow for a better understanding of how science works. "Scientific development is like Darwinian evolution, a process driven from behind rather than pulled toward some fixed goal to which it grows ever closer" (T. S. Kuhn 1992 *The Trouble with the Historical Philosophy of Science*, cited after [7] (p. 139)).

One way in which history of science can benefit scientists is also by letting go of some illusions: "the delight of demystifying science of its more pretentious claims to rationality and orderly procedure, the delight of revealing for once, or seeming to reveal, what scientists "really do"—which turns out to include coffee breaks and gossip" [8] (p. 26).

Maienschein and colleagues distinguish at least five general categories in which the history of science matters for scientists:

- "Self-improvement, illuminating science and making it better
- Efficiency, avoiding and learning from past mistakes
- Perspective, providing judgment and clarity and therefore making science better
- Imagination, offering a wider repertoire of ideas to choose from
- Education, improving public understanding of science and scientific literacy" [3] (p. 342).

There is some awareness of the importance of history in contemporary research and teaching practice, as a welcome example from a recent textbook shows: "we have continued to present the field in a historical context, with the intent of sensitizing and inspiring students (and others) to the realities of how research progress unfolds and how ideas develop and attain maturity—or not. We have refrained wherever possible from unadulterated dogma and from presenting the field [...] as anywhere near total clarification. While we are aware of presenting viewpoints that are sometimes controversial and even conflicting, we trust that readers, especially students, are not unduly confused or frustrated by our reluctance to always provide the final word, as it were. Rather, it is our hope that such controversies and complexities will inspire further studies" [9] (pp. xxv–xxvi).

Historical considerations seem to be the common element in most philosophical discussions about molecular biology (cf. the extensive reference list on the history of molecular biology in [10]). Taken together, there has been a rather large interest in the history of molecular biology, although it has subsided a bit recently. (It appears paradoxical, as S. de Chadarevian observes that "steady advances in the field of molecular biology" coincide with "the fading interest in its history" [11] (p. 464).) However, there are still a number of chapters left to process. The aim of this paper, however, is neither a complete overview nor a particularly original compilation of this history. One may even question whether that is possible at all: ("The history of the past can be described—most often wrongly—but I do not think it can be explained; or perhaps better, there are too many different explanations that are equally valid and equally useless" [12] (p. 239, my translation).) "Unfortunately

I have concluded that there cannot be a history of natural science that would be more profound than, let's say, a history of fashion" [12] (p. 131, my translation). In order to still make good use of history, I think it is best to openly acknowledge one's necessarily limited perspective. "In truth, history always appears to me to be rather subjective, arbitrary and selective. And whoever engages in historical reflections would be maybe well advised to drop any claims to objectivity and instead at least give one's criteria for their arbitrary and subjective selection of topics and their treatment and to disclose what one's aims are" [13] (p. XI, my translation).

To be transparent about my aims here: I want to argue that studying the history—not only of one's own discipline—enriches the everyday business of systematic researchers. In doing so, one should not only consider the already established and easily accessible resources but also those that might have received less attention. Who knows what kind of treasures are still buried in the archives? The value that the history of molecular biology has to offer—in addition to the general reasons outlined above—also lies in the fact that biology is a dynamic and rapidly evolving domain of research. In order not to float aimlessly alongside the latest trends without any orientation, studying some history provides an important perspective. Finally, there is the title question of this paper: where to define the place of molecular biology as a scientific discipline?

3. "Molecular Biology"—Pleonasm or Denotation for a Discipline of Its Own?

"The biggest difficulty in discussing the history of molecular biology is in knowing just what the term means, and what lines of demarcation can be drawn between it and other fields of biology. Of course biology is multidisciplinary, and its area of interest lies at the interface between several established fields [...] and the question arises what distinctive element there is in the approach of molecular biologists to justify a special name" [14] (p. 5).

Before tackling the multi-faceted question about molecular biology's status as a scientific discipline, let us start with the most basic question: what is molecular biology?

3.1. What Is Molecular Biology?

"What is x?" questions are often the most difficult to answer and often how philosophical debates get started. When asked by family or friends what one's area of scientific work centers around, one often has to first clarify that "microbiology" and "molecular biology" are different terms and different fields. With that out of the way, you might then have to explain to your increasingly confused conversation partner that it is a kind of mixture of biology, physics, chemistry, genetics, etc., that studies the processes of living organisms on the molecular level. You may go on to highlight, for example, the particular protein or pathway your lab is working on—and how it might in the future contribute to understanding a certain disease. Frequently, that will be acknowledged with some interest and skepticism. Sooner or later, however, the inevitable question on an afternoon walk will come up: "which kind of plant or animal is that? Aren't you a biologist? You should know!" Unless by lucky chance you can answer this question, your opposite will feel confirmed in their skepticism. "This one wants to be a biologist? They don't even know the animals/plants (cf. [15] (p. 14, my translation))! No surprise, if they study everything mixed together without *really* knowing any of these scientific disciplines".

A very different experience will unfold when you have to explain to the same person something about CRISPR genetics that was mentioned on the news. However, the impression will remain: molecular biology appears to be neither fish nor fowl. Is this impression entirely unwarranted? Or does it hit a meaningful weak spot? Or, perhaps, might it actually be one of molecular biology's assets? A look at the history and development of molecular biology and the origin of its name will allow us to answer these questions and shed light on some of its central theoretical features.

The main question of this paper is whether molecular biology is a scientific discipline of its own. Since the answer seemingly could not be anything other than "obviously, yes—duh", the question might appear ridiculous and absurd. Surely there are departments, study programs, scientific journals and societies bearing that name. Whether or not that is indeed a good criterion is a separate question. However, a closer inspection of such study programs and the term "life sciences" in general reveals some interesting observations. The usage of the plural form "scienc*es*" seems to "suggest that a number of (still?) heterogeneous areas are assembled under this umbrella term" [16] (p. 115, my translation). Additionally, looking into the curricula of study programs does not contribute to the impression of molecular biology as a coherent and stringent discipline, as you will find courses on chemistry, physics, genetics, mathematics, cell biology, biochemistry, biophysics, even biophysical chemistry, bioinformatics, molecular medicine, etc.

Thus, one could easily dispute that molecular biology is a discipline, as a subject with its own methods and issues. Others, on the other hand, might equally feel inclined to say that "every piece of genetic or other biological research contains some part of molecular biology, because ultimately everything can be traced back to the changes of molecules" [5] (pp. 43–44, my translation). However, Nachtigall eventually concludes that "not for any epistemological reasons, but due to purely practical aspects, biology is a closed discipline" [5] (p. 49, my translation), and further: "Biological questions are often characterized by the fact that in order to answer them a diversity of different methods have to be combined in multiple ways. This fact makes modern biological research so difficult" [5] (p. 50, my translation).

Thus, the question "whether the life sciences constitute a *logical whole*" [17] (p. 181, my translation, original emphasis) deserves more attention. In any event, it seems to be a fact when Sarkar finds that "[t]oday, for good or for bad, depending on one's point of view, most of biology is molecular biology" [18] (p. 1). However, is that really the case for all of biology? How does molecular biology relate to the rest of biology? How does it relate to other scientific disciplines such as physics and chemistry? How did this relationship evolve? What are the possibilities and dangers of such relationships? What can philosophy of science learn from these developments—and contribute to them? These are a number of questions to which a historical and philosophical analysis of molecular biology can contribute. In return, molecular biology poses a prime study object for the history and philosophy of science to study topics of, e.g., theory change and dynamics. *Interdisciplinarity* is another important aspect. From an inter- and transdisciplinary point of view, it might not be a flaw at all that the status of molecular biology as a scientific discipline can be called into question. To the contrary, it might be a sign of healthy interdisciplinary practice.

If we were to take the name of a field of research as a main indicator for its status as a discipline and considering all the practical and organizational reasons for doing so, it clearly cannot be the only factor. Along the same lines, it seems questionable whether the brackets holding molecular biology together have been consistent enough to allow its disciplinary identity through time: "molecular biology had a strong disciplinary identity perhaps only in the fifteen years or so following the discovery of the structure of DNA. [...] In saying that molecular biology had a strong disciplinary identity we mean that the name became, for a time at least, associated with recognisable clusters of concepts, projects, tools, institutions, and individuals" [19] (p. 62).

Perhaps then it will be better to call molecular biology an "ultra-discipline" [20] (p. 4) or "super-discipline" [21] (p. 79)—something overarching and uniting parts of other disciplines—or to simply accept a less narrow sense of "discipline", in which we do not require strict boundaries. In any event, it is clear that we have to reconsider the concepts of and requirements for the identity and development of scientific disciplines (for a case study about molecular evolution, see [22]). This is especially true for molecular biology, for which R. M. Burian has the following fitting claim: "there is not, I claim, a central theory that binds together the many distinct (sub)disciplines of molecular biology. What serves to unify the distinct sorts of work in molecular biology together is *not* a central theory, but the use of an immense battery of techniques and a general approach to explaining—and altering—organismic function by reference to, and use of, an *omnium gatherum* of detailed molecular mechanisms.

In brief, molecular biology is less like Newtonian mechanics than it is like auto mechanics.

Despite the lack of an overarching theory, a Newtonian or quantum mechanics of its very own, molecular biology has become a unifying discipline in virtue of the powers of its techniques, its ability to extrapolate from the molecular to higher levels, and its synthesis of problems of form and function at the molecular level. This synthesis of form and function is a central, ill-understood, and historically important feature of molecular biology. Among other things it explains the degree of truth—and of error—in Erwin Chargaff's and Seymour Cohen's mordant witticism that molecular biology is the practise of biochemistry without a license" [21] (pp. 67–68, original emphasis).

Finally, it is insightful to emphasize that the question of molecular biology as a discipline might sometimes have been confused with the question of molecular biology *as a theory*. (The "central dogma" as a potential candidate for such a (theoretical) role is discussed later. Its assumptions "may well be taken to form the theoretical core of molecular biology (to the extent–and this is a matter of controversy–that molecular biology has any theory)" [23] (p. 187).) Unlike other areas of science, however, the notion of "theory"—understood as a deductively closed set of sentences—might not be suited for most areas of biology (for a seminal study on theory structure and change in molecular biology, see [24]). If not through a central theory, how could something be considered a scientific discipline? We will address this question next.

3.2. What Is a Scientific Discipline?

Following B. Gräfrath's encyclopedia entry, *scientific discipline* is the "denotation for an area of science that can be demarcated by its subject matter, method or epistemic interest" [25] (p. 237, my translation). Now, is molecular biology such an area of science that can be demarcated by its subject matter, method, or epistemological interest?

The subject matter encompasses all processes and entities of the living world, with special regard to their molecular features. What would set molecular biology apart from the rest of biology other than this *special regard*? Moreover, would it suffice as a demarcation criterion (especially against biochemistry)? Other areas of biological research with different aims often pay an equal special regard to molecular features. On the other hand, molecular biology is not only interested in and limited to molecular features; the macroscopic, cellular, and physiological levels also play an important role in molecular biology all the time. Without going further down the rabbit hole of the notions of levels and their relations [26], we can see that subject matter alone does not allow for a satisfactory demarcation of molecular biology.

Perhaps we fare better with the method, consisting of the "immense battery of techniques" Burian observed above to provide a constitutive element for molecular biology through its methods. Among the methods that might put some shape to the body of molecular biology are indeed a whole arsenal of techniques. Without claiming to give a remotely complete list, these include: PCR, cloning, sequencing, gel electrophoresis, X-ray crystallography, fluorescence in situ hybridization, Southern blots (and all other cardinal direction variations), and many more. Although it does warrant specialized training, education, and awarding academic degrees, this bandwidth will most likely not work to firmly demarcate molecular biology as its own discipline either. (It is an interesting observation (thanks to an anonymous reviewer) that the name "molecular biology" has increasingly disappeared from study programs and textbooks, where some of the most influential ones are titled "Molecular Biology of the Cell" [27] and "Molecular Cell Biology" [28]).

Finally, to our third and last ingredient. The epistemic interest might in fact offer the best justification for the term "molecular" in this denotation for an area of biology. What falls into the epistemic interest, however, is less a question of curricula and academic teaching but rather depends on the research interests of individual researchers. In today's molecular biology, the interests of people working in the field are perhaps even more diverse than what we normally consider to be within the scope of a given discipline. For that reason, I will leave the discussion about the current status and future trajectory of molecular biology for another section (Section 4). There, we will also address the question regarding whether the epistemic interest still allows us to denote molecular biology as its own scientific discipline. For now, we will stick with the epistemic interest as the most promising candidate for uniting researchers within a scientific discipline and investigate the epistemic interest of those researchers who, according to a large consensus, have contributed to the emergence of molecular biology as a scientific discipline in the first half and middle of the twentieth century.

3.3. The Many Origins of Molecular Biology

Rarely have I encountered a phrase as frequently as "origin(s) of molecular biology" during my research on this topic. Even when it is not directly part of a book's or paper's title, the content of hundreds of pieces of writing could easily have this phrase in their title. (It would be easy to provide dozens of references as proof, but for reasons of readability and not blowing up the references section, I refrain from doing so.) This immense interest in the early days and beginning of molecular biology is striking. Not only because it is an interesting topic that deservedly receives a lot of attention. It is particularly striking because each time the story is being told differently—at least from a different perspective or with a different emphasis. Obviously, the reason for this variability is that there is no single origin of molecular biology, there are many. Neither is it just the development of another branch of science, nor the wedging off from another, larger discipline, nor the planned and orchestrated convergence or merging of other separate disciplines. It is a powerful synergy of several tendencies, disciplines, persons, and their development that gradually emerged in the scientific landscape of the twentieth century.

The geneticist H. J. Muller, frustrated with the limitations of X-ray crystallography for viral substances, publicly called upon others: "The geneticist himself is helpless to analyze these properties further. Here the physicist, as well as the chemist, must step in. Who will volunteer to do so?" (cited after [29] (p. 400)). Directly or indirectly, this call reached the right ears. The interdisciplinary efforts and synergetic emergence that ensued can hardly be found in other areas of science; nor can anything comparable easily be found in the history of science. Sometimes, these debates might not have been about the actual history and scientific content but about intellectual territory and the division of financial resources: "The bitterness of many of these discussions reflects the fact that the stakes go far beyond the history of molecular biology, relating to the orientation of current research and its financing" [30] (p. 99). "[T]he discussions surrounding the disciplinary status of molecular biology are as virulent as ever indicating that the stakes are high. What is at issue are intellectual territories and ways of producing knowledge, the distribution of resources and the recruitment of new generations of researchers" [20] (p. 5). This can cast doubt and fuel the debates on the status of molecular biology as a scientific discipline. It also calls into question the classical conceptions of disciplines and their demarcation and development, including other areas of science.

However one wants to frame the relationship between biology and molecular biology, the denotation for both is relatively young. This might be less surprising for molecular biology, but also the name "biology" for the general study and science of life as a demarcated area of science arrived as late as the 19th century (Cf. [31–34]). Molecular biology fully emerged in the second half of the twentieth century, even though lots of groundwork already happened in the 1930s and 1940s. The origin of the term "molecular biology" warrants special attention. W. Weaver, director for natural sciences in the Rockefeller Foundation (for a historical study on the role of this institution see [35]) proposed the name in 1938 and sketched the field as follows (see [36]; cf. also [37–39]): "Among the studies to which the Foundation is giving support is a series in a relatively new field, which may be called molecular biology, in which delicate modern techniques are being used to investigate even more minute details of certain life processes" (cited after [36] (p. 582)).

The usage and scope of this new term were not left uncontested. (This is exemplified in the debate "Molecular biology or ultrastructural biology?" between C. H. Waddington and W. T. Astbury in *Nature* [40–42].) The name, nevertheless, stuck—not least because of practical reasons, as F. H. C. Crick described: "I myself was forced to call myself a molecular biologist because when inquiring clergymen asked me what I did, I got tired of explaining that I was a mixture of crystallographer, biophysicist, biochemist, and geneticist, an explanation which in any case they found too hard to grasp" (cited after [43] (p. 390)).

Crick played a very important role in the history of molecular biology (see also Section 3.5.1). For now, however, let us focus on someone who has been frequently credited as the "founding father" or "scientific hero" of molecular biology—and one of its major critics.

3.4. Erwin Schrödinger and Erwin Chargaff: Hero and Anti-Hero in the Foundational Myth of Molecular Biology

The title heroes of this section have something in common. Not only their first name "Erwin" and their Austrian background. Both play an important role in the history of molecular biology. The biochemist Erwin Chargaff contributed pioneering work to the biochemical origins of molecular biology. He is now probably best known for his observation on the ratios of bases in DNA, known as "Chargaff's rules". In his later days, however, he became an outspoken critic of the routes that science had taken (to the worse, in his opinion). Despite his sometimes-cynical criticism, he is a witty and educated author who is still worth reading today. Reading the texts of this "anti-hero of molecular biology" [44] (p. 97) pays off despite—or even because of this heavy criticism— for biologists and philosophers. Whereas Chargaff became an explicit critic of science, the physicist and Nobel laureate Erwin Schrödinger is held in the highest esteem. In particular, there is a glorified picture about how he contributed to molecular biology. To this day, there are publications and symposia on Schrödinger and his texts (for example [45–47]).

The relationship between physics and biology is a special one for several reasons, cf. [33]. Of particular interest for the history of science has been the role that physics played in the emergence of molecular biology (cf. [29] and the literature cited on p. 389). A classical question from the philosophy of science is whether biology could be *reduced* to physics. (This reductionism debate remains a battle ground to this day [48]. Regardless of this debate, the relationship between physics and biology has become a much more fruitful enterprise, looking at how their modes of explanation can be integrated, see for example [49]).

Even more prominently than physics, it is the *physicists* who played an essential role in carrying over their approaches, techniques and views from the field of physics to tackling biological phenomena. Along this line, of significant influence is Schrödinger's little book *What is life?* [50]. It has been said to "mark the birth of molecular biology" [17] (p. 79). (G. S. Stent calls it "Uncle Tom's Cabin of the revolution in biology that, when the dust had cleared, left molecular biology as its legacy" [43] (p. 392).) In addition to Schrödinger, also Niels Bohr's [51] and Max Delbrück's writings [52] have sometimes been claimed to have had a similar effect. However, Schrödinger clearly left the biggest impact and reputation. His little book *What is life?*, based on public lectures he gave in Dublin in 1943, carries the subtitle "The physical basis of the living cell". With the constant excuse that he himself is no biologist, he summarizes the textbook knowledge of the time and fuels the expectation that the mechanisms of the living cell will turn out to be based on physical principles. The hope to discover new physical laws is born from some of his formulations (for more nuance, see [53], forthcoming).

With regard to the unmanageable amount of accumulated knowledge, Schrödinger sees the universality of knowledge in danger: "[I]t has become next to impossible for a single mind fully to command more than a small specialized portion of it.

I can see no other escape from this dilemma (lest our true aim be lost for ever) than that some of us should venture to embark on a synthesis of facts and theories, albeit with second-hand and incomplete knowledge of some of them—and at the risk of making fools of ourselves" [50] (p. 1).

What has now been the reason for the great popularity and attractiveness of this book? M. Morange puts it as follows: "Schrödinger presented the new results of genetics in a lively, compelling way—much better than the biologists had. Fifty years later, the book has lost none of its seductiveness: its clarity and simplicity make it a pleasure to read.

A modern molecular biologist would feel quite at home reading Schrödinger's book" [30] (p. 74). (I can confirm this from personal experience, reading it for the first time as an undergraduate student in one sitting.) F. H. C. Crick said about Schrödinger's book: "Its main point was one that only a physicist would feel it necessary to make, but the book [...] conveyed in an exciting way the idea that, in biology, molecular explanations were just around the corner. This had been said before, but Schrödinger's book was very timely" (cited after [29] (p. 404)).

F. Jacob's analysis of the time and spirit make clear what made Schrödinger's book so "timely": "After the Second World War, many young physicists were shocked by the military use made of atomic energy. Some of them, moreover, were dissatisfied with the direction taken by experiments in nuclear physics, by their slowness and by the complexity inherent in the use of large machines. They saw in this the end of a science, and looked around for other activities. Some turned to biology with a mixture of anxiety and hope: anxiety, because all they usually knew about living organisms consisted of vague recollections of zoology and botany acquired at school; hope because some of their most celebrated elders pointed to biology as a science full of promise. Niels Bohr saw it as the source of new laws of physics awaiting discovery. Schrödinger also prophesied a new and exhilarating era for biology, particularly in the field of heredity. Just to hear one of the leaders in quantum mechanics asking "What is life?" and then describing heredity in terms of molecular structures, interatomic bonds and thermodynamic stability was enough to fire the enthusiasm of certain young physicists and to bestow some sort of legitimacy on biology. Their ambition and interest were limited to a single problem: the physical basis of genetic information" [54] (pp. 259–260).

What truth is there now to all these mythical tales—are they more than mere foundational myths? Indeed, many pioneers of molecular biology have credited Schrödinger's book as having played a major role in their turn to biology. ("Among those who acknowledged their debt to Schrödinger's book are M. Delbruck [sic], G. Stent, J. D. Watson, F. Crick, M. F. Wilkins, and S. Benzer" [55] (p. 1071). As was J. D. Watson influenced: "from the moment I read Schrödinger's 'What is Life,' [...] I became polarized toward finding out the secret of the gene" [56] (p. 239).) However, the actual role and influence of such eminent authorities have been viewed much less heroic (cf. [44,55,57–60]; S. Sarkar, despite all criticism, gives a positive outline: "Even if all these criticisms of *What is Life?* were fair, it would still remain to Schrödinger's credit that he had introduced the idea of the genetic code well before 1953, when the structure of DNA was discovered" [60] (p. 633)). "Perhaps this is a case of the new science's constructing its own historiography, furnishing itself with such prestigious founders as Niels Bohr and Erwin Schrödinger" [30] (p. 75). "It was characteristic of this early phase of homemade history of science by scientists that it privileged groups and schools that one could easily identify. In particular, it concentrated on the role that physics—theoretical physics—played in the foundation of the new direction in biology. Niels Bohr's paper 'Light and life' as well as Erwin Schrödinger's book *What is life?* played a central role in this foundational myth" [61] (p. 8).

Even if it was indeed a *foundational myth* (cf. [21,62]), it is hard to brush off the impact of more or less famous physicists on molecular biology: "Physicists played an important role in this change in the form of biological knowledge, by the way they conceived and carried out their experiments. In following Delbrück and asking simple questions of biological objects, they obliged these objects to reply in the same simple language. [...] The most important contribution of the physicists was perhaps simply to have been convinced (with a certain dose of naiveté) and to have convinced the biologists that the secret of life was not an eternal mystery, but was within reach" [30] (p. 101).

Despite all that, however, it must not be overlooked that a physical perspective did not and does not exhaust the methodological approaches in biology. This is now also widely accepted by philosophers of science: "It [biology] knows questions and modes of explanation that do not exist anywhere else. A philosophy of science that is exclusively oriented towards physics cannot do justice to biology and thus the entirety of science" [15], my translation.

A potential obstacle in this regard is likely the fact that biology is less suited for idealizations and formalized considerations. Unlike physics, where *ideal gas* assumptions are common, to conceptualize an *ideal organism* does not seem appropriate for all kinds of reasons. The term alone appears to summon the wraiths of vitalism and eugenics. However, there is a deeper problem. Biological regularities are substantially different from physical laws. The latter often claim unrestricted validity for all times and spaces. Biology, on the other hand, deals with a restricted subject area, where generalized claims are regularities and tendencies that are full of exceptions and limitations. By comparing biology to the ideals of physics, one could be tempted to ask with J. J. C. Smart: "Can Biology Be an Exact Science?" [63]. This is also among the reasons why biological theories are rarely expressed in neat mathematical form as those in physics. (Although there have been attempts to formalize and even axiomatize biology [64,65]).

In place of some important equations or axioms that would provide the theoretical foundations of molecular biology, there is a peculiar scheme often referred to as the *central dogma of molecular biology*.

3.5. The So-Called "Central Dogma of Molecular Biology"

Another option for establishing molecular biology's disciplinary identity might be the so-called "central dogma of molecular biology". After a brief historical view on the unfortunate term "dogma", I want to investigate whether there is anything "dogmatic" about molecular biology in its current form.

3.5.1. Crick's Legacy

We owe the scheme behind and term of the "central dogma" to F. H. C. Crick [66,67], in addition to many other more or less problematic insights. When hearing about the "central dogma of molecular biology" for the first time, I thought this was a sort of terminological pun, playing around with terms usually used in theological discussions (cf. [68] (p. 92): "This shorter scheme is called the central dogma, a jokingly ecclesiastical reference by Francis Crick to a summary of bulk cellular information flow."). Perhaps already containing a grain of truth, some criticism—or something more—might be indicated by this odd choice of words. For that reason, a historical clarification of how the term "central dogma" had been introduced is helpful (see also [8] (pp. 337–338)).

"I called this idea the central dogma, for two reasons, I suspect. I had already used the obvious word hypothesis in the sequence hypothesis, and in addition I wanted to suggest that this new assumption was more central and more powerful. As it turned out, the use of the word dogma caused almost more trouble than it was worth. Many years later Jacques Monod pointed out to me that I did not appear to understand the correct use of the word dogma, which is a belief *that cannot be doubted*. I did apprehend this in a vague sort of way but since I thought that *all* religious beliefs were without foundation, I used the word the way I myself thought about it, not as most of the world does, and simply applied it to a grand hypothesis that, however plausible, had little direct experimental support" [69] (p. 109, original emphasis).

However, something does not simply become a dogma (in its usual meaning) by calling it one. Rather, it takes people who believe in it without question and blindly defend it against any sort of doubt or criticism. Is that something that happened with the so-called *central dogma of molecular biology*? Although often reproduced in textbooks without much context, it is very clear that the underlying assumptions can and have been called into question, criticized, modified and changed multiple times (for example, [70–72]; looking at those challenges for the central dogma in detail, however, is beyond the scope of this paper). Overall, these lively debates show the openness of molecular biology, rather than its dogmatism.

3.5.2. What Place do Dogmas Have in Molecular Biology?

Stent [43] distinguishes three phases of molecular biology: a romantic phase (1938–1952, centering around Max Delbrück and the *phage group*), a dogmatic phase (1953–1963, dominated by Watson, Crick and the so-called *central dogma of molecular biology*) and an academic phase (1963 until today?). What characterizes today's phase of molecular biology? Is it new dogmas? Whereas dogmas have their place in religion, in other areas—and especially science—being dogmatic is a sign of decline and something to be avoided. Does molecular biology have any dogmatic or religious traits? Chargaff calls it "cult-like" [12] (p. 94, my translation), talks of "molecular fundamentalists" [12] (p. 107, my translation) and a "molecular kabbalah" [12] (p. 176). Hausmann talks of "newly converted disciples of molecular biology" [13] (p. 53, my translation). Delbrück calls DNA "unmoved mover" [73] (p. 55).

Fortunately, the life sciences do not appear to be trapped in such a dogmatic predicament. To the contrary, it is the strength of the scientific method that ideas and concepts are up to debate at all times (at least in principle). With sufficient justification, once well-established assumptions will be refuted. There is something known as the academic "mainstream" that one cannot easily object to, and the incentives and publication pressure have made science inherently conservative [74]. However, understood as something that a true believer must not doubt, dogmas have no place in science. (The term "dogma" does repeatedly reappear in contemporary debate when uncritically long-held beliefs are called into question, e.g., [75]).

3.5.3. What Remains of Molecular Biology When the Dogmas Have fallen?

The title of this subsection is a direct reference to P. Bieri's paper [76], in which he asks: "what remains of analytic philosophy when the dogmas have fallen?" (my translation). As different as these two fields—molecular biology and analytic philosophy—might appear on first (perhaps also second and third) impression, and given that Bieri is comparing two very different academic traditions, his intention in this paper is not too different from mine here. (There are some parallels and complementarity that incentivized me to pursue both studying molecular biology and philosophy in the first place.) In both instances, it is a nuisance when *analytic philosophers* look down on other *continental philosophers* or when *molecular biologists* are disdainful towards colleagues working in *ecology*.

If we take the central dogma and the ensuing research on the relations between RNA, DNA and proteins as indicative for a central part of molecular biology's research program, we can perhaps best put it with R. Creath like this: "every dogma has its day" [77]. I would even like to adapt his summary to our case of molecular biology here: "For all the difficulties we have noted, [the "central dogma"] is by no means a failure. Instead, it is one of the landmarks of [biologic] achievements of the twentieth century. This is not just because it has been historically influential. Its importance is much deeper than that [...].

While [there exist] several non-trivial difficulties with [Crick's] version of that research program, his strategy as a whole has by no means been refuted. It is doubtful, moreover that it is even possible to refute a strategy [...].

In any case, [Crick's] program is not utterly without antecedents [...] [molecular biology] makes sufficient improvement on those antecedents that, as long as we consider [the "central dogma"] as a program rather than a doctrine, we can hope that some way can be found for it to triumph over the difficulties we have noted.

In the meantime we have [several] programs of [biological] research [...] As programs of research perhaps neither should be reviled as dogma" (modified from: [77] (pp. 384–385)).

Taken together, we can see that the central dogma is also not suited for ensuring molecular biology's disciplinary identity in a strict sense. On the contrary, it might be seen as indicative of an area of science without any such strict boundaries. The same observation is underpinned by the diversity of different methods, techniques, figures, and intricate historical interweaving of sciences and scientists we have seen so far. One might thus wonder whether molecular biology might just simply be a mixture of other areas of science.

3.6. Is Molecular Biology Just a Mash-Up of Other Different Scientific Disciplines?

Does all this mixing of methods and techniques make molecular biology "a conglomerate, in which biochemistry, structural research, biophysics, genetics, microbiology, immunology, virology and whatever else is mixed together" [2] (p. 34, my translation) or even a "science goulash"? ("[W]hoever violates these boundary lines [between the individual sciences] makes a goulash out of the sciences and becomes, for example, a molecular biologist" [2] (p. 72, my translation)). And what about the rest of (non-molecular) biology? The appearance of "classical" biology has certainly changed through its "molecularization" or "molecular revolution". However, it might rather be the concept of a demarcated scientific discipline that needs to be called into question than "mourning" other branches of biology that have not disappeared.

From all we have established so far, it is safe to say that regardless of its status as its own scientific discipline, molecular biology has always been a fruitful synergy of different approaches and techniques—and will continue to be so. It is exactly the openness and changes that drive science forward. Which label we put on certain things might be important for organizing departments and curricula, but insisting on strict disciplinary boundaries appears to be obsolete in today's scientific landscape.

Instead of pitting different fields against each other, perhaps the best path forward is to acknowledge the interdisciplinary character of molecular biology. "In truth the chief feature of molecular biology has been its interdisciplinarity" [78] (p. 511). Along the same lines, it is not surprising and testifies to the interdisciplinary attitude that in a bioinformatics textbook [79] the following is given as a motto: "If nature does not discriminate between physics, chemistry and biology, why should [...] the sciences that study it do so?" [80], (p. 3, my translation).

Now that we have established interdisciplinarity as an intrinsic feature of molecular biology—held together by a common epistemic interest that tries to understand biological phenomena by studying the interactions of molecules in and between cells—we can proceed to answer the title question.

3.7. Answering the Question: Is "Molecular Biology" a Pleonasm or Denotation for a Discipline of Its Own?

My answer is this: molecular biology has always been more of an interdisciplinary framework and thus not a separate scientific discipline of its own (understood in a *narrow* sense). However, for practical and organizational reasons, for academic research and teaching, and for reasons of a common epistemic interest, it is indeed a subject area that might justifiably be called a discipline in a *wider* sense, acknowledging the vagueness of applying the term loosely.

The truth, once again, lies somewhere in the middle it appears. *Neither* is "molecular biology" a pleonasm that adds nothing to the semantic content of "biology", *nor* is it a clear-cut and well-delineated denotation for a discipline of its own. With this answer and a distinction between a wider and more narrow conception of what counts as a scientific discipline, I find myself close to R. Olby's position: (Similarly also: "molecular biology (broadly defined as the set of disciplines looking for molecular explanations and using molecular tools like DNA sequencing and PCR: developmental biology, molecular genetics, genomics, immunology, molecular neurology, etc.)" [81] (p. 2).) "It is this association between structural investigation and genetic mechanisms that gave to the molecular biology of the 1950s its alleged novelty, and justified the claim that here was a new discipline formed out of the fusion of specialism, and quite unlike anything that had preceded it. This *narrow* conception is in contrast to the *broad* conception of the subject held by those responsible for introducing the term 'molecular biology' in the 1930s and 1940s" [78] (pp. 503–504, original emphasis).

In sum, it is justified to call molecular biology a scientific discipline due to its underlying epistemic interests to investigate biological phenomena through the molecular interactions within and between cells, the battery of techniques required to do so, and the efforts to advance them, irrespective of any demarcation between the life sciences.

Even in uncontroversial cases that count as scientific disciplines (e.g., mathematics, geology, Egyptology, etc.), they all have a formative history. This applies to other areas and periods of science as well. (This also includes other areas of biology. An interesting comparison could be made to the "modern synthesis" of evolutionary biology and its recent extensions and modifications. Interesting parallels might exist to the inception of this branch of science, according to E. Mayr [82] and more recent scholarship [83]. However, this would be beyond the scope of this paper.) Nevertheless, the tension of the topic at hand is best exemplified in molecular biology: "The [disciplines] that are institutionally [. . .] anchored today are historically grown organizational structures that are not due to the "nature" of things" [25] (p. 237, my translation). Trying to define and defend strict demarcation lines between disciplines is equally unhelpful as putting on blinders and specializing in increasingly smaller and narrower domains. "Excessive specialization can impair the perception of problems that do not fall within the framework of a single [scientific discipline] [. . .]. Working across academic subjects, i.e., intradisciplinary or even interdisciplinary or transdisciplinary research, counteracts this" [25] (p. 238, my translation).

I want to close here with a quotation that nicely summarizes our results so far, as well as pointing out open questions—especially with regards to what makes up the epistemic interest that holds molecular biology together: "Molecular biology, as I hope to show, must be regarded as the result of an extraordinarily complex development that can by no means be described in an adequate fashion through, for example, the fusion of already existing biological disciplines, such as microbiology, genetics, or biochemistry. Nor is it simply another biological discipline supplementing the historically established canon of biological disciplines" [7] (p. 33). "It is therefore not simply a tautological joke when Francis Crick proposed, on "doubtful grounds," as he admitted with an ironically self-deprecating gesture, that molecular biology "can be defined as anything that interests molecular biologists" [84]" [7] (p. 33). (An interesting suggestion has been made by one of the reviewers, namely, to characterize molecular biology *negatively*, i.e., by referring to anything that does *not* interest molecular biologists. These might include biological phenomena without chemical details, physiological aspects of entire organisms, ecological interactions, etc. I agree that this might often provide clearer disciplinary boundaries in practice).

With a more rigorous analysis of its disciplinary past and trajectory in hand, however, we can very well characterize molecular biology as a scientific discipline via its underlying epistemic interest.

4. Molecular Biology: An Interdisciplinary Past and an Open Future

With this result and answer to the title question, we might also ask whether this still applies to molecular biology today and what we can expect for its future. What can we speculate about the future of this area of research called molecular biology? Any predictions and prognosis might be futile, as it is the nature of science not to follow a directional path towards a pre-defined goal. (It is also interesting to revisit Crick's predictions for "molecular biology in the year 2000" [84].) "What we can guess today will not be realized. Change is bound to occur anyway, but the future will be different from what we believe. This is especially true in science. The search for knowledge is an endless process and one can never tell how it is going to turn out. Unpredictability is in the nature of the scientific enterprise. If what is to be found is really new, then it is by definition unknown in advance. There is no way of telling where a particular line of research will lead" [85] (p. 67, cited after [7] (p. 182)).

One thing appears to me of great importance here. It is not something that only concerns molecular biology but is of particular relevance—that is the large amounts of data that pose their own conceptual problems for the life sciences [86,87]. All the genome, sequence and protein databases and the excessive number of papers being published provide a "rapidly evolving body of knowledge, the accumulation of innumerable incalculable facts that overwhelm the individual's ability to think and remember" [2] (p. 115, my translation)—thus calling for new methods to deal with them.

In light of these challenges, some turn towards new horizons and claim to see such new approaches—and new disciplines—in the emergence of systems biology and synthetic biology. "New generations", "scientific revolutions", "paradigm changes", the "end" or "death" of molecular biology, or even a "new era" of biology have been announced multiple times (I am also omitting citations here to keep the references to a manageable size). Given the result we have obtained as an answer to the title question, however, we can be cautious in following such claims. When discussing the alleged "death of molecular biology", M. Morange offers the following claim: "Molecular biology was never a discipline [in a narrow sense] and the apparent signs of death as a discipline—absence of new chairs and journals—is meaningless. It was created as a research area, which is still fully active" [88] (p. 40).

Besides all that, however, I argue that the framework of molecular biology as a discipline (in a wider sense) offers enough space and possibilities for new developments to be included in its epistemic interest. Only if that were not the case, one could hold dogmatism against molecular biology. Even if science (for a number of structural and sociopolitical reasons) has become more conservative [74] and less "disruptive" [89], the accusation of dogmatism would not be justified (see Section 3.5.2). Regardless of single narrow-minded individuals, there are no principal obstacles to trying new approaches in molecular biology. For example, including more of a macroscopic perspective has—despite its reductionist tendencies in the past—never been totally excluded from molecular biology. This is even acknowledged by proponents of such an "evolution of molecular biology into systems biology": "It is, however, incorrect to state that integrative thinking is new to molecular biology" [90] (p. 1249).

"[S]ystems and synthetic biology, [. . .] are, ultimately, more a continuation and culmination of long-standing research programmes in biology than revolutionary new approaches" [91] (p. S42). Before prematurely claiming scientific "revolutions" or "paradigm shifts", biology would do well to remember its past. "Although the development of systems biology and synthetic biology cannot be compared to the rise of molecular biology in the 1950s, the changes that these two new fields make to the way that biologists work and to what they consider as proof of value, as well as their possible contribution to reducing the gap between functional biology and evolutionary biology, are important for the discipline as a whole" [92] (p. S53).

Putting on a new disciplinary label would hardly bring any significant innovation or improvement to biology. Instead, I think it would be beneficial to use the term "biology" in a more inclusive, interdisciplinary semantic meaning, as it has already been the case for "molecular biology" from the start. Perhaps the expression "life sciences" is the most inclusive one that embraces the plurality of (sub)disciplines at work. Rigid disciplinary boundaries and habitual adherence to assumptions and principles, of which no one knows exactly when—and for what reason—they became established, have certainly promoted scientific progress less often than those phases where new approaches, methods, and technologies were used to open new scientific horizons. These are lessons one can take from the history and philosophy of molecular biology.

The obscurity and large sizes of scientific projects can be overwhelming and call for particular emphasis on interdisciplinary collaborations. This cannot just be a temporary meeting of representatives of different disciplines in the same room or the publication of articles between the same two book covers without really having reached any understanding and afterwards everyone proceeding in their field as before. To reach a proper understanding and progress, "science needs philosophy" [93].

My aim in this paper was to show that the history and philosophy of science offer important lessons and can contribute to a proper understanding of the disciplinary status of molecular biology as well as its current situation and future developments. I argued that "molecular biology" is neither a pleonasm nor a discipline in a narrow sense but that we should acknowledge its interdisciplinary past and open future. One essential insight in that respect is "that we turn away from the perspective of a more or less well-defined disciplinary matrix of twentieth century biology" [7] (p. 34). That is all the more valid for twenty-first century biology.

In addition, learning from the past about one's area of research—irrespective of any labels seemingly demarcating separate disciplines—can be informative for contemporary researchers forging their own path, which often might not follow any clear demarcation lines either. Life scientists of the twenty-first century will be well advised to be open to new technologies and experimental techniques from different directions instead of defending any disciplinary territory. In fact, it may be exactly this openness that facilitated molecular biology as a scientific discipline in the first place—properly understood as an epistemic interest to study biological phenomena at the molecular and cellular level—while implementing a diverse set of techniques that allow us to manipulate and study the relevant macromolecules at play. In my opinion, this is what qualified molecular biology as a discipline in the past, still does at present, and will continue to do so in the future.

Funding: Open Access Funding by the University of Vienna.

Acknowledgments: I wish to thank the reviewers and editors or their work and critical comments. This project has received support from the European Research Council (ERC) under the European Union's Horizon 2020 research and innovation programme (grant agreement No. 818772).

Conflicts of Interest: The author declares no conflict of interest.

References

1. Keynes, M.; Cox, T. William Bateson, the Rediscoverer of Mendel. *J. R. Soc. Med.* **2008**, *101*, 104. [CrossRef]
2. Chargaff, E. *Unbegreifliches Geheimnis: Wissenschaft Als Kampf Für Und Gegen Die Natur*, 2nd ed.; Klett-Cotta: Stuttgart, Germany, 1981; ISBN 978-3-12-901621-3.
3. Maienschein, J.; Laubichler, M.; Loettgers, A. How Can History of Science Matter to Scientists? *Isis* **2008**, *99*, 341–349. [CrossRef]
4. Wuketits, F.M. *Biologische Erkenntnis: Grundlagen und Probleme*; UTB; Fischer: Stuttgart, Germany, 1983; ISBN 978-3-437-20296-4.
5. Nachtigall, W. *Biologische Forschung: Aspekte, Argumente, Aussagen*; Quelle & Meyer: Heidelberg, Germany, 1972; ISBN 978-3-494-00689-5.
6. Brush, S.G. Should the History of Science Be Rated X?: The Way Scientists Behave (According to Historians) Might Not Be a Good Model for Students. *Science* **1974**, *183*, 1164–1172. [CrossRef]
7. Rheinberger, H.-J. *Towards a History of Epistemic Things: Synthesizing Proteins in the Test Tube*; Stanford University Press: Redwood City, CA, USA, 1997.
8. Judson, H.F. *The Eighth Day of Creation: Makers of the Revolution in Biology*; Simon and Schuster: New York, NY, USA, 1979.
9. Friedberg, E.C.; Walker, G.C.; Siede, W.; Wood, R.D.; Schultz, R.A.; Ellenberger, T. *DNA Repair and Mutagenesis*; ASM Press: Washington, DC, USA, 2005; ISBN 978-1-68367-188-6.
10. Tabery, J.; Piotrowska, M.; Darden, L. Molecular Biology. In *The Stanford Encyclopedia of Philosophy*; Zalta, E.N., Ed.; Stanford University: Stanford, CA, USA, 2016.
11. de Chadarevian, S. Of Some Paradoxes in the Historiography of Molecular Biology. *Ber Wiss.* **2022**, *45*, 462–467. [CrossRef]
12. Chargaff, E. *Über das Lebendige: Ausgewählte Essays*; Klett-Cotta: Stuttgart, Germany, 1993; ISBN 978-3-608-95976-5.
13. Hausmann, R. *Und Wollten Versuchen, das Leben zu Verstehen: Betrachtungen zur Geschichte der Molekularbiologie*; Wiss. Buchges. [Abt. Verl.]: Darmstadt, Germany, 1995; ISBN 978-3-534-11575-4.
14. Kendrew, J.C. Some Remarks on the History of Molecular Biology. *Biochem. Soc. Symp.* **1970**, *30*, 5–10. [PubMed]
15. Vollmer, G. Die Wissenschaft Vom Leben: Das Bild Der Biologie in Der Öffentlichkeit. In *Biophilosophie*; Vollmer, G., Ed.; Reclam: Stuttgart, Germany, 1995; pp. 5–32.
16. Janich, P.; Weingarten, M. *Wissenschaftstheorie der Biologie: Methodische Wissenschaftstheorie und Die Begründung der Wissenschaften*; UTB für Wissenschaft Uni-Taschenbücher Philosophie, Biologie; Fink: München, Germany, 1999; ISBN 978-3-8252-2033-4.

17. Wuketits, F.M. *Wissenschaftstheoretische Probleme Der Modernen Biologie*; Duncker & Humblot: Berlin, Germany, 1978; ISBN 978-3-428-04279-1.
18. Sarkar, S. *Molecular Models of Life: Philosophical Papers on Molecular Biology*; MIT Press: Cambridge, MA, USA; London, UK, 2005.
19. Powell, A.; Dupré, J. From Molecules to Systems: The Importance of Looking Both Ways. *Stud. Hist. Philos. Biol. Biomed. Sci.* **2009**, *40*, 54–64. [CrossRef] [PubMed]
20. de Chadarevian, S.; Rheinberger, H.-J. Introduction. *Stud. Hist. Philos. Sci. Part C Stud. Hist. Philos. Biol. Biomed. Sci.* **2009**, *40*, 4–5. [CrossRef]
21. Burian, R.M. Underappreciated Pathways Toward Molecular Genetics as Illustrated by Jean Brachet's Cytochemical Embryology. In *The Philosophy and History of Molecular Biology: New Perspectives*; Sarkar, S., Ed.; Kluwer Academic Publishers: Dordrecht, The Netherlands; Boston, MA, USA; London, UK, 1996; pp. 67–85.
22. Suárez-Díaz, E. Molecular Evolution: Concepts and the Origin of Disciplines. *Stud. Hist. Philos. Sci. Part C Stud. Hist. Philos. Biol. Biomed. Sci.* **2009**, *40*, 43–53. [CrossRef]
23. Sarkar, S. Biological Information: A Skeptical Look at Some Central Dogmas of Molecular Biology. In *The Philosophy and History of Molecular Biology: New Perspectives*; Sarkar, S., Ed.; Kluver: Dordrecht, The Netherlands, 1996; pp. 187–231.
24. Culp, S.; Kitcher, P. Theory Structure and Theory Change in Contemporary Molecular Biology. *Br. J. Philos. Sci.* **1989**, *40*, 459–483. [CrossRef]
25. Gräfrath, B. Disziplin, Wissenschaftliche. In *Enzyklopädie Philosophie und Wissenschaftstheorie*; Metzler: Stuttgart, Germany, 2005; Band II; pp. 237–238.
26. Brooks, D.S.; DiFrisco, J.; Wimsatt, W.C. (Eds.) *Levels of Organization in the Biological Sciences*; MIT Press: Cambridge, MA, USA; London, UK, 2021.
27. Alberts, B.; Johnson, A.; Lewis, J.; Morgan, D.; Raff, M.; Roberts, K.; Walter, P. (Eds.) *Molecular Biology of the Cell*, 6th ed.; Garland Science: New York, NY, USA, 2015.
28. Lodish, H.F. *Molecular Cell Biology*, 9th ed.; Macmillan Learning: Austin, TX, USA, 2021; ISBN 978-1-319-20852-3.
29. Keller, E.F. Physics and the Emergence of Molecular Biology: A History of Cognitive and Political Synergy. *J. Hist. Biol.* **1990**, *23*, 389–409. [CrossRef]
30. Morange, M. *A History of Molecular Biology*; Harvard University Press: Cambridge, MA, USA, 1998.
31. Wolters, G. Biologie. In *Enzyklopädie Philosophie und Wissenschaftstheorie*; Metzler: Stuttgart, Germany, 2005; Band I; pp. 474–477.
32. McLaughlin, P. Naming Biology. *J. Hist. Biol.* **2002**, *35*, 1–4. [CrossRef]
33. Poon, W.C.K. Interdisciplinary Reflections: The Case of Physics and Biology. *Stud. Hist. Philos. Sci. Part C Stud. Hist. Philos. Biol. Biomed. Sci.* **2011**, *42*, 115–118. [CrossRef]
34. Sander, K. Zur Geschichte Der Biologie. In *Biologie 1: Systeme des Lebendigen*; Todt, D., Ed.; Fischer-Taschenbuch-Verlag: Frankfur am Main, Germany, 1976; pp. 23–44.
35. Kay, L.E. *The Molecular Vision of Life: Caltech, The Rockefeller Foundation and the Rise of the New Biology*; Oxford University Press: Oxford, UK, 1993.
36. Weaver, W. Molecular Biology: Origin of the Term. *Science* **1970**, *170*, 581–582. [CrossRef] [PubMed]
37. Kendrew, J.C. (Ed.) *The Encyclopedia of Molecular Biology*; Reprinted as a Paperback; Blackwell Science: Oxford, UK, 1995; ISBN 978-0-86542-621-4.
38. Rheinberger, H.-J. Kurze Geschichte Der Molekularbiologie. In *Geschichte der Biologie*; Jahn, I., Ed.; Fischer: Jena, Germany, 1998; pp. 642–663.
39. Kendrew, J.C. Molecular Biology. In *The Encyclopedia of Molecular Biology*; Kendrew, J.C., Ed.; Blackwell Science: Oxford, UK; Wien, Austria, 1994; p. 664.
40. Waddington, C.H. Molecular Biology or Ultrastructural Biology? *Nature* **1961**, *190*, 184. [CrossRef]
41. Waddington, C.H. Molecular Biology or Ultrastructural Biology? *Nature* **1961**, *190*, 1124–1125. [CrossRef]
42. Astbury, W.T. Molecular Biology or Ultrastructural Biology? *Nature* **1961**, *190*, 1124. [CrossRef] [PubMed]
43. Stent, G.S. That Was the Molecular Biology That Was. *Science* **1968**, *160*, 390–395. [CrossRef]
44. Abir-Am, P.G. Themes, Genres and Orders of Legitimation in the Consolidation of New Scientific Disciplines: Deconstructing the Historiography of Molecular Biology. *Hist. Sci.* **1985**, *23*, 73–117. [CrossRef]
45. Götschl, J. *Erwin Schrödinger's World View: The Dynamics of Knowledge and Reality*; Springer: Dordrecht, The Netherlands, 1992; ISBN 978-94-011-2428-7.
46. Phillips, R. Schrödinger's What Is Life? At 75. *Cell Syst.* **2021**, *12*, 465–476. [CrossRef]
47. Ball, P. Schrödinger's Cat among Biology's Pigeons: 75 Years of What Is Life? *Nature* **2018**, *560*, 548–550. [CrossRef]
48. Van Regenmortel, M.H.V. Reductionism and Complexity in Molecular Biology. *EMBO Rep.* **2004**, *5*, 1016–1020. [CrossRef] [PubMed]
49. Morange, M. Recent Opportunities for an Increasing Role for Physical Explanations in Biology. *Stud. Hist. Philos. Sci. Part C Stud. Hist. Philos. Biol. Biomed. Sci.* **2011**, *42*, 139–144. [CrossRef]
50. Schrödinger, E. *What Is Life? The Physical Aspect of the Living Cell; With, Mind and Matter; & Autobiographical Sketches*; Cambridge University Press: Cambridge, UK; New York, NY, USA, 1944; ISBN 978-0-521-42708-1.
51. Bohr, N. Light and Life. *Nature* **1933**, *131*, 421–423, 457–459. [CrossRef]

52. Delbrück, M. *A Physicist Looks at Biology*; Reprinted 1966 in: Phage and the Origins of Molecular Biology; Cold Spring Harbor Laboratory Press: Cold Spring Harbor, New York, USA, 1949.
53. Nicholson, D.J. *(Forthcoming) What Is Life? Revisited*; Cambridge University Press: Cambridge, UK, 2023; ISBN 978-1-00-912418-8.
54. Jacob, F. *The Logic of Life: A History of Heredity*; New Princeton Science Library, Ed.; Princeton Science Library; Princeton University Press: Princeton, NJ, USA, 2022; ISBN 978-0-691-18284-1.
55. Dronamraju, K.R. Erwin Schrödinger and the Origins of Molecular Biology. *Genetics* **1999**, *153*, 1071–1076. [CrossRef]
56. Watson, J.D. Growing up in the Phage Group. In *Phage and the Origins of Molecular Biology*; Cairns, J., Stent, G.S., Watson, J.D., Eds.; Cold Spring Harbor Laboratory Press: Plainview, NY, USA, 1966; pp. 239–245.
57. Olby, R. Schrödinger's Problem: What Is Life? *J. Hist. Biol.* **1971**, *4*, 119–148. [CrossRef]
58. Yoxen, E.J. Where Does Schroedinger's "What Is Life?" Belong in the History of Molecular Biology? *Hist. Sci.* **1979**, *17*, 17–52. [CrossRef]
59. Crow, J.F. Erwin Schrödinger and the Hornless Cattle Problem. *Genetics* **1992**, *130*, 237–239. [CrossRef]
60. Sarkar, S. What Is Life? Revisited. *BioScience* **1991**, *41*, 631–634. [CrossRef]
61. Rheinberger, H.-J. Recent Science and Its Exploration: The Case of Molecular Biology. *Stud. Hist. Philos. Sci. Part C Stud. Hist. Philos. Biol. Biomed. Sci.* **2009**, *40*, 6–12. [CrossRef]
62. Abir-Am, P.G. How Scientists View Their Heroes: Some Remarks on the Mechanism of Myth Construction. *J. Hist. Biol.* **1982**, *15*, 281–315. [CrossRef] [PubMed]
63. Smart, J.J.C. Can Biology Be an Exact Science? *Synthese* **1959**, *11*, 359–368. [CrossRef]
64. Woodger, J.H. *The Axiomatic Method in Biology*, 1st ed.; Cambridge University Press: Cambridge, UK, 1937.
65. Woodger, J.H. Biology and the Axiomatic Method. *Ann. N. Y. Acad. Sci.* **1962**, *96*, 1093–1116. [CrossRef]
66. Crick, F.H. On Protein Synthesis. *Symp. Soc. Exp. Biol.* **1958**, *12*, 138–163. [PubMed]
67. Crick, F.H. Central Dogma of Molecular Biology. *Nature* **1970**, *227*, 561–563. [CrossRef] [PubMed]
68. Yarus, M. *Life from an RNA World: The Ancestor Within*; Harvard University Press: Cambridge, MA, USA; London, UK, 2010; ISBN 978-0-674-06071-5.
69. Crick, F.H. *What Mad Pursuit: A Personal View of Scientific Discovery*; Basic Books: New York, NY, USA, 1988.
70. Shapiro, J.A. Revisiting the Central Dogma in the 21st Century. *Ann. N. Y. Acad. Sci.* **2009**, *1178*, 6–28. [CrossRef]
71. Koonin, E.V. Does the Central Dogma Still Stand? *Biol. Direct* **2012**, *7*, 27. [CrossRef]
72. Ille, A.M.; Lamont, H.; Mathews, M.B. The Central Dogma Revisited: Insights from Protein Synthesis, CRISPR, and Beyond. *WIREs RNA* **2022**, *13*, e1718. [CrossRef]
73. Delbrück, M. Aristotle-Totle-Totle. In *Of Microbes and Life*; Monod, J., Borek, E., Eds.; Columbia University Press: New York, NY, USA, 1971; pp. 50–55.
74. O'Connor, C. The Natural Selection of Conservative Science. *Stud. Hist. Philos. Sci. Part A* **2019**, *76*, 24–29. [CrossRef]
75. Ledford, H. Dogma-Defying Bacteria Package DNA in Unusual Ways. *Nature* **2023**, *614*, 401. [CrossRef]
76. Bieri, P. Was Bleibt von Der Analytischen Philosophie? *Dtsch. Z. Für Philos.* **2007**, *55*, 333–344. [CrossRef]
77. Creath, R. Every Dogma Has Its Day. *Erkenntnis* **1991**, *35*, 347–389. [CrossRef]
78. Olby, R. The Molecular Revolution in Biology. In *Companion to the History of Modern Science*; Olby, R.C., Cantor, G.N., Cristie, J.R.R., Hodge, M.J.S., Eds.; Routledge: Oxford, UK, 1990; pp. 503–520.
79. Hütt, M.-T.; Dehnert, M. *Methoden der Bioinformatik: Eine Einführung*; Springer-Lehrbuch; Springer: Berlin/Heidelberg, Germany, 2006; ISBN 978-3-540-25687-8.
80. Mittelstraß, J. Transdisziplinarität—Eine Chance Für Wissenschaft Und Philosophie. *Phys. Blätter* **1999**, *55*, 3. [CrossRef]
81. Barberousse, A.; Morange, M.; Pradeu, T. Introduction. In *Mapping the Future of Biology: Evolving Concepts and Theories*; Barberousse, A., Morange, M., Pradeu, T., Eds.; Boston Studies in the Philosophy of Science; Springer: Dordrecht, The Netherlands, 2009; ISBN 978-1-4020-9635-8.
82. Mayr, E.; Provine, W.B. (Eds.) *The Evolutionary Synthesis: Perspectives on the Unification of Biology*; Harvard University Press: Cambridge, MA, USA, 1980; ISBN 978-0-674-86537-2.
83. Laland, K.N.; Uller, T.; Feldman, M.W.; Sterelny, K.; Müller, G.B.; Moczek, A.; Jablonka, E.; Odling-Smee, J. The Extended Evolutionary Synthesis: Its Structure, Assumptions and Predictions. *Proc. R. Soc. B* **2015**, *282*, 20151019. [CrossRef]
84. Crick, F. Molecular Biology in the Year 2000. *Nature* **1970**, *228*, 613–615. [CrossRef]
85. Jacob, F. *The Possible and the Actual*; The Jessie and John Danz lectures; University of Washington Press: Seattle, DC, USA, 1982; ISBN 978-0-295-95888-0.
86. Leonelli, S. *Data-Centric Biology: A Philosophical Study*; The University of Chicago Press: Chicago, IL, USA; London, UK, 2016; ISBN 978-0-226-41633-5.
87. Leonelli, S. The Challenges of Big Data Biology. *eLife* **2019**, *8*, e47381. [CrossRef]
88. Morange, M. The Death of Molecular Biology? *Hist. Philos. Life Sci.* **2008**, *30*, 31–42.
89. Park, M.; Leahey, E.; Funk, R.J. Papers and Patents Are Becoming Less Disruptive over Time. *Nature* **2023**, *613*, 138–144. [CrossRef]
90. Westerhoff, H.V.; Palsson, B.O. The Evolution of Molecular Biology into Systems Biology. *Nat. Biotechnol.* **2004**, *22*, 1249–1252. [CrossRef] [PubMed]
91. Potthast, T. Paradigm Shifts versus Fashion Shifts? *EMBO Rep.* **2009**, *10*, S42–S45. [CrossRef] [PubMed]

92. Morange, M. A New Revolution? The Place of Systems Biology and Synthetic Biology in the History of Biology. *EMBO Rep.* **2009**, *10*, S50–S53. [CrossRef] [PubMed]
93. Laplane, L.; Mantovani, P.; Adolphs, R.; Chang, H.; Mantovani, A.; McFall-Ngai, M.; Rovelli, C.; Sober, E.; Pradeu, T. Opinion: Why Science Needs Philosophy. *Proc. Natl. Acad. Sci. USA* **2019**, *116*, 3948–3952. [CrossRef]

biomolecules

Review

The History and Science of the Major Birch Pollen Allergen Bet v 1

Heimo Breiteneder * and Dietrich Kraft

Division of Medical Biotechnology, Department of Pathophysiology and Allergy Research, Center of Pathophysiology, Infectiology and Immunology, Medical University of Vienna, 1090 Vienna, Austria; dietrich.kraft@aon.at
* Correspondence: heimo.breiteneder@meduniwien.ac.at

Abstract: The term allergy was coined in 1906 by the Austrian scientist and pediatrician Clemens Freiherr von Pirquet. In 1976, Dietrich Kraft became the head of the Allergy and Immunology Research Group at the Department of General and Experimental Pathology of the University of Vienna. In 1983, Kraft proposed to replace natural extracts used in allergy diagnostic tests and vaccines with recombinant allergen molecules and persuaded Michael Breitenbach to contribute his expertise in molecular cloning as one of the mentors of this project. Thus, the foundation for the Vienna School of Molecular Allergology was laid. With the recruitment of Heimo Breiteneder as a young molecular biology researcher, the work began in earnest, resulting in the publication of the cloning of the first plant allergen Bet v 1 in 1989. Bet v 1 has become the subject of a very large number of basic scientific as well as clinical studies. Bet v 1 is also the founding member of the large Bet v 1-like superfamily of proteins with members—based on the ancient conserved Bet v 1 fold—being present in all three domains of life, i.e., archaea, bacteria and eukaryotes. This suggests that the Bet v 1 fold most likely already existed in the last universal common ancestor. The biological function of this protein was probably related to lipid binding. However, during evolution, a functional diversity within the Bet v 1-like superfamily was established. The superfamily comprises 25 families, one of which is the Bet v 1 family, which in turn is composed of 11 subfamilies. One of these, the PR-10-like subfamily of proteins, contains almost all of the Bet v 1 homologous allergens from pollen and plant foods. Structural and functional comparisons of Bet v 1 and its non-allergenic homologs of the superfamily will pave the way for a deeper understanding of the allergic sensitization process.

Keywords: Bet v 1; major birch pollen allergen; molecular allergology; PR-10-like protein family; allergen ligands

Citation: Breiteneder, H.; Kraft, D. The History and Science of the Major Birch Pollen Allergen Bet v 1. *Biomolecules* **2023**, *13*, 1151. https://doi.org/10.3390/biom13071151

Academic Editors: Mark Rinnerthaler and Markus Ralser

Received: 19 June 2023
Revised: 14 July 2023
Accepted: 17 July 2023
Published: 19 July 2023

1. Events That Led to the Discovery of the Major Birch Pollen Allergen Bet v 1

Clemens von Pirquet, a true legend of immunology [1], observed that antibodies were not only part of protective immune responses but could also cause diseases. He coined the word "allergy," derived from ancient Greek (allos = other and ergon = work), to generally describe a change in the reactive capability of the immune system. He presented his findings in 1906 in the Münchener Medizinische Wochenschrift, thus introducing the term "allergy" to the medical terminology.

In 1976, Dietrich Kraft [2] became the head of the Allergy and Immunology Research Group at the Department of General and Experimental Pathology, then of the Medical Faculty of the University of Vienna, which would become the Medical University of Vienna in 2004. In autumn 1983, Kraft turned his full focus to allergic diseases. Working in an outpatient allergy clinic in Vienna together with his colleague Herwig Ebner, he came to the insight that allergy test solutions and hyposensitization solutions could only be standardized based on pure recombinant molecules. Kraft elaborated on this idea during a lengthy stay in hospital due to a severe bicycle accident. Consequently, on 12 December

1983, he called Michael Breitenbach on the phone, as a meeting in person was impossible due to the unusually heavy snowfall that day. During this one hour call, Kraft managed to persuade Michael Breitenbach to contribute his essential expertise in molecular cloning to this newly hatched project. Thus, the keystone for the founding of the Vienna School of Molecular Allergology was put into place. Kraft also recruited Otto Scheiner and Helmut Rumpold, who brought additional expertise, i.e., immunohistochemistry and medical laboratory techniques, to the project.

From 1983 to 1984, investing in the cloning and production of recombinant allergens was of no interest for funding agencies in Austria. In a similar manner, pharmaceutical companies active in the allergy field in countries including Denmark, Sweden, Germany, and the USA did not support Kraft's plans, deeming them to be too exotic. During a banquet held by the Austrian Society of Allergology and Immunology (ÖGAI), Kraft ended up sitting next to Jörg Mayrhofer, to whom he laid open his vision for recombinant allergens. Mayrhofer, then owner of the Schutzengel Apotheke (Guardian Angel Pharmacy) in Linz, Austria, showed an interest in Kraft's plans as a potential business opportunity. During a follow-up visit to Linz, Jörg Mayrhofer and his father, Theodor Mayrhofer, agreed to fund Kraft's ambitious plans. Consequently, in 1984, the company Biomay was founded that would finance all the formative years of Kraft's team of molecular allergology.

As around 5% of the Austrian population suffer from pollinosis induced by birch pollen in early spring, and as there is only one major allergen present in birch pollen extract, this allergen was chosen as the target. Following the rules of allergen nomenclature laid down by the WHO/IUIS Allergen Nomenclature Sub-Committee [3] and the then-valid taxonomical name for the white birch, *Betula verrucosa*, this allergen was designated as Bet v 1. Michael Breitenbach selected Heimo Breiteneder as the person who would be responsible for the cloning of the cDNA for Bet v 1. Breiteneder had just finished his thesis on the genome organization of the cyanelles of *Cyanophora paradoxa* [4] under the guidance of Wolfgang Löffelhardt at the Institute of General Biochemistry of the University of Vienna.

On 2 May 1985, the work on the cloning of the Bet v 1 cDNA started in earnest. At first, the isolation of RNA from birch tissues needed to be established. This was done from roots of birch seedlings, and from leaves and inflorescences of mature trees. Finally, pure pollen from birch trees was obtained, RNA was isolated and poly(A)$^+$ mRNA was enriched. This mRNA preparation was translated in vitro in a cell-free wheat germ system, and the proteins synthesized were separated by SDS-PAGE and transferred to nitrocellulose. When the blots were incubated with sera from patients allergic to birch pollen, IgE binding to a 17-kD protein, presumably Bet v 1, was observed, indicating the presence of allergen-coding mRNA in the RNA preparations [5]. Based on this finding, the *E. coli* phage λgt11 was selected to produce a cDNA expression library from birch pollen poly(A)$^+$ mRNA. Michael Breitenbach was also instrumental in obtaining further knowhow from Arnold Bito and Klaus Richter at the Academy of Sciences in Salzburg on working with phage λgt11. The expression library was going to be screened, in a manner identical to immunoblots—with sera from birch pollen allergic patients—a combination of techniques that had not been previously used.

Phage λgt11 clones containing Bet v 1-encoding cDNAs, and thus producing recombinant Bet v 1 in infected *E. coli* cells, were first identified on 3 July 1988. The full cDNA sequence of Bet v 1 was published in the EMBO Journal in 1989 [6], representing the most abundant isoform in birch pollen, called Bet v 1.0101. Thus, Bet v 1 became the first cloned plant allergen and also the first allergenic PR-10-like protein that was published worldwide. Its sequence was similar to the N-terminal peptide sequences of the pollen allergens of hazel, alder and hornbeam, trees belonging to the order of Fagales and, therefore, closely related to birch [7]. What came as a surprise was that Bet v 1 also displayed a 55% sequence identity with a pea disease resistance response gene. Pea and birch being taxonomically quite distant, this was the first indication of a gene family whose founding member must have existed a long time ago. In 1991, the cDNA sequence and recombinant protein of the alder pollen allergen Aln g 1 was published [8], followed in 1993 by the major allergen of hazel pollen, Cor a 1 [9]. This publication describes the identification of four cDNA clones whose open reading frames coded for different isoforms of the major hazel pollen allergen. Interestingly, recombinant proteins of these four isoforms displayed differing IgE binding capacities. In 1995, the cDNAs and recombinant proteins of the major allergens of apple, Mal d 1 [10], and of celery, Api g 1 [11], joined the ranks of allergenic PR-10-like proteins, thus firmly establishing the presence of members of this gene family in the plant kingdom.

2. The PR-10-like Family of Allergenic Proteins

In 1980, van Lonn and colleagues defined pathogenesis-related proteins (PR proteins) as "proteins encoded by the host plant but induced only in pathological or related situations" and suggested the first groupings of such PR proteins into families [12]. Currently, 17 such families are described whose members are key components of the plant innate immune system [13]. Plants use PR proteins in inducible defense responses to combat various biotic and abiotic stresses. In contrast, Bet v 1, which is related by sequence to the PR-10 family of proteins, is constitutively expressed in pollen at rather high concentrations. Hence, the correct term for Bet v 1 and its homologous proteins is PR-10-like proteins. The Bet v 1 homologs are slightly acidic small (154–163 amino acid residues) and predominantly cytoplasmic proteins with molecular masses of around 17 kDa. In general, these proteins are susceptible to gastric digestion [14] and heat processing results in a loss of the native protein fold via denaturation, oligomerization and precipitation along with a subsequent reduction in IgE recognition [15]. Some exceptions exist; for example, the allergenicity of the Bet v 1 homolog from carrot, Dau c 1, is not destroyed by cooking [16].

Birch flowers in spring, and its pollen is one of the most common causes of IgE-mediated allergies in Northern and Central Europe as well as in North America [17]. The major sensitizing allergen present in birch pollen is Bet v 1 which is regarded as a marker allergen for a primary sensitization to pollen of birch and other Fagales trees (e.g., alder, hazel, hornbeam, beech, oak). However, allergic reactions to Fagales pollen can be initiated independently by PR-10-like proteins from pollen of all members of the Betulaceae and Fagaceae families (Table 1). However, 25% of the IgE-binding epitopes of PR-10-like allergens from the pollen of trees of the Betuloideae and Coryloideae families are unique to these subfamilies, while pollen allergens from the Fagaceae are generally cross-reactive [18].

Table 1. Bet v 1-homologous pollen allergens of trees of the order Fagales.

Plant Family	Subfamily	Allergen Source	Allergen	References	UniProt/PDB
Betulaceae	Betuloideae	Birch (*Betula pendula*)	Bet v 1.0101	[6]	P15494/4A88
		Alder (*Alnus glutinosa*)	Aln g 1.0101	[8]	P38948
	Coryloideae	Hazel (*Corylus avellana*)	Cor a 1.0101	[9]	Q09407
		Hornbeam (*Carpinus betulus*)	Car b 1.0101	[19]	P38949
		Hop-hornbeam (*Ostrya carpinifolia*)	Ost c 1.0101	[18]	E2GL17
Fagaceae	Fagoideae	Beech (*Fagus silvatica*)	Fag s 1.0101	[18]	B7TWE6/6ALK
	Quercoideae	Oak (*Quercus alba*)	Que a 1.0201	[20]	B6RQS1
		Sawtooth oak (*Quercus acutissima*)	Que ac 1.0101	[21]	GenBank QOL10866.1
		Holly oak (*Quercus ilex*)	Que i 1.0101	[22]	A0A7D0TA82
		Mongolian oak (*Quercus mongolica*)	Que m 1.0101	[23]	GenBank AUH28179
	Castanoideae	Chestnut (*Castanea sativa*)	Cas s 1.0101	[18]	B7TWE3

Bet v 1 and its homologs in pollen are considered important inducers of birch pollen-associated plant food allergies [17]. The most frequently observed clinical entity, the oral allergy syndrome (OAS) [24], is caused by IgE antibodies that cross-react between Bet v 1 and its homologs in fruits, nuts, seeds and vegetables. Homologs of Bet v 1 have been identified in a wide range of plant foods (Table 2). Hence, Bet v 1-allergic patients are at risk to even react to novel foods without prior exposure.

Table 2. Plant food allergenic PR-10-like proteins.

Plant Family	Allergen Source	Allergen	References	UniProt/PDB
Actinidiaceae	Golden kiwifruit (*Actinidia chinensis*)	Act c 8.0101	[25]	D1YSM4
	Green kiwifruit (*Actinidia deliciosa*)	Act d 8.0101 Act d 11.0101	[25] [26]	D1YSM5 P85524/4IGV
Anacardiaceae	Mango (*Mangifera indica*)	Man i 2.0101	[27]	GenBank UYO79702.1
Apiaceae	Celery (*Apium graveolens*)	Api g 1.0101	[11]	P49372/2BK0
	Carrot (*Daucus carota*)	Dau c 1.0103	[28]	O04298/2WQL
Cannabaceae	Indian hemp (*Cannabis sativa*)	Can s 5.0101	[29]	I6XT51
Corylaceae	Hazelnut (*Corylus avellana*)	Cor a 1.0401	[30]	Q9SWR4/6GQ9, 6Y3H
Fabaceae	Peanut (*Arachis hypogaea*)	Ara h 8.0101	[31]	Q6VT83/4M9B
	Soybean (*Glycine max*)	Gly m 4.0101	[32]	P26987/2K7H
	Mung bean (*Vigna radiata*)	Vig r 1.0101 Vig r 6.0101	[33] [34]	Q2VU97 A0A1S3THR8/2FLH, 3C0V
Juglandaceae	English walnut (*Juglans regia*)	Jug r 5.0101	[35]	GenBank Acc. No. KX034087.1
Rosaceae	Strawberry (*Fragaria x ananassa*)	Fra a 1.0101	[36]	Q5ULZ4/6ST8
	Apple (*Malus domestica*)	Mal d 1.0101	[10]	P43211/5MMU
	Apricot (*Prunus armeniaca*)	Pru ar 1.0101	unpublished	O50001
	Cherry (*Prunus avium*)	Pru av 1.0101	[37]	O24248/1E09, 1H20
	Almond (*Prunus dulcis*)	Pru du 1.0101	[38]	B6CQS9
	Peach (*Prunus persica*)	Pru p 1.0101	[39]	Q2I6V8/6Z98
	Pear (*Pyrus communis*)	Pyr c 1.0101	[40]	O65200
	Raspberry (*Rubus idaeus*)	Rub 1 1.0101	[41]	Q0Z8U9
Solanaceae	Tomato (*Solanum lycopersicum*)	Sola l 4.0101	[42]	K4CWC5

In contrast to Bet v 1, Bet v 1-related food allergens are in general unable to act as primary sensitizers of predisposed individuals. The major reason for the very low ability of most plant food Bet v 1 homologs to induce sensitization is their high susceptibility to gastric digestion [14]. In addition to the not life-threatening symptoms of the OAS (e.g., itching, redness and tearing of the eyes, itch in the nose and oropharynx, sneezing, runny or stuffy nose, wheezing), severe reactions to Gly m 4, the Bet v 1 homolog from soybean, have been observed in a subpopulation of Bet v 1-allergic individuals [24,43].

3. Natural Ligands of PR-10-like Allergenic Proteins and Their Biological Functions

The three-dimensional structure of the major birch pollen allergen Bet v 1 was the first experimentally determined structure of a clinically important major inhalant allergen [44]. The so-called Bet v 1 fold (Figure 1) consists of a seven-stranded anti-parallel beta-sheet that wraps around a 25 residue-long C-terminal alpha-helix. The beta-sheet and the C-terminal part of the long a-helix are separated be two consecutive alpha-helices. The main structural feature of the Bet v 1 fold is a long forked cavity that penetrates the whole protein and that is solvent accessible via three openings to the protein's surface. The volume of the cavity is about 1500 Å^3 and its surface is predominantly hydrophobic. This hydrophobic cavity has the ability to bind a variety of experimental and physiologic ligands [45,46]. A detailed description of ligand classes interacting with PR-10 allergens is provided in the 2020 review by Aglas and colleagues [47]. McBride and colleagues chose various polyphenols, fatty acids, phenols, plant hormones, and one plant and one animal sterol to routinely screen for ligand binding characteristics of PR-10-like allergens [48]. However, studies on the physiologic natural ligands of PR-10-like allergens are still limited.

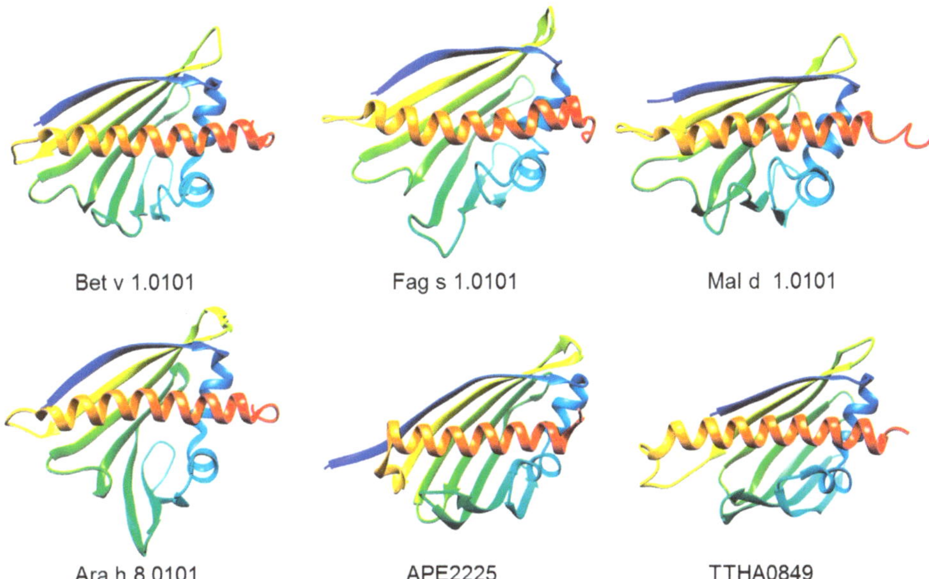

Figure 1. Ribbon representations of birch pollen Bet v 1 (PDB 4A88) and homologs from beech pollen (Fag s 1; PDB 6ALK), apple (Mal d 1; PDB 5MMU), peanut (Ara h 8; PDB 4M9B), the hyperthermophilic archaeon *Aeropyrum pernix* (APE2225; 2NS9) and the extremely thermophilic bacterium *Thermus thermophilus* (TTHA0849; PDB 2D4R) rainbow-colored from blue at the N-terminus to red at the C-terminus. The 3D images were created with the molecular modeling system UCSF ChimeraX (https://www.rbvi.ucsf.edu/chimerax/, accessed on 9 June 2023).

The first physiologic ligand of a PR-10-like protein was determined for Bet v 1. Quercetin-3-*O*-sophoroside (Q3OS), a glycosylated flavonol, was found as a natural ligand bound to Bet v 1 isolated from mature birch pollen [49]. Flavonoids contribute to pigment formation in flowering plants, are involved in plant hormone signaling, facilitate pollen tube formation and protect pollen DNA from UV radiation [50,51]. Flavonoids are stored in pollen as glycosylated precursors and, during the rehydration of the pollen grain, are processed into their active form by pollen glycosyltransferases. At this time point, Q3OS would need to be displaced from the Bet v 1-Q3OS complex to become accessible for deglycosylation. In solution, Bet v 1.0101 is conformationally heterogeneous and cannot be represented by a single structure [52]. NMR relaxation data suggest that structural dynamics are fundamental for ligand access to the protein interior. Complex formation then leads to a significant rigidification of the protein, along with a compaction of its three-dimensional structure, which is observed in the crystalized protein. Ligand binding stabilizes the conformation of Bet v 1, resulting in an increased melting point as well as drastically increased resistance towards endo-/lysosomal proteolysis [53]. The increased stability could hinder optimal proteolytic processing of the allergen, which would favor the development of a Th2 immune response. Analyses of human IgE binding on Bet v 1 in mediator release assays revealed that ligand-bound allergen-induced degranulation at lower concentrations. However, in basophil activation tests with human basophils, ligand-binding did not show this effect [53]. Phytoprostane E1 (PPE1) was identified as another physiologic ligand of Bet v 1 [54]. PPE1 interacts with Bet v 1, increasing its stability to proteolytic degradation. Pollen-derived PPE1 interacts with Bet v 1 with high affinity, increasing its stability and attenuating its degradation by processing by lysosomal cathepsin S. PPE1 also inhibited lysosomal cathepsins by blocking their catalytic cysteines. In addition, processing of Bet v 1.0101 and a hypoallergenic isoform differed distinctly, resulting in low- or high-density class II MHC loading and subsequently in Th2 and Th1 polarization, respectively [55]. Binding of glycosylated flavonoids to Bet v 1 isoforms is governed by the sugar moiety, and various isoforms show individual and highly specific binding behaviors for the different ligands [56]. It is tempting to speculate that the binding of phytoprostanes is also isoform-specific, resulting in the observed differences in Bet v 1 isoform allergenicity.

Natural Cor a 1 was extracted from mature hazel pollen, and its ligand was identified as quercetin-3-O-(2-O-β-D-glucopyranosyl)-β-D-galactopyranoside (Q3O-(Glc)-Gal) [57]. Similarly to nBet v 1, nCor a 1 is composed of different isoallergens and variants. The authors of this study were able to confirm the presence of the variants Cor a 1.0103 and Cor a 1.0104 on the basis of variant-specific tryptic peptides. Furthermore, four known Cor a 1 variants previously detected in hazel pollen and one isoform, Cor a 1.0401, detected only in hazel nuts [30] were analyzed as recombinant proteins for binding the ligand Q3O-(Glc)-Gal. Interestingly, Q3O-(Glc)-Gal was identified as a natural ligand of Cor a 1.0401 [57]. It was further demonstrated that Cor a 1.0401 and Bet v 1.0101 exhibited highly selective binding for their specific ligand but not for the respective ligand of the other allergen. A hypothesis of the role of Q3O-(Glc)-Gal in the unusual reproductive biology of hazel is presented by the authors of the study. The ligand is released from pollen, bound by Cor a 1.0401 present in the stigma, and hence in the nut, to prevent premature deglycosylation, and then only released after several months to be converted into quercetin to assist the formation of the secondary pollen tube.

Emanuelsson and co-workers reported that fruits of colorless white strawberry cultivars showed very low levels of Fra a 1 expression in contrast to red-colored fruits, as well as a downregulation of several enzymes in the pathway for the biosynthesis of flavonoids to which the red color pelargonidin belongs [58]. Casañal and coworkers presented crystallographic structures of Fra a 1 isoforms in complex with the naturally occurring flavonoid catechin, providing for the first time, a molecular basis for the function of these proteins in flavonoid biosynthesis [59]. The authors conclude that Fra a 1 proteins could act as transporters or "chemical chaperones" binding to flavonoid intermediates and making them available to processing enzymes, or they function as cytosolic transporters of flavonoids from the endoplasmic reticulum to other cellular membranes.

4. The Bet v 1-like Superfamily of Proteins

The determination of the Bet v 1 structure [44] paved the way for the search and definition of the Bet v 1-like superfamily. Radauer and colleagues used psc++, an improved version of the ProSup structural alignment program [60] to search the Protein Data Bank for structural homologs of Bet v 1 [61]. The resulting structures were then classified into eleven families, of which the Bet v 1 family was one. Today the so called Bet_v_1_like set, which corresponds to the original Pfam Clan CL0209, encompasses 25 families (https://www.ebi.ac.uk/interpro/set/pfam/CL0209/; accessed on 19 June 2023). The underscores in the allergen designation were added by the InterPro database. This designation is not in accordance with the official designation that was assigned by the WHO/IUIS Allergen Nomenclature Sub-Committee (http://allergen.org/viewallergen.php?aid=129, accessed on 19 June 2023). The most widely distributed families of the Bet v 1 superfamily were the polyketide cyclase family and the AHA1 (activator of Hsp90 ATPase homolog 1) family. Members of both families are found in bacteria, archaea and eukaryotes. The ring hydroxylases and the CoxG (named after the *cox g* gene encoding a subunit of the carbon monoxide dehydrogenase) families are widely distributed in bacteria and archaea and the StART (steroidogenic acute regulatory protein-related lipid transfer) family in bacteria and eukaryotes. The phosphatidylinositol transfer proteins are found only in eukaryotes, and members of the Bet v 1 family are exclusively present in plants.

The Bet v 1 family was classified into 11 subfamilies, including 9 from plants and 2 from bacteria [61]. The largest subfamily is the dicot PR-10 subfamily, which contains genuine PR proteins whose expression is upregulated upon pathogen infection, wounding or by abiotic stress, and PR-10-like proteins whose expression is developmentally regulated. The vast majority of allergens are found in the PR-10 subfamily. Allergenic members of other subfamilies were only rarely described and include the mung bean allergen Vig r 6 [34], a member of the subfamily of the cytokinin-specific binding proteins, and the kiwi allergen Act d 11 [26], a member of the major latex protein/ripening-related protein subfamily. So far, this limits the allergens to 3 of 11 subfamilies and to 1 of 25 families of the Bet v 1-like superfamily. The reason that allergenic members of the Bet v 1-like superfamily were found almost exclusively in the PR-10 subfamily is probably linked to the specific ligands they harbor (see Section 3 above), to other biologically active matrix components of the specific plant tissues and the ease or absence of exposure.

A truly ancient protein possessing the Bet v 1 fold must have existed as the last universal common ancestor of this superfamily, giving rise to proteins with diverse functions by insertion of additional structural elements or by fusion to other domains. During evolution, sequence similarities decreased to very low values, making sequence-based searches for Bet v 1 homologs unfeasible. The APE2225 (PDB 2NS9), a CoxG family member from the archaeon *Aeropyrum pernix* K1, has a fold identical to Bet v 1 (Figure 1) but only 14% sequence identity. *A. pernix* K1 is an aerobic hyperthermophilic archaeon isolated from a coastal hydrothermal vent at Kodakara-Jima Island, Japan [62]. This archaeon grows optimally at 90 to 95 °C, pH 7.0, and a salinity of 3.5%, which indicates a high stability of its Bet v 1 fold, a feature that was lost once the Bet v 1 predecessor gene moved into land plants. Recently, a bacterial Bet v 1-related protein, possibly a member of the polyketide cyclase family, TTHA0849 (Figure 1) [63] from *Thermus thermophilus* served as a non-allergenic scaffold to create chimeric proteins by grafting individual epitope-sized, contiguous surface patches of the allergen Bet v 1 onto its surface [64]. These chimeras were then used to determine patient-specific patterns of epitope recognition by IgE antibodies. The norcoclaurine synthases (NCS) form another subfamily of the Bet v 1 family [61]. The NCS from the plant meadow rue (*Thalictrum flavum*), a structural homolog of Bet v 1, does not bind Bet v 1-reactive IgE and was therefore used as another scaffold for grafting of a Bet v 1-specific IgE epitope [65].

5. The Yeast Connection

Yeast represents a highly useful genetic model system due to its easily manipulated genome and easily manageable methods for the functional analysis of gene products. Thus, yeast is easily suited to identify biological functions of new and unknown eukaryotic protein-coding genes. We have therefore searched for Bet v 1 homologs in the yeast genome and found a family of sterol transfer proteins recently described in the literature.

Sterols play a key role in regulating the fluidity and barrier function of plasma membranes in eukaryotic cells [66]. Sterol transport proteins distribute sterols from their point of synthesis in the endoplasmic reticulum (ER) to their exact subcellular localization. The ability to solubilize lipids into aqueous solutions is a general property of steroidogenic acute regulatory protein-related lipid transfer (StART) -like domains. Gatta and colleagues have identified a family of StART domain containing proteins that are anchored at membrane contact sites and that transport sterols from the ER to the plasma membrane [67]. The yeast *Saccharomyces cerevisiae* possesses six such proteins with StARkin domains [67–69] of which four (Lam1–Lam4) are anchored to the ER membrane. The StARkin domain (the name is derived from kin of steroidogenic acute regulatory protein) is an α/β helix-grip-fold structure with a deep hydrophobic pocket [70]. The major birch pollen allergen Bet v 1 represents the first described StARkin domain [44]. StARkin domains were used to define a large superfamily of proteins, which is also referred to as the Bet v 1-like superfamily [61] or START domain superfamily [71]. The name StARkin was proposed as a more inclusive name for this superfamily [72].

The name Lam is derived from "lipid transport proteins anchored at membrane contact sites". They are integral membrane proteins anchored into the ER membrane by a C-terminal transmembrane helix [72]. The yeast proteins Lam1 and Lam3 have a single StARkin domain, whereas Lam2 and Lam4 each have two StARkin domains. Jentsch and colleagues carried out a structure-function analysis of the second StARkin domain of Lam4 (Lam4 SD2), showing that Lam4 SD2 undergoes conformational changes upon binding sterol and that it catalyzes sterol transport between vesicles in vitro [73]. Crystal structures of Lam4 SD2 with and without bound 25-hydroxycholesterol were also obtained. Tong and colleagues obtained crystal structures of both StARkin domains of Lam2 and Lam4 in the apo form and of Lam2 SD2 in complex with ergosterol [74].

To identify proteins with similar structures to Bet v 1.0101, we used the online available Vector Alignment Search Tool+ (https://structure.ncbi.nlm.nih.gov/Structure/VAST/vast.shtml, accessed on 13 July 2023) [75,76]. As a bait, NMR data (PDB: 6R3C) of the allergen were chosen [57]. Figure 2 shows ribbon representations of Bet v 1.0101 and the StARkin domains of Lam2 and Lam4, whereby Lam2 SD2 is only available as a swapped dimer. Figure 3 shows an overlay of Bet v 1.0101 and the first StARkin domain of Lam4 (PDB 5YQJ) [74], one of the results of the VAST+ analysis.

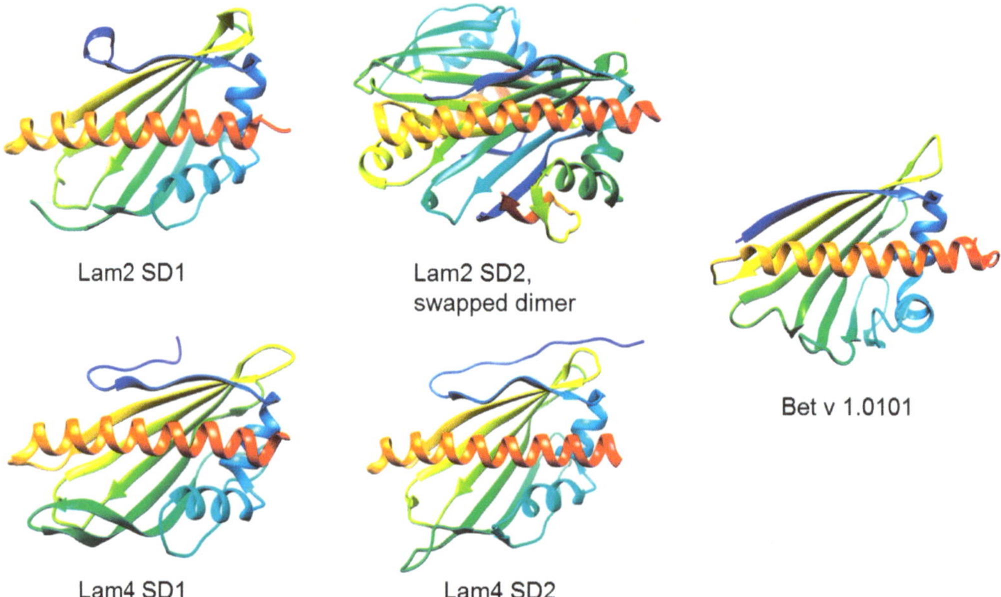

Figure 2. Ribbon representations of Bet v 1 (PDB 4A88) homologous structures from the yeast *Saccharomyces cerevisiae*. The StARkin domains of Lam2 (StARkin domain 1, PDB 5YQI; StARkin domain 2 as a swapped dimer, PDB 5YQQ) and of Lam4 (StARkin domain 1, PDB 5YQJ; StARkin domain 2, PDB 5YQP) are shown. The structures are rainbow-colored from blue at the N-terminus to red at the C-terminus. The 3D images were created with the molecular modeling system UCSF ChimeraX (https://www.rbvi.ucsf.edu/chimerax/, accessed on 19 June 2023).

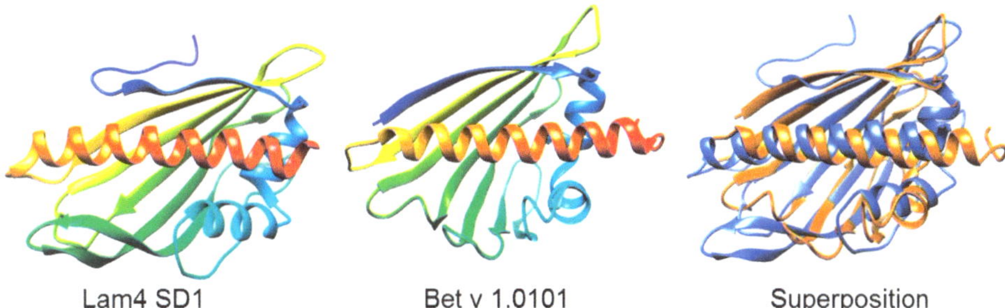

Figure 3. Superposition of the ribbon structure of Bet v 1.0101 (PDB 4A88, aa2-160) in orange and the structure of the StARkin domain 1 of the yeast Lam4 protein (PDB 5YQJ, aa749-929). The 3D images were created with the molecular modeling system UCSF ChimeraX (https://www.rbvi.ucsf.edu/chimerax/, accessed on 19 June 2023).

6. Conclusions

The research on Bet v 1 has spanned five decades. It started in the 1980s with an allergy-driven focus and has expanded into numerous and highly diversified studies of the functions of the members of the Bet v 1-like/StARkin superfamily of proteins. The first described StARkin domain was Bet v 1. Identification of ligands of Bet v 1 and its homologs offer the first insights into the varied functions of these proteins. However, information on ligands and functions of the various isoforms of Bet v 1 in birch pollen is still missing, as it is unclear why there are so many isoforms present. The Bet v 1 scaffold is a very old and versatile one and has been used for a wide variety of functions. We do not believe that the Bet v 1 homologs present in birch or in other organisms are the result of convergent evolution. Rather, it is a very useful scaffold that was conserved and used in many different ways. The Bet v 1-like/StARkin superfamily keeps growing, and new members are still being discovered.

Author Contributions: Conceptualization, H.B. and D.K.; methodology, H.B.; software, H.B..; writing—original draft preparation, H.B. and D.K.; writing—review and editing, H.B.; funding acquisition, H.B. All authors have read and agreed to the published version of the manuscript.

Funding: This research received no external funding.

Acknowledgments: Author H.B. gratefully acknowledges the general support of the Danube Allergy Research Cluster Project P-06 funded by the Country of Lower Austria. The authors thank Mark Rinnerthaler for the analysis of Bet v 1 homologous structures in yeast.

Conflicts of Interest: The authors declare no conflict of interest.

References

1. Breiteneder, H.; Hendler, P.N.; Kraft, D. Legends of allergy and immunology: Clemens von Pirquet. *Allergy* **2020**, *75*, 1276–1277. [CrossRef]
2. Breiteneder, H. Legends of allergy/immunology: Dietrich Kraft. *Allergy* **2019**, *74*, 1591–1593. [CrossRef]
3. King, T.P.; Hoffman, D.; Lowenstein, H.; Marsh, D.G.; Platts-Mills, T.A.; Thomas, W. Allergen nomenclature. *Allergy* **1995**, *50*, 765–774. [CrossRef] [PubMed]
4. Breiteneder, H.; Seiser, C.; Loffelhardt, W.; Michalowski, C.; Bohnert, H.J. Physical map and protein gene map of cyanelle DNA from the second known isolate of *Cyanophora paradoxa* (Kies-strain). *Curr. Genet.* **1988**, *13*, 199–206. [CrossRef] [PubMed]
5. Breiteneder, H.; Hassfeld, W.; Pettenburger, K.; Jarolim, E.; Breitenbach, M.; Rumpold, H.; Kraft, D.; Scheiner, O. Isolation and characterization of messenger RNA from male inflorescences and pollen of the white birch (*Betula verrucosa*). *Int. Arch. Allergy Appl. Immunol.* **1988**, *87*, 19–24. [CrossRef] [PubMed]
6. Breiteneder, H.; Pettenburger, K.; Bito, A.; Valenta, R.; Kraft, D.; Rumpold, H.; Scheiner, O.; Breitenbach, M. The gene coding for the major birch pollen allergen Betv1, is highly homologous to a pea disease resistance response gene. *EMBO J.* **1989**, *8*, 1935–1938. [CrossRef] [PubMed]

7. Valenta, R.; Breiteneder, H.; Petternburger, K.; Breitenbach, M.; Rumpold, H.; Kraft, D.; Scheiner, O. Homology of the major birch-pollen allergen, *Bet v* I, with the major pollen allergens of alder, hazel, and hornbeam at the nucleic acid level as determined by cross-hybridization. *J. Allergy Clin. Immunol.* **1991**, *87*, 677–682. [CrossRef] [PubMed]

8. Breiteneder, H.; Ferreira, F.; Reikerstorfer, A.; Duchene, M.; Valenta, R.; Hoffmann-Sommergruber, K.; Ebner, C.; Breitenbach, M.; Kraft, D.; Scheiner, O. Complementary DNA cloning and expression in *Escherichia coli* of Aln g I, the major allergen in pollen of alder (*Alnus glutinosa*). *J. Allergy Clin. Immunol.* **1992**, *90*, 909–917. [CrossRef]

9. Breiteneder, H.; Ferreira, F.; Hoffmann-Sommergruber, K.; Ebner, C.; Breitenbach, M.; Rumpold, H.; Kraft, D.; Scheiner, O. Four recombinant isoforms of *Cor a* I, the major allergen of hazel pollen, show different IgE-binding properties. *Eur. J. Biochem.* **1993**, *212*, 355–362. [CrossRef]

10. Vanek-Krebitz, M.; Hoffmann-Sommergruber, K.; Laimer da Camara Machado, M.; Susani, M.; Ebner, C.; Kraft, D.; Scheiner, O.; Breiteneder, H. Cloning and sequencing of Mal d 1, the major allergen from apple (*Malus domestica*), and its immunological relationship to Bet v 1, the major birch pollen allergen. *Biochem. Biophys. Res. Commun.* **1995**, *214*, 538–551. [CrossRef]

11. Breiteneder, H.; Hoffmann-Sommergruber, K.; O'Riordain, G.; Susani, M.; Ahorn, H.; Ebner, C.; Kraft, D.; Scheiner, O. Molecular characterization of Api g 1, the major allergen of celery (*Apium graveolens*), and its immunological and structural relationships to a group of 17-kDa tree pollen allergens. *Eur. J. Biochem.* **1995**, *233*, 484–489. [CrossRef]

12. Antoniw, J.F.; Ritter, C.E.; Pierpoint, W.S.; Vanloon, L.C. Comparison of 3 pathogenesis-related proteins from plants of 2 cultivars of tobacco infected with TMV. *J. Gen. Virol.* **1980**, *47*, 79–87. [CrossRef]

13. Ali, S.; Ganai, B.A.; Kamili, A.N.; Bhat, A.A.; Mir, Z.A.; Bhat, J.A.; Tyagi, A.; Islam, S.T.; Mushtaq, M.; Yadav, P.; et al. Pathogenesis-related proteins and peptides as promising tools for engineering plants with multiple stress tolerance. *Microbiol. Res.* **2018**, *212–213*, 29–37. [CrossRef] [PubMed]

14. Sancho, A.I.; Wangorsch, A.; Jensen, B.M.; Watson, A.; Alexeev, Y.; Johnson, P.E.; Mackie, A.R.; Neubauer, A.; Reese, G.; Ballmer-Weber, B.; et al. Responsiveness of the major birch allergen Bet v 1 scaffold to the gastric environment: Impact on structure and allergenic activity. *Mol. Nutr. Food Res.* **2011**, *55*, 1690–1699. [CrossRef] [PubMed]

15. Rib-Schmidt, C.; Riedl, P.; Meisinger, V.; Schwaben, L.; Schulenborg, T.; Reuter, A.; Schiller, D.; Seutter von Loetzen, C.; Rosch, P. pH and heat resistance of the major celery allergen Api g 1. *Mol. Nutr. Food Res.* **2018**, *62*, e1700886. [CrossRef]

16. Jacob, T.; Vogel, L.; Reuter, A.; Wangorsch, A.; Kring, C.; Mahler, V.; Wohrl, B.M. Food processing does not abolish the allergenicity of the carrot allergen Dau c 1: Influence of pH, temperature, and the food matrix. *Mol. Nutr. Food Res.* **2020**, *64*, e2000334. [CrossRef] [PubMed]

17. Dramburg, S.; Hilger, C.; Santos, A.F.; de Las Vecillas, L.; Aalberse, R.C.; Acevedo, N.; Aglas, L.; Altmann, F.; Arruda, K.L.; Asero, R.; et al. EAACI Molecular Allergology User's Guide 2.0. *Pediatr. Allergy Immunol.* **2023**, *34* (Suppl. S28), e13854. [CrossRef] [PubMed]

18. Hauser, M.; Asam, C.; Himly, M.; Palazzo, P.; Voltolini, S.; Montanari, C.; Briza, P.; Bernardi, M.L.; Mari, A.; Ferreira, F.; et al. Bet v 1-like pollen allergens of multiple Fagales species can sensitize atopic individuals. *Clin. Exp. Allergy* **2011**, *41*, 1804–1814. [CrossRef] [PubMed]

19. Larsen, J.N.; Stroman, P.; Ipsen, H. PCR based cloning and sequencing of isogenes encoding the tree pollen major allergen Car b I from *Carpinus betulus*, hornbeam. *Mol. Immunol.* **1992**, *29*, 703–711. [CrossRef] [PubMed]

20. Wallner, M.; Erler, A.; Hauser, M.; Klinglmayr, E.; Gadermaier, G.; Vogel, L.; Mari, A.; Bohle, B.; Briza, P.; Ferreira, F. Immunologic characterization of isoforms of Car b 1 and Que a 1, the major hornbeam and oak pollen allergens. *Allergy* **2009**, *64*, 452–460. [CrossRef] [PubMed]

21. Jeong, K.Y.; Lee, J.; Sang, M.K.; Lee, Y.S.; Park, K.H.; Lee, J.H.; Park, J.W. Sensitization profile to sawtooth oak component allergens and their clinical implications. *J. Clin. Lab. Anal.* **2021**, *35*, e23825. [CrossRef] [PubMed]

22. Pedrosa, M.; Guerrero-Sanchez, V.M.; Canales-Bueno, N.; Loli-Ausejo, D.; Castillejo, M.A.; Quirce, S.; Jorrin-Novo, J.V.; Rodriguez-Perez, R. Quercus ilex pollen allergen, Que i 1, responsible for pollen food allergy syndrome caused by fruits in Spanish allergic patients. *Clin. Exp. Allergy* **2020**, *50*, 815–823. [CrossRef] [PubMed]

23. Lee, J.Y.; Yang, M.; Jeong, K.Y.; Sim, D.W.; Park, J.H.; Park, K.H.; Lee, J.H.; Park, J.W. Characterization of a major allergen from Mongolian oak, *Quercus mongolica*, a dominant species of oak in Korea. *Int. Arch. Allergy Immunol.* **2017**, *174*, 77–85. [CrossRef] [PubMed]

24. Vieths, S.; Scheurer, S.; Ballmer-Weber, B. Current understanding of cross-reactivity of food allergens and pollen. *Ann. N. Y. Acad. Sci.* **2002**, *964*, 47–68. [CrossRef]

25. Oberhuber, C.; Bulley, S.M.; Ballmer-Weber, B.K.; Bublin, M.; Gaier, S.; DeWitt, A.M.; Briza, P.; Hofstetter, G.; Lidholm, J.; Vieths, S.; et al. Characterization of Bet v 1-related allergens from kiwifruit relevant for patients with combined kiwifruit and birch pollen allergy. *Mol. Nutr. Food Res.* **2008**, *52* (Suppl. S2), S230–S240. [CrossRef]

26. D'Avino, R.; Bernardi, M.L.; Wallner, M.; Palazzo, P.; Camardella, L.; Tuppo, L.; Alessandri, C.; Breiteneder, H.; Ferreira, F.; Ciardiello, M.A.; et al. Kiwifruit Act d 11 is the first member of the ripening-related protein family identified as an allergen. *Allergy* **2011**, *66*, 870–877. [CrossRef]

27. Zhao, L.; Xie, H.; Wang, X.; Liu, M.; Ma, T.; Fu, W.; Wang, Y.; Wu, D.; Feng, Y.; Liu, Y.; et al. Molecular characterization of allergens and component-resolved diagnosis of IgE-mediated mango fruit allergy. *Allergy* **2023**. [CrossRef]

28. Hoffmann-Sommergruber, K.; O'Riordain, G.; Ahorn, H.; Ebner, C.; Laimer Da Camara Machado, M.; Puhringer, H.; Scheiner, O.; Breiteneder, H. Molecular characterization of Dau c 1, the Bet v 1 homologous protein from carrot and its cross-reactivity with Bet v 1 and Api g 1. *Clin. Exp. Allergy* **1999**, *29*, 840–847. [CrossRef]

29. Ebo, D.G.; Decuyper, I.I.; Rihs, H.P.; Mertens, C.; Van Gasse, A.L.; van der Poorten, M.M.; De Puysseleyr, L.; Faber, M.A.; Hagendorens, M.M.; Bridts, C.H.; et al. IgE-binding and mast cell-activating capacity of the homologue of the major birch pollen allergen and profilin from Cannabis sativa. *J. Allergy Clin. Immunol. Pract.* **2021**, *9*, 2509–2512.e3. [CrossRef]

30. Luttkopf, D.; Muller, U.; Skov, P.S.; Ballmer-Weber, B.K.; Wuthrich, B.; Skamstrup Hansen, K.; Poulsen, L.K.; Kastner, M.; Haustein, D.; Vieths, S. Comparison of four variants of a major allergen in hazelnut (*Corylus avellana*) Cor a 1.04 with the major hazel pollen allergen Cor a 1.01. *Mol. Immunol.* **2002**, *38*, 515–525. [CrossRef]

31. Mittag, D.; Akkerdaas, J.; Ballmer-Weber, B.K.; Vogel, L.; Wensing, M.; Becker, W.M.; Koppelman, S.J.; Knulst, A.C.; Helbling, A.; Hefle, S.L.; et al. Ara h 8, a Bet v 1-homologen allergen from peanut, is a major allergen in patients with combined birch pollen and peanut allergy. *J. Allergy Clin. Immunol.* **2004**, *114*, 1410–1417. [CrossRef]

32. Crowell, D.N.; John, M.E.; Russell, D.; Amasino, R.M. Characterization of a stress-induced, developmentally regulated gene family from soybean. *Plant Mol. Biol.* **1992**, *18*, 459–466. [CrossRef]

33. Mittag, D.; Vieths, S.; Vogel, L.; Wagner-Loew, D.; Starke, A.; Hunziker, P.; Becker, W.M.; Ballmer-Weber, B.K. Birch pollen-related food allergy to legumes: Identification and characterization of the Bet v 1 homologue in mungbean (*Vigna radiata*), Vig r 1. *Clin. Exp. Allergy* **2005**, *35*, 1049–1055. [CrossRef] [PubMed]

34. Guhsl, E.E.; Hofstetter, G.; Hemmer, W.; Ebner, C.; Vieths, S.; Vogel, L.; Breiteneder, H.; Radauer, C. Vig r 6, the cytokinin-specific binding protein from mung bean (Vigna radiata) sprouts, cross-reacts with Bet v 1-related allergens and binds IgE from birch pollen allergic patients' sera. *Mol. Nutr. Food Res.* **2014**, *58*, 625–634. [CrossRef] [PubMed]

35. Wangorsch, A.; Jamin, A.; Lidholm, J.; Grani, N.; Lang, C.; Ballmer-Weber, B.; Vieths, S.; Scheurer, S. Identification and implication of an allergenic PR-10 protein from walnut in birch pollen associated walnut allergy. *Mol. Nutr. Food Res.* **2017**, *61*, 1600902. [CrossRef] [PubMed]

36. Musidlowska-Persson, A.; Alm, R.; Emanuelsson, C. Cloning and sequencing of the Bet v 1-homologen allergen Fra a 1 in strawberry (*Fragaria ananassa*) shows the presence of an intron and little variability in amino acid sequence. *Mol. Immunol.* **2007**, *44*, 1245–1252. [CrossRef]

37. Scheurer, S.; Pastorello, E.A.; Wangorsch, A.; Kastner, M.; Haustein, D.; Vieths, S. Recombinant allergens Pru av 1 and Pru av 4 and a newly identified lipid transfer protein in the in vitro diagnosis of cherry allergy. *J. Allergy Clin. Immunol.* **2001**, *107*, 724–731. [CrossRef]

38. Kabasser, S.; Crvenjak, N.; Schmalz, S.; Kalic, T.; Hafner, C.; Dubiela, P.; Kucharczyk, A.; Bazan-Socha, S.; Lukaszyk, M.; Breiteneder, H.; et al. Pru du 1, the Bet v 1-homologue from almond, is a major allergen in patients with birch pollen associated almond allergy. *Clin. Transl. Allergy* **2022**, *12*, e12177. [CrossRef] [PubMed]

39. Wisniewski, M.; Bassett, C.; Arora, R. Distribution and partial characterization of seasonally expressed proteins in different aged shoots and roots of 'Loring' peach (*Prunus persica*). *Tree Physiol.* **2004**, *24*, 339–345. [CrossRef]

40. Karamloo, F.; Scheurer, S.; Wangorsch, A.; May, S.; Haustein, D.; Vieths, S. Pyr c 1, the major allergen from pear (*Pyrus communis*), is a new member of the Bet v 1 allergen family. *J. Chromatogr. B Biomed. Sci. Appl.* **2001**, *756*, 281–293. [CrossRef]

41. Marzban, G.; Herndl, A.; Kolarich, D.; Maghuly, F.; Mansfeld, A.; Hemmer, W.; Katinger, H.; Laimer, M. Identification of four IgE-reactive proteins in raspberry (*Rubus ideaeus* L.). *Mol. Nutr. Food Res.* **2008**, *52*, 1497–1506. [CrossRef] [PubMed]

42. Wangorsch, A.; Jamin, A.; Foetisch, K.; Malczyk, A.; Reuter, A.; Vierecke, S.; Schulke, S.; Bartel, D.; Mahler, V.; Lidholm, J.; et al. Identification of Sola l 4 as Bet v 1 homologous pathogenesis related-10 allergen in tomato fruits. *Mol. Nutr. Food Res.* **2015**, *59*, 582–592. [CrossRef] [PubMed]

43. Kleine-Tebbe, J.; Vogel, L.; Crowell, D.N.; Haustein, U.F.; Vieths, S. Severe oral allergy syndrome and anaphylactic reactions caused by a Bet v 1- related PR-10 protein in soybean, SAM22. *J. Allergy Clin. Immunol.* **2002**, *110*, 797–804. [CrossRef] [PubMed]

44. Gajhede, M.; Osmark, P.; Poulsen, F.M.; Ipsen, H.; Larsen, J.N.; Joost van Neerven, R.J.; Schou, C.; Lowenstein, H.; Spangfort, M.D. X-ray and NMR structure of Bet v 1, the origin of birch pollen allergy. *Nat. Struct. Biol.* **1996**, *3*, 1040–1045. [CrossRef]

45. Mogensen, J.E.; Wimmer, R.; Larsen, J.N.; Spangfort, M.D.; Otzen, D.E. The major birch allergen, Bet v 1, shows affinity for a broad spectrum of physiological ligands. *J. Biol. Chem.* **2002**, *277*, 23684–23692. [CrossRef]

46. Kofler, S.; Asam, C.; Eckhard, U.; Wallner, M.; Ferreira, F.; Brandstetter, H. Crystallographically mapped ligand binding differs in high and low IgE binding isoforms of birch pollen allergen bet v 1. *J. Mol. Biol.* **2012**, *422*, 109–123. [CrossRef]

47. Aglas, L.; Soh, W.T.; Kraiem, A.; Wenger, M.; Brandstetter, H.; Ferreira, F. Ligand Binding of PR-10 Proteins with a Particular Focus on the Bet v 1 Allergen Family. *Curr. Allergy Asthma Rep.* **2020**, *20*, 25. [CrossRef]

48. McBride, J.K.; Cheng, H.; Maleki, S.J.; Hurlburt, B.K. Purification and characterization of pathogenesis related class 10 panallergens. *Foods* **2019**, *8*, 609. [CrossRef]

49. Seutter von Loetzen, C.; Hoffmann, T.; Hartl, M.J.; Schweimer, K.; Schwab, W.; Rosch, P.; Hartl-Spiegelhauer, O. Secret of the major birch pollen allergen Bet v 1: Identification of the physiological ligand. *Biochem. J.* **2014**, *457*, 379–390. [CrossRef]

50. Brunetti, C.; Fini, A.; Sebastiani, F.; Gori, A.; Tattini, M. Modulation of Phytohormone Signaling: A Primary Function of Flavonoids in Plant-Environment Interactions. *Front. Plant Sci.* **2018**, *9*, 1042. [CrossRef]

51. Agati, G.; Brunetti, C.; Di Ferdinando, M.; Ferrini, F.; Pollastri, S.; Tattini, M. Functional roles of flavonoids in photoprotection: New evidence, lessons from the past. *Plant Physiol. Biochem.* **2013**, *72*, 35–45. [CrossRef] [PubMed]

52. Grutsch, S.; Fuchs, J.E.; Freier, R.; Kofler, S.; Bibi, M.; Asam, C.; Wallner, M.; Ferreira, F.; Brandstetter, H.; Liedl, K.R.; et al. Ligand binding modulates the structural dynamics and compactness of the major birch pollen allergen. *Biophys. J.* **2014**, *107*, 2972–2981. [CrossRef] [PubMed]

53. Asam, C.; Batista, A.L.; Moraes, A.H.; de Paula, V.S.; Almeida, F.C.; Aglas, L.; Kitzmuller, C.; Bohle, B.; Ebner, C.; Ferreira, F.; et al. Bet v 1—A Trojan horse for small ligands boosting allergic sensitization? *Clin. Exp. Allergy* **2014**, *44*, 1083–1093. [CrossRef] [PubMed]
54. Soh, W.T.; Aglas, L.; Mueller, G.A.; Gilles, S.; Weiss, R.; Scheiblhofer, S.; Huber, S.; Scheidt, T.; Thompson, P.M.; Briza, P.; et al. Multiple roles of Bet v 1 ligands in allergen stabilization and modulation of endosomal protease activity. *Allergy* **2019**, *74*, 2382–2393. [CrossRef]
55. Freier, R.; Dall, E.; Brandstetter, H. Protease recognition sites in Bet v 1a are cryptic, explaining its slow processing relevant to its allergenicity. *Sci. Rep.* **2015**, *5*, 12707. [CrossRef]
56. Seutter von Loetzen, C.; Jacob, T.; Hartl-Spiegelhauer, O.; Vogel, L.; Schiller, D.; Sporlein-Guttler, C.; Schobert, R.; Vieths, S.; Hartl, M.J.; Rosch, P. Ligand recognition of the major birch pollen allergen Bet v 1 is isoform dependent. *PLoS ONE* **2015**, *10*, e0128677. [CrossRef]
57. Jacob, T.; von Loetzen, C.S.; Reuter, A.; Lacher, U.; Schiller, D.; Schobert, R.; Mahler, V.; Vieths, S.; Rosch, P.; Schweimer, K.; et al. Identification of a natural ligand of the hazel allergen Cor a 1. *Sci. Rep.* **2019**, *9*, 8714. [CrossRef] [PubMed]
58. Hjerno, K.; Alm, R.; Canback, B.; Matthiesen, R.; Trajkovski, K.; Bjork, L.; Roepstorff, P.; Emanuelsson, C. Down-regulation of the strawberry Bet v 1-homologous allergen in concert with the flavonoid biosynthesis pathway in colorless strawberry mutant. *Proteomics* **2006**, *6*, 1574–1587. [CrossRef] [PubMed]
59. Casanal, A.; Zander, U.; Munoz, C.; Dupeux, F.; Luque, I.; Botella, M.A.; Schwab, W.; Valpuesta, V.; Marquez, J.A. The strawberry pathogenesis-related 10 (PR-10) Fra a proteins control flavonoid biosynthesis by binding to metabolic intermediates. *J. Biol. Chem.* **2013**, *288*, 35322–35332. [CrossRef]
60. Lackner, P.; Koppensteiner, W.A.; Sippl, M.J.; Domingues, F.S. ProSup: A refined tool for protein structure alignment. *Protein Eng.* **2000**, *13*, 745–752. [CrossRef]
61. Radauer, C.; Lackner, P.; Breiteneder, H. The Bet v 1 fold: An ancient, versatile scaffold for binding of large, hydrophobic ligands. *BMC Evol. Biol.* **2008**, *8*, 286. [CrossRef] [PubMed]
62. Sako, Y.; Nomura, N.; Uchida, A.; Ishida, Y.; Morii, H.; Koga, Y.; Hoaki, T.; Maruyama, T. Aeropyrum pernix gen. nov., sp. nov., a novel aerobic hyperthermophilic archaeon growing at temperatures up to 100 degrees C. *Int. J. Syst. Bacteriol.* **1996**, *46*, 1070–1077. [CrossRef] [PubMed]
63. Nakabayashi, M.; Shibata, N.; Komori, H.; Ueda, Y.; Iino, H.; Ebihara, A.; Kuramitsu, S.; Higuchi, Y. Structure of a conserved hypothetical protein, TTHA0849 from Thermus thermophilus HB8, at 2.4 A resolution: A putative member of the StAR-related lipid-transfer (START) domain superfamily. *Acta Crystallogr. Sect. F Struct. Biol. Cryst. Commun.* **2005**, *61*, 1027–1031. [CrossRef] [PubMed]
64. Schmalz, S.; Mayr, V.; Shosherova, A.; Gepp, B.; Ackerbauer, D.; Sturm, G.; Bohle, B.; Breiteneder, H.; Radauer, C. Isotype-specific binding patterns of serum antibodies to multiple conformational epitopes of Bet v 1. *J. Allergy Clin. Immunol.* **2022**, *149*, 1786–1794.e1712. [CrossRef]
65. Berkner, H.; Seutter von Loetzen, C.; Hartl, M.; Randow, S.; Gubesch, M.; Vogel, L.; Husslik, F.; Reuter, A.; Lidholm, J.; Ballmer-Weber, B.; et al. Enlarging the toolbox for allergen epitope definition with an allergen-type model protein. *PLoS ONE* **2014**, *9*, e111691. [CrossRef]
66. Holthuis, J.C.; Menon, A.K. Lipid landscapes and pipelines in membrane homeostasis. *Nature* **2014**, *510*, 48–57. [CrossRef]
67. Gatta, A.T.; Wong, L.H.; Sere, Y.Y.; Calderon-Norena, D.M.; Cockcroft, S.; Menon, A.K.; Levine, T.P. A new family of StART domain proteins at membrane contact sites has a role in ER-PM sterol transport. *eLife* **2015**, *4*, e07253. [CrossRef]
68. Elbaz-Alon, Y.; Eisenberg-Bord, M.; Shinder, V.; Stiller, S.B.; Shimoni, E.; Wiedemann, N.; Geiger, T.; Schuldiner, M. Lam6 regulates the extent of contacts between organelles. *Cell Rep.* **2015**, *12*, 7–14. [CrossRef]
69. Murley, A.; Sarsam, R.D.; Toulmay, A.; Yamada, J.; Prinz, W.A.; Nunnari, J. Ltc1 is an ER-localized sterol transporter and a component of ER-mitochondria and ER-vacuole contacts. *J. Cell Biol.* **2015**, *209*, 539–548. [CrossRef]
70. Dresden, C.E.; Ashraf, Q.; Husbands, A.Y. Diverse regulatory mechanisms of StARkin domains in land plants and mammals. *Curr. Opin. Plant Biol.* **2021**, *64*, 102148. [CrossRef]
71. Iyer, L.M.; Koonin, E.V.; Aravind, L. Adaptations of the helix-grip fold for ligand binding and catalysis in the START domain superfamily. *Proteins* **2001**, *43*, 134–144. [CrossRef] [PubMed]
72. Wong, L.H.; Levine, T.P. Lipid transfer proteins do their thing anchored at membrane contact sites . . . but what is their thing? *Biochem. Soc. Trans.* **2016**, *44*, 517–527. [CrossRef]
73. Jentsch, J.A.; Kiburu, I.; Pandey, K.; Timme, M.; Ramlall, T.; Levkau, B.; Wu, J.; Eliezer, D.; Boudker, O.; Menon, A.K. Structural basis of sterol binding and transport by a yeast StARkin domain. *J. Biol. Chem.* **2018**, *293*, 5522–5531. [CrossRef]
74. Tong, J.; Manik, M.K.; Im, Y.J. Structural basis of sterol recognition and nonvesicular transport by lipid transfer proteins anchored at membrane contact sites. *Proc. Natl. Acad. Sci. USA* **2018**, *115*, E856–E865. [CrossRef] [PubMed]
75. Gibrat, J.F.; Madej, T.; Bryant, S.H. Surprising similarities in structure comparison. *Curr. Opin. Struct. Biol.* **1996**, *6*, 377–385. [CrossRef]
76. Madej, T.; Marchler-Bauer, A.; Lanczycki, C.; Zhang, D.; Bryant, S.H. Biological assembly comparison with VAST. *Methods Mol. Biol.* **2020**, *2112*, 175–186. [CrossRef] [PubMed]

 *biomolecules*

Opinion

Senopathies—Diseases Associated with Cellular Senescence

Oleh Lushchak [1,2,3], Markus Schosserer [4,5] and Johannes Grillari [1,5,6,*]

1   Ludwig Boltzmann Institute for Traumatology, The Research Center in Cooperation with AUVA, 1200 Vienna, Austria; oleh.lushchak@pnu.edu.ua
2   Department of Biochemistry and Biotechnology, Precarpathian National University, 76000 Ivano-Frankivsk, Ukraine
3   Research and Development University, 76018 Ivano-Frankivsk, Ukraine
4   Institute of Medical Genetics, Center for Pathobiochemistry and Genetics, Medical University of Vienna, 1090 Vienna, Austria; markus.schosserer@meduniwien.ac.at
5   Austrian Cluster for Tissue Regeneration, 1200 Vienna, Austria
6   Institute of Molecular Biotechnology, Department of Biotechnology, University of Natural Resources and Life Sciences, 1190 Vienna, Austria
*   Correspondence: johannes.grillari@trauma.lbg.ac.at

Abstract: Cellular senescence describes a stable cell cycle arrest state with a characteristic phenotype. Senescent cells accumulate in the human body during normal aging, limiting the lifespan and promoting aging-related, but also several non-related, pathologies. We propose to refer to all diseases whose pathogenesis or progression is associated with cellular senescence as "senopathies". Targeting senescent cells with senolytics or senomorphics is likely to mitigate these pathologies. Examples of senopathies include cardiovascular, metabolic, musculoskeletal, liver, kidney, and lung diseases and neurodegeneration. For all these pathologies, animal studies provide clear mechanistic evidence for a connection between senescent cell accumulation and disease progression. The major persisting challenge in developing novel senotherapies is the heterogeneity of senescence phenotypes, causing a lack of universal biomarkers and difficulties in discriminating senescent from non-senescent cells.

Keywords: cellular senescence; aging; senopathy; senolytic; senomorphic; senotherapy; geroscience; senescence-associated secretory phenotype (SASP); pathology; age-related disease

Citation: Lushchak, O.; Schosserer, M.; Grillari, J. Senopathies—Diseases Associated with Cellular Senescence. *Biomolecules* **2023**, *13*, 966. https://doi.org/10.3390/biom13060966

Academic Editors: Mark Rinnerthaler and Markus Ralser

Received: 8 May 2023
Revised: 5 June 2023
Accepted: 6 June 2023
Published: 8 June 2023

1. Introduction

Cellular senescence is a state of stable cell cycle arrest. It is, in vivo, most commonly induced in response to various physiological and pathological conditions through DNA damage signaling [1–4] or the expression of oncogenes [5,6]. Leonard Hayflick discovered cellular senescence in 1965 as a phenomenon when human fetal fibroblasts stopped dividing but maintained viability and metabolic activity after prolonged cultivation [7]. In these early days, the permanent cell cycle arrest and characteristic morphological changes associated with senescence were considered mere artifacts of in vitro cultivation of primary cells [8,9]. Several decades later, the progressive shortening of telomeres was shown to be associated with the proliferation limit of cells in culture, known now as replicative senescence [10]. After that, an increasing body of evidence emerged, demonstrating that senescent cells (SCs) also accumulate in vivo, preferentially in aged individuals, and specifically at sites of aging-related pathologies [11–13]. Ultimately, elegant experiments in mice showed that the accumulation of SCs during aging limits the healthy murine lifespan [14,15]. These discoveries initiated ongoing endeavors to identify novel "senotherapies", which comprise pharmacological compounds to eliminate SCs specifically ("senolytics") or to mitigate or modify their senescence-associated secretory phenotype (SASP) ("senomorphics") [16–21].

The SASP is considered the main trigger of tissue damage caused by SCs and is composed of pro-inflammatory cytokines, chemokines, and matrix-remodeling factors [22]. In addition, extracellular vesicles, miRNAs [23–26], and pro-inflammatory lipids [27] have been identified as SASP members. Recent evidence suggests that SCs attract innate immune cells by their SASP and activate dendritic and CD8 T-cells by increased MHC class I expression and self-antigen presentation [28]. However, SCs are also highly efficient in resisting immune clearance, partially explaining the increased SC burden in aged individuals. Cellular senescence is also associated with increased cell plasticity and a concomitant loss of cell identity. Senescent cells promote the de-differentiation of non-senescent neighboring cells, causing further tissue damage [29]. This relatively new concept is supported by a study demonstrating that Alzheimer's disease (AD) patients induced neurons simultaneously expressed markers of de-differentiation and cellular senescence [30]. On the contrary, SCs play essential roles in embryogenesis, wound healing, and tissue regeneration [31,32]. Thus, while the transient presence of SCs is beneficial for maintaining tissue homeostasis and preventing tumorigenesis early in life, the accumulation of these cells drives age-related disease and promotes chronic inflammation [31].

Similarly, the contribution of cellular senescence to a wide range of pathological conditions is highly complex, combining beneficial and detrimental effects [1,33]. Many, but not all, of these are associated with aging, such as cardiovascular pathologies [12,34], neurodegeneration [35], and ocular diseases [36]. However, it is often unclear if the accumulation of SCs is a cause or consequence of the respective pathological condition [37]. Cystic fibrosis [38] and acute infections, such as SARS-CoV-2 [39], are not considered age-related pathologies but are both characterized and worsened by SC presence and accumulation.

We propose to refer to all diseases whose pathogenesis or progression is associated with cellular senescence as "senopathies" and aim to provide examples and mechanistic explanations in this opinion paper. While, on the one hand, precision medicine identifies more and more subtypes of disease entities, senescence might, on the other hand, be a common denominator and therapeutic (co-) target for a variety of very different disease entities. Here, we do not cover the connection between cellular senescence, cancer, and senolytic cancer therapy. This topic has already been extensively reviewed elsewhere [40–42] and is recently extended by the idea of a one-two-punch strategy against cancer [43].

2. Biomarkers of Cellular Senescence

In 1995, Dimri and colleagues demonstrated that cellular senescence occurs in cultures of human cells and aging skin by developing a first biomarker, such as the activity of β-galactosidase at pH 6.0, referred to as SA-β-Gal [11]. Increased SA-β-Gal due to elevated lysosomal content with altered activity is widely adopted as a biomarker both in vitro and in vivo. Later studies showed that SCs are distinct from quiescent and differentiated cells with specific morphological and metabolic signatures, reorganized chromatin, altered gene expression, and expression of the SASP. Before highlighting biomarkers of SCs based on these characteristic features, it is essential to emphasize that not all senescent cells display the full panel of all senescence biomarkers. Moreover, some senescence markers are also present in apoptotic and quiescent cells. Thus, only complex phenotyping using several biomarkers might lead to correct conclusions.

In general, SCs are characterized by a stable cell cycle arrest primarily due to DNA damage signaling in response to the shortening of telomeres, exposure to growth factors and inflammatory cytokines, high levels of reactive oxygen species (ROS), and mitochondrial dysfunction. These cells are typically larger, displaying a flattened shape and extensive vacuolization. Disrupted integrity of the nuclear envelope is observed due to a loss of Lamin B1 protein, whose decreased expression is currently one of the most useful senescence biomarkers [44]. Senescent cells accumulate many dysfunctional mitochondria producing increased levels of ROS, further causing DNA damage [45].

Chromatin reorganization in SCs provokes the formation of senescence-associated heterochromatin foci (SAHFs) to induce the silencing of genes that promote proliferation, including E2F target genes. SAHFs are characterized by extensive DAPI staining and immunoreactivity to macroH2A, heterochromatin protein 1 (HP1), and lysine 9 di-or-tri-methylated histone H3 (H3K9me2/3) [46].

Moreover, SCs display a persistent DNA damage response (DDR) and contain nuclear foci called "DNA segments with chromatin alterations reinforcing senescence" (DNA-SCARS). DNA-SCARS in uncapped telomeres are called "telomere dysfunction-induced foci" (TIF). Another indicator of the DDR is the phosphorylated form of H2A.X or γ-H2A.X, a variant of a histone required for checkpoint-mediated cell cycle arrest and DNA repair of double-stranded DNA breaks. Regulatory protein p53, as a mediator of the DDR, promotes the activation of p21$^{WAF1/CIP1}$ (CDKN1A) and p16^{INK4A} (CDKN2A) to inhibit the activity of cyclin-dependent kinase [3]. Both protein and mRNA levels for p53, p16^{INK4A}, and p21$^{WAF1/CIP1}$, are extensively used to detect cellular senescence in vitro and in vivo [3,37,47].

Senescent cells maintain their metabolic activity and secrete inflammatory cytokines (IL-6, IL-8, TNF-α), growth factors such as TGFβ, chemokines, specific micro-RNAs, and matrix metalloproteinases, collectively known as the SASP [47]. The SASP modulates the surrounding environment, exerting its pathophysiological effects [47,48]. Given the nature of the SASP, senescence signaling is often linked to inflammation. Remarkably, some SASP factors, such as colony-stimulating factor 1 (CSF1), CCL2, and IL-8, recruit the immune system to promote the self-clearance of SCs [4]. This process seems critical to control the levels of SCs in a given setting and prevent chronic inflammatory responses.

Manifestation of each senescence hallmark is context-dependent and varies according to the trigger, cell or tissue type, and time from induction of the senescence program [49]. Changes at the levels of transcripts or proteins in response to multiple senescence inducers in different cell types demonstrate variability within specific pathological contexts and the requirement for a multi-marker system to identify SCs with precise accuracy [50]. A three-step multi-marker workflow was suggested to include (1) SA-β-gal or lipofuscin accumulation; (2) co-staining with markers frequently present (p16^{INK4A}, p21$^{WAF1/CIP1}$) or absent (proliferation markers, Lamin B1); and (3) identification of factors predicted to be altered in specific senescence contexts (SASP, DNA damage, and PI3K/FOXO/mTOR) [3]. Similarly, recent endeavors led to the development of the SenMayo marker panel, which identifies SCs based on their mRNA expression signatures across different tissues with high accuracy [51].

3. Selected Examples of Senopathies and Their Treatment with Senotherapies

Aging and related functional declines are strong drivers for many diseases and can be treated pharmacologically [52–55]. Aging-related senopathies are characterized by the accumulation of SCs over time, causing tissue and organ dysfunctions and the development of pathophysiological changes [56,57]. Targeting SCs with senolytics; however, may improve the functionality of various organ systems and prevent disease progression [18,20,58–60]. Similar mechanistic relationships are more difficult to establish for senomorphics because these compounds target the SASP, not SCs as such. Since most SASP components are commonly present in inflammatory diseases, it is difficult to determine if a therapeutically effective compound's mode of action is via its senomorphic or more general anti-inflammatory properties. Thus, readouts, such as the most common SASP factors IL-6 or IL-8, are insufficient to establish a clear mechanistic connection between SCs and a specific disease. Several compounds with senomorphic activity in vitro, such as rapamycin, metformin, aspirin, statins and NF-κB-, p38MAPK-, JAK/STAT-, and ATM-inhibitors, are effective in potential senopathies [61].

Several studies connecting SC accumulation to disease progression used the *p16-3MR* [32] or the *INK-ATTAC* [14] mouse model. Both models contain a transgene that permits the visualization and inducible elimination of p16^{INK4A}-expressing cells.

3.1. Cardiovascular Diseases

Improved cardiovascular function, increased regeneration, and prevention of hypertrophy and fibrosis were shown under the treatment of old animals with navitoclax or a combination of dasatinib and quercetin (D+Q) [16,62]. These drugs significantly decreased the SC burden, shown by reduced SA-β-gal staining, telomere-associated DNA damage foci (TAF), and p16^{INK4A} mRNA. Navitoclax effectively eliminated SCs from advanced atherosclerotic lesions in LDL receptor knockout mice exposed to a high-fat diet. Thereby, navitoclax prevented elastic fiber degradation and fibrous cap thinning through increased metalloprotease production [63].

3.2. Metabolic Disorders

Ruxolitinib treatment enhanced adipogenesis, reduced fat tissue loss and lipotoxicity, and improved insulin sensitivity of old animals by reducing the amounts of SA-β-gal positive cells and transcription of IL-6, p21$^{WAF1/CIP1}$, and p16^{INK4A} [64]. Even more so, diabetes was prevented in the *NOD* mouse model of type 1 diabetes, pointing toward diabetes as a potential senopathy [65]. Alleviated anxiety-related behavior and decreased inflammation in *db/db* diabetes-prone animals fed a high-fat diet were observed after treatment with D+Q [66]. A subset of insulin-producing pancreatic β-cells acquires a SASP during the natural history of T1D in humans. Senescent β-cells upregulated the pro-survival mediator Bcl-2 in non-obese diabetic mice. Thompson and colleagues used ABT-737 and ABT-199, inhibitors of Bcl-2, to eliminate senescent β-cells sufficiently to prevent diabetes [65]. Metformin is an approved drug for type 2 diabetes in humans, but also positively affects other age-related disorders [67] and extends the healthy lifespan of male mice [68]. Although several senescence markers in different cell types are down-regulated upon exposure to this drug [69,70], the precise mechanisms of metformin action remain elusive [61].

3.3. Musculoskeletal Diseases

Significant reduction in SCs and SASP factors in old mice by treatment with Ruxolitinib or D+Q prevented loss of bone mass and increased their strength, effectively preventing osteoporosis and frailty [71]. Moreover, UBX0101 inhibited erosion of articular cartilage by targeting transcription of SASP-associated genes [72] and thus points towards osteoarthritis as senopathy as well. Aspirin, for which senomorphic properties were described [73], is also effective in arthritis and osteoporosis [74]. The novel small molecule NF-κB-inhibitor SR12343 reduced senescence markers and improved muscle pathologies in the *Zmpste24$^{-/-}$* progeroid mouse model [75].

3.4. Liver and Kidney Diseases

Age-related liver steatosis and fibrosis were effectively treated with A-1331852 or D+Q by elimination of SCs with decreased production of the SASP, both in *INK-ATTAC* and in the *Mdr2$^{-/-}$* mouse model of primary sclerosing cholangitis [76,77]. Moreover, treatment with FOXO4-DRI improved kidney function by affecting amounts of Lamin B1 positive cells and expression of IL-6 [21].

3.5. Neurodegeneration

Senopathies have detrimental effects on the nervous system by affecting both its structure and function. One prominent example of a neurodegenerative condition likely promoted by the accumulation of SCs and characterized by the expression of tau is AD. Studies in *SAMP8* mice developing pathophysiological alterations similar to human AD and in the *tau$_{NFT}$-Mapt$^{0/0}$* model acquiring tau-related pathologies at a faster rate show that treatments with fisetin or D+Q decreased inflammation and stress with restored synaptic function, cognitive deficits, and cerebral blood flow [78,79]. Moreover, navitoclax prevented cognitive function decline, gliosis, and hyper-phosphorylation of tau in *INK-ATTAC PS19* mice, which express high levels of human tau, specifically in neurons [80].

3.6. Lung Diseases

An accumulation of senescent cells induces fibrotic pulmonary disease, especially in idiopathic pulmonary fibrosis [81]. D+Q improved lung function in bleomycin-treated mice by affecting the mRNA levels of Mcpl, 11-6, and Mmpl2 [81]. Moreover, signatures of pulmonary fibrosis induced by irradiation were successfully treated with navitoclax [82]. In this model, the elimination of SCs and decreased inflammation were shown by reduced amounts of SA-β-gal positive cells and mRNA levels of p16^{INK4A}, Bcl-2, IL-la, and IL-1β. It would be interesting if also chronic obstructive pulmonary disease (COPD), in which senescent cells are known to accumulate, were alleviated by senolytic treatment [83].

In Table 1, we summarize the effects of senolytic treatment on cardiovascular diseases, detrimental changes in bones, cartilage, liver, kidney, brain, and fat tissues, as well as treatment of atherosclerosis, neurodegeneration, fibrotic pulmonary disease, and diabetes, clearly qualifying these diseases as senopathies.

Table 1. Examples of medical conditions with evidence for an implication of senescent cells. Italic font indicates mRNA. "Age" reflects the age at analysis. * Initial age of mice before the start of the treatment is not specified; mo: month; WT: wildtype; p16-3MR: p16^{INK4A} trimodality reporter; HFD: high-fat diet; INK-ATTAC: p16^{INK4A}-induced apoptosis through targeted activation of caspase; SAMP8: Senescence Accelerated Mouse-Prone 8; PS19: Tau P301S (PS19Tg); NOD: Non-Obese Diabetic Mouse; db/db: BKS.Cg-Dock7^m+/+Lepr^{db}J; D+Q: Dasatinib and Quercetin; SASP: senescence-associated secretory phenotype; TAF: telomere-associated DNA damage repair foci; AECII: alveolar epithelial type II cell; SA-β-gal: senescence-associated β-galactosidase activity.

Organ/System/ Diseases	Age/Genotype	Senolytic Treatment	Evidence for Targeting SCs	Effects	Reference
Cardiovascular system	24-mo-old WT	D+Q	SA-β-gal, *p16^{INK4A}*	Improved cardiovascular function	[16]: Zhu et al., 2015
	23-mo-old WT	navitoclax	TAF, *p16^{INK4A}*	Reduced hypertrophy and fibrosis, increased regeneration	[62]: Anderson et al., 2019
	8-mo-old p16-3MR; Ldlr$^{-/-}$ HFD	navitoclax	SA-β-gal, *p16^{INK4A}*, *pl9^{ARF}*, *p21$^{WAF1/CIP1}$*, *IL-1a, Mmp3, Mmpl3*	Reduced plaque number and size, smaller fatty streaks, and protection from plaques rupture	[63]: Childs et al., 2016
Fat tissue	24-mo-old WT	ruxolitinib	SA-β-gal, *IL-6*, *p21$^{WAF1/CIP1}$*, *p16^{INK4A}*	Enhanced adipogenesis, reduced loss of fat tissue, reduced lipotoxicity, and enhanced insulin sensitivity	[64]: Xu et al., 2015
	min. 4-mo-old * INK-ATTAC HFD	D+Q	SA-β-gal, *p16^{INK4A}*, TAF, SASP factors in plasma	Alleviated anxiety-related behavior and decreased inflammation	[66]: Ogrodnik et al., 2019
Bones and cartilages	20-mo-old WT	D+Q	*p16^{INK4A}*, bone fluorescence, distention of satellites osteocytes	Higher bone mass and strength, better bone microarchitecture, reduced osteoporosis, and frailty	[71]: Farr et al., 2017
	22-mo-old WT	ruxolitinib			
	19-mo-old p16-3MR	UBX0101	*p16^{INK4A}*, *p21$^{WAF1/CIP1}$*, *11-6, Mmpl3*	Inhibited articular cartilage erosion	[72]: Jeon et al., 2017
	5-mo-old Zmpste24$^{-/-}$	SR12343	SA-β-gal, *p16^{INK4A}*, *p21$^{WAF1/CIP1}$*, *IL-6*	Inhibited fibrosis formation (trichrome staining), improved myogenesis	[75]: Zhang et al., 2021
Liver and kidney	4-mo-old Mdr$^{-/-}$	A-1331852	*p16^{INK4A}*-positive cholangiocytes and SASP levels	Reduction in liver fibrosis and amounts of growth factors and cytokines	[77]: Moncsek et al., 2018
	24-mo-old INK-ATTAC	D+Q	Karyomegalic and TAF-positive hepatocytes	Reduced hepatic steatosis	[76]: Ogrodnik et al., 2017
	24-28-mo-old p16-3MR	FOXO4-DRI	Lamin B1-positive cells, *IL-6*	Improved kidney function	[21]: Baar et al., 2017

Table 1. *Cont.*

Organ/System/ Diseases	Age/Genotype	Senolytic Treatment	Evidence for Targeting SCs	Effects	Reference
Brain and neurodegeneration	10-mo-old SAMP8	fisetin		Reduced cognitive deficits, restored synaptic function, stress, and inflammation	[78]: Currais et al., 2018
	23-mo-old tau$_{NFT}$-Mapt$^{0/0}$	D+Q	$p16^{INK4A}$, $p21^{WAF1/CIP}$, $Cxcl1$, IL-$1a$	Neuroprotective effects, improved brain structure, and cerebral blood flow	[79]: Musi et al., 2018
	6-8-mo-old INK-ATTAC; PS19	navitoclax	X-Gal crystals	Prevented gliosis and hyper-phosphorylation of tau, preserved cognitive function	[80]: Bussian et al., 2018
Fibrotic pulmonary disease	4-9-mo-old INK-ATTAC bleomycin-treated	D+Q	$p16^{INK4A}$, $Mcpl$, 11-6, $Mmpl2$	Improvement of lung function measured by plethysmography, increased body weight, and improved exercise capacity	[81]: Schafer et al., 2017
	8-9-mo-old WT irradiated	navitoclax	SA-β-gal positive AECIIs, $p16^{INK4A}$, Bcl-2, IL-$1a$, IL-$1b$	Reversed pulmonary fibrosis	[82]: Pan et al., 2017
Diabetes	9-mo-old NOD	ABT-737 ABT-199	SA-β-gal, $p16^{INK4A}$, $p21^{WAF1/CIP}$, $Igfbp3$, IL-6	Prevention of type-1 diabetes by inhibition of b-cell destruction	[65]: Thompson et al., 2019
	min. 2-mo-old * db/db mice	D+Q	SA-β-gal, $p16^{INK4A}$, TAF	Alleviated anxiety-related behavior	[66]: Ogrodnik et al., 2019

4. Conclusions

While we are moving towards precision medicine with more disease entities identified by molecular patterning than before by clinical phenotypic classification alone [84], one common denominator of several very different and heterogeneous conditions emerges: cellular senescence is present and accumulates in a plethora of mainly, but not exclusively, age-associated diseases and can be targeted by senotherapies as a single treatment strategy. Therefore, we postulate the term "senopathy" to define such diseases and conditions. One caveat, however, is the so far not well-described and understood heterogeneity of senescence depending on different senescence triggers and specific cell types. At this point, developing single senolytic or senomorphic drugs that specifically target all senescent cells in an organism is not well-conceivable.

Author Contributions: O.L.: Conceptualization, Investigation, Writing—original draft, Writing—review and editing. M.S.: Conceptualization, Writing—original draft, Writing—review and editing. J.G.: Conceptualization, Writing—original draft, Writing—review and editing, Supervision, Project administration, Funding acquisition. All authors have read and agreed to the published version of the manuscript.

Funding: We kindly acknowledge support by the Austrian Science Fund (FWF) and 'Herzfelder'sche Familienstiftung: P35268-B and FG 2400-B. OL is supported by a FEBS Ukrainian Short-Term Fellowship. This research was conducted while MS was a Hevolution/AFAR New Investigator Awardee in Aging Biology and Geroscience Research. The APC was funded by Open Access Funding by the Austrian Science Fund (FWF): P35268-B.

Institutional Review Board Statement: Not applicable.

Informed Consent Statement: Not applicable.

Data Availability Statement: Not applicable.

Conflicts of Interest: J.G. is co-founder and shareholder of Rockfish Bio AG. O.L. and M.S. declare no conflict of interest.

References

1. Borghesan, M.; Hoogaars, W.M.H.; Varela-Eirin, M.; Talma, N.; Demaria, M. A Senescence-Centric View of Aging: Implications for Longevity and Disease. *Trends Cell Biol.* **2020**, *30*, 777–791. [CrossRef] [PubMed]
2. Van Deursen, J.M. The Role of Senescent Cells in Ageing. *Nature* **2014**, *509*, 439–446. [CrossRef] [PubMed]
3. Gorgoulis, V.; Adams, P.D.; Alimonti, A.; Bennett, D.C.; Bischof, O.; Bishop, C.; Campisi, J.; Collado, M.; Evangelou, K.; Ferbeyre, G.; et al. Cellular Senescence: Defining a Path Forward. *Cell* **2019**, *179*, 813–827. [CrossRef]
4. Muñoz-Espín, D.; Serrano, M. Cellular Senescence: From Physiology to Pathology. *Nat. Rev. Mol. Cell Biol.* **2014**, *15*, 482–496. [CrossRef]
5. Baker, D.J.; Alimirah, F.; van Deursen, J.M.; Campisi, J.; Hildesheim, J. Oncogenic Senescence: A Multi-Functional Perspective. *Oncotarget* **2017**, *8*, 27661–27672. [CrossRef]
6. Toussaint, O.; Royer, V.; Salmon, M.; Remacle, J. Stress-Induced Premature Senescence and Tissue Ageing. *Biochem. Pharmacol.* **2002**, *64*, 1007–1009. [CrossRef]
7. Hayflick, L. The Limited in Vitro Lifetime of Human Diploid Cell Strains. *Exp. Cell Res.* **1965**, *37*, 614–636. [CrossRef]
8. Hayflick, L.; Moorhead, P.S. The Serial Cultivation of Human Diploid Cell Strains. *Exp. Cell Res.* **1961**, *25*, 585–621. [CrossRef] [PubMed]
9. Shay, J.W.; Wright, W.E. Hayflick, His Limit, and Cellular Ageing. *Nat. Rev. Mol. Cell Biol.* **2000**, *1*, 72–76. [CrossRef]
10. Harley, C.B.; Futcher, A.B.; Greider, C.W. Telomeres Shorten during Ageing of Human Fibroblasts. *Nature* **1990**, *345*, 458–460. [CrossRef]
11. Dimri, G.P.; Lee, X.; Basile, G.; Acosta, M.; Scott, G.; Roskelley, C.; Medrano, E.E.; Linskens, M.; Rubelj, I.; Pereira-Smith, O. A Biomarker That Identifies Senescent Human Cells in Culture and in Aging Skin In Vivo. *Proc. Natl. Acad. Sci. USA* **1995**, *92*, 9363–9367. [CrossRef]
12. Erusalimsky, J.D.; Kurz, D.J. Cellular Senescence In Vivo: Its Relevance in Ageing and Cardiovascular Disease. *Exp. Gerontol.* **2005**, *40*, 634–642. [CrossRef]
13. Herbig, U.; Ferreira, M.; Condel, L.; Carey, D.; Sedivy, J.M. Cellular Senescence in Aging Primates. *Science* **2006**, *311*, 1257. [CrossRef]
14. Baker, D.J.; Wijshake, T.; Tchkonia, T.; LeBrasseur, N.K.; Childs, B.G.; van de Sluis, B.; Kirkland, J.L.; van Deursen, J.M. Clearance of P16Ink4a-Positive Senescent Cells Delays Ageing-Associated Disorders. *Nature* **2011**, *479*, 232–236. [CrossRef] [PubMed]
15. Baker, D.J.; Childs, B.G.; Durik, M.; Wijers, M.E.; Sieben, C.J.; Zhong, J.; Saltness, R.A.; Jeganathan, K.B.; Verzosa, G.C.; Pezeshki, A.; et al. Naturally Occuring P16INK4a-Positive Cells Shorten Healthy Lifespan. *Nature* **2016**, *530*, 184–189. [CrossRef]
16. Zhu, Y.; Tchkonia, T.; Pirtskhalava, T.; Gower, A.C.; Ding, H.; Giorgadze, N.; Palmer, A.K.; Ikeno, Y.; Hubbard, G.B.; Hara, S.P.O.; et al. The 'Achilles' Heel of Senescent Cells: From Transcriptome to Senolytic Drugs. *Aging Cell* **2015**, *14*, 644–658. [CrossRef]
17. Xu, M.; Pirtskhalava, T.; Farr, J.N.; Weigand, B.M.; Palmer, A.K.; Weivoda, M.M.; Inman, C.L.; Ogrodnik, M.B.; Hachfeld, C.M.; Fraser, D.G.; et al. Senolytics Improve Physical Function and Increase Lifespan in Old Age. *Nat. Med.* **2018**, *24*, 1246–1256. [CrossRef] [PubMed]
18. Robbins, P.D.; Jurk, D.; Khosla, S.; Kirkland, J.L.; LeBrasseur, N.K.; Miller, J.D.; Passos, J.F.; Pignolo, R.J.; Tchkonia, T.; Niedernhofer, L.J. Senolytic Drugs: Reducing Senescent Cell Viability to Extend Health Span. *Annu. Rev. Pharmacol. Toxicol.* **2021**, *61*, 779–803. [CrossRef]
19. Tchkonia, T.; Zhu, Y.; van Deursen, J.; Campisi, J.; Kirkland, J.L. Cellular Senescence and the Senescent Secretory Phenotype: Therapeutic Opportunities. *J. Clin. Investig.* **2013**, *123*, 966–972. [CrossRef] [PubMed]
20. Birch, J.; Gil, J. Senescence and the SASP: Many Therapeutic Avenues. *Genes Dev.* **2020**, *34*, 1565–1576. [CrossRef]
21. Baar, M.P.; Brandt, R.M.C.; Putavet, D.A.; Klein, J.D.D.; Derks, K.W.J.; Bourgeois, B.R.M.; Stryeck, S.; Rijksen, Y.; van Willigenburg, H.; Feijtel, D.A.; et al. Targeted Apoptosis of Senescent Cells Restores Tissue Homeostasis in Response to Chemotoxicity and Aging. *Cell* **2017**, *169*, 132–147.e16. [CrossRef] [PubMed]
22. Coppé, J.-P.; Patil, C.K.; Rodier, F.; Sun, Y.; Muñoz, D.P.; Goldstein, J.; Nelson, P.S.; Desprez, P.-Y.; Campisi, J. Senescence-Associated Secretory Phenotypes Reveal Cell-Nonautonomous Functions of Oncogenic RAS and the P53 Tumor Suppressor. *PLoS Biol.* **2008**, *6*, 2853–2868. [CrossRef]
23. Weilner, S.; Schraml, E.; Redl, H.; Grillari-Voglauer, R.; Grillari, J. Secretion of Microvesicular MiRNAs in Cellular and Organismal Aging. *Exp. Gerontol.* **2013**, *48*, 626–633. [CrossRef]
24. Terlecki-Zaniewicz, L.; Lämmermann, I.; Latreille, J.; Bobbili, M.R.; Pils, V.; Schosserer, M.; Weinmüllner, R.; Dellago, H.; Skalicky, S.; Pum, D.; et al. Small Extracellular Vesicles and Their MiRNA Cargo Are Anti-Apoptotic Members of the Senescence-Associated Secretory Phenotype. *Aging* **2018**, *10*, 1103–1132. [CrossRef] [PubMed]
25. Fafián-Labora, J.A.; O'Loghlen, A. NF-KB/IKK Activation by Small Extracellular Vesicles within the SASP. *Aging Cell* **2021**, *20*, e13426. [CrossRef]
26. Tanaka, Y.; Takahashi, A. Senescence-Associated Extracellular Vesicle Release Plays a Role in Senescence-Associated Secretory Phenotype (SASP) in Age-Associated Diseases. *J. Biochem.* **2021**, *169*, 147–153. [CrossRef]
27. Lopes-Paciencia, S.; Saint-Germain, E.; Rowell, M.-C.; Ruiz, A.F.; Kalegari, P.; Ferbeyre, G. The Senescence-Associated Secretory Phenotype and Its Regulation. *Cytokine* **2019**, *117*, 15–22. [CrossRef]

28. Marin, I.; Boix, O.; Garcia-Garijo, A.; Sirois, I.; Caballe, A.; Zarzuela, E.; Ruano, I.; Stephan-Otto Attolini, C.; Prats, N.; Lopez-Dominguez, J.A.; et al. Cellular Senescence Is Immunogenic and Promotes Anti-Tumor Immunity. *Cancer Discov.* **2022**, *13*, 410–431. [CrossRef] [PubMed]
29. Ring, N.A.R.; Valdivieso, K.; Grillari, J.; Redl, H.; Ogrodnik, M. The Role of Senescence in Cellular Plasticity: Lessons from Regeneration and Development and Implications for Age-Related Diseases. *Dev. Cell* **2022**, *57*, 1083–1101. [CrossRef]
30. Mertens, J.; Herdy, J.R.; Traxler, L.; Schafer, S.T.; Schlachetzki, J.C.M.; Böhnke, L.; Reid, D.A.; Lee, H.; Zangwill, D.; Fernandes, D.P.; et al. Age-Dependent Instability of Mature Neuronal Fate in Induced Neurons from Alzheimer's Patients. *Cell Stem Cell* **2021**, *28*, 1533–1548.e6. [CrossRef]
31. Schosserer, M. The Role and Biology of Senescent Cells in Ageing-Related Tissue Damage and Repair. *Mech. Ageing Dev.* **2022**, *202*, 111629. [CrossRef] [PubMed]
32. Demaria, M.; Ohtani, N.; Youssef, S.A.; Rodier, F.; Toussaint, W.; Mitchell, J.R.; Laberge, R.M.; Vijg, J.; VanSteeg, H.; Dollé, M.E.T.; et al. An Essential Role for Senescent Cells in Optimal Wound Healing through Secretion of PDGF-AA. *Dev. Cell* **2014**, *31*, 722–733. [CrossRef] [PubMed]
33. He, S.; Sharpless, N.E. Senescence in Health and Disease. *Cell* **2017**, *169*, 1000–1011. [CrossRef] [PubMed]
34. Mehdizadeh, M.; Aguilar, M.; Thorin, E.; Ferbeyre, G.; Nattel, S. The Role of Cellular Senescence in Cardiac Disease: Basic Biology and Clinical Relevance. *Nat. Rev. Cardiol.* **2022**, *19*, 250–264. [CrossRef]
35. Kritsilis, M.; Rizou, V.S.; Koutsoudaki, P.N.; Evangelou, K.; Gorgoulis, V.G.; Papadopoulos, D. Ageing, Cellular Senescence and Neurodegenerative Disease. *Int. J. Mol. Sci.* **2018**, *19*, 2937. [CrossRef]
36. Sreekumar, P.G.; Hinton, D.R.; Kannan, R. The Emerging Role of Senescence in Ocular Disease. *Oxidative Med. Cell. Longev.* **2020**, *2020*, e2583601. [CrossRef]
37. Childs, B.G.; Durik, M.; Baker, D.J.; van Deursen, J.M. Cellular Senescence in Aging and Age-Related Disease: From Mechanisms to Therapy. *Nat. Med.* **2015**, *21*, 1424–1435. [CrossRef]
38. Fischer, B.M.; Wong, J.K.; Degan, S.; Kummarapurugu, A.B.; Zheng, S.; Haridass, P.; Voynow, J.A. Increased Expression of Senescence Markers in Cystic Fibrosis Airways. *Am. J. Physiol. Lung Cell Mol. Physiol.* **2013**, *304*, L394–L400. [CrossRef]
39. Schmitt, C.A.; Tchkonia, T.; Niedernhofer, L.J.; Robbins, P.D.; Kirkland, J.L.; Lee, S. COVID-19 and Cellular Senescence. *Nat. Rev. Immunol.* **2022**, *23*, 251–263. [CrossRef]
40. Schosserer, M.; Grillari, J.; Breitenbach, M. The Dual Role of Cellular Senescence in Developing Tumors and Their Response to Cancer Therapy. *Front. Oncol.* **2017**, *7*, 278. [CrossRef]
41. Campisi, J.; Andersen, J.K.; Kapahi, P.; Melov, S. Cellular Senescence: A Link between Cancer and Age-Related Degenerative Disease? *Semin. Cancer Biol.* **2011**, *21*, 354–359. [CrossRef] [PubMed]
42. Loaiza, N.; Demaria, M. Cellular Senescence and Tumor Promotion: Is Aging the Key? *Biochim. Biophys. Acta Rev. Cancer* **2016**, *1865*, 155–167. [CrossRef] [PubMed]
43. Wang, L.; Lankhorst, L.; Bernards, R. Exploiting Senescence for the Treatment of Cancer. *Nat. Rev. Cancer* **2022**, *22*, 340–355. [CrossRef] [PubMed]
44. Matias, I.; Diniz, L.P.; Damico, I.V.; Araujo, A.P.B.; Neves, L.D.S.; Vargas, G.; Leite, R.E.P.; Suemoto, C.K.; Nitrini, R.; Jacob-Filho, W.; et al. Loss of Lamin-B1 and Defective Nuclear Morphology Are Hallmarks of Astrocyte Senescence In Vitro and in the Aging Human Hippocampus. *Aging Cell* **2022**, *21*, e13521. [CrossRef] [PubMed]
45. Giorgi, C.; Marchi, S.; Simoes, I.C.M.; Ren, Z.; Morciano, G.; Perrone, M.; Patalas-Krawczyk, P.; Borchard, S.; Jędrak, P.; Pierzynowska, K.; et al. Mitochondria and Reactive Oxygen Species in Aging and Age-Related Diseases. In *International Review of Cell and Molecular Biology*; Elsevier: Amsterdam, The Netherlands, 2018; Volume 340, pp. 209–344. ISBN 978-0-12-815736-7.
46. Aird, K.M.; Zhang, R. Detection of Senescence-Associated Heterochromatin Foci (SAHF). In *Cell Senescence*; Galluzzi, L., Vitale, I., Kepp, O., Kroemer, G., Eds.; Methods in Molecular Biology Series; Humana Press: Totowa, NJ, USA, 2013; Volume 965, pp. 185–196. ISBN 978-1-62703-238-4.
47. Coppé, J.-P.; Desprez, P.-Y.; Krtolica, A.; Campisi, J. The Senescence-Associated Secretory Phenotype: The Dark Side of Tumor Suppression. *Annu. Rev. Pathol.* **2010**, *5*, 99–118. [CrossRef]
48. Hearps, A.C.; Martin, G.E.; Angelovich, T.A.; Cheng, W.-J.; Maisa, A.; Landay, A.L.; Jaworowski, A.; Crowe, S.M. Aging Is Associated with Chronic Innate Immune Activation and Dysregulation of Monocyte Phenotype and Function. *Aging Cell* **2012**, *11*, 867–875. [CrossRef]
49. Calcinotto, A.; Kohli, J.; Zagato, E.; Pellegrini, L.; Demaria, M.; Alimonti, A. Cellular Senescence: Aging, Cancer, and Injury. *Physiol. Rev.* **2019**, *99*, 1047–1078. [CrossRef]
50. Basisty, N.; Kale, A.; Jeon, O.H.; Kuehnemann, C.; Payne, T.; Rao, C.; Holtz, A.; Shah, S.; Sharma, V.; Ferrucci, L.; et al. A Proteomic Atlas of Senescence-Associated Secretomes for Aging Biomarker Development. *PLoS Biol.* **2020**, *18*, e3000599. [CrossRef]
51. Saul, D.; Kosinsky, R.L.; Atkinson, E.J.; Doolittle, M.L.; Zhang, X.; LeBrasseur, N.K.; Pignolo, R.J.; Robbins, P.D.; Niedernhofer, L.J.; Ikeno, Y.; et al. A New Gene Set Identifies Senescent Cells and Predicts Senescence-Associated Pathways across Tissues. *Nat. Commun.* **2022**, *13*, 4827. [CrossRef]
52. Piskovatska, V.; Strilbytska, O.; Koliada, A.; Vaiserman, A.; Lushchak, O. Health Benefits of Anti-Aging Drugs. *Subcell. Biochem.* **2019**, *91*, 339–392. [CrossRef]
53. Lushchak, O.; Strilbytska, O.; Koliada, A.; Zayachkivska, A.; Burdyliuk, N.; Yurkevych, I.; Storey, K.B.; Vaiserman, A. Nanodelivery of Phytobioactive Compounds for Treating Aging-Associated Disorders. *Geroscience* **2020**, *42*, 117–139. [CrossRef]

54. Vaiserman, A.; Koliada, A.; Lushchak, O. Neuroinflammation in Pathogenesis of Alzheimer's Disease: Phytochemicals as Potential Therapeutics. *Mech. Ageing Dev.* **2020**, *189*, 111259. [CrossRef]
55. Vaiserman, A.; Koliada, A.; Lushchak, O.; Castillo, M.J. Repurposing Drugs to Fight Aging: The Difficult Path from Bench to Bedside. *Med. Res. Rev.* **2021**, *41*, 1676–1700. [CrossRef] [PubMed]
56. Khosla, S.; Farr, J.N.; Tchkonia, T.; Kirkland, J.L. The Role of Cellular Senescence in Ageing and Endocrine Disease. *Nat. Rev. Endocrinol.* **2020**, *16*, 263–275. [CrossRef] [PubMed]
57. Doolittle, M.L.; Monroe, D.G.; Farr, J.N.; Khosla, S. The Role of Senolytics in Osteoporosis and Other Skeletal Pathologies. *Mech. Ageing Dev.* **2021**, *199*, 111565. [CrossRef]
58. Sikora, E.; Bielak-Zmijewska, A.; Mosieniak, G. Targeting Normal and Cancer Senescent Cells as a Strategy of Senotherapy. *Ageing Res. Rev.* **2019**, *55*, 100941. [CrossRef] [PubMed]
59. Al-Naggar, I.M.A.; Kuchel, G.A.; Xu, M. Senolytics: Targeting Senescent Cells for Age-Associated Diseases. *Curr. Mol. Biol. Rep.* **2020**, *6*, 161–172. [CrossRef]
60. Owens, W.A.; Walaszczyk, A.; Spyridopoulos, I.; Dookun, E.; Richardson, G.D. Senescence and Senolytics in Cardiovascular Disease: Promise and Potential Pitfalls. *Mech. Ageing Dev.* **2021**, *198*, 111540. [CrossRef]
61. Zhang, L.; Pitcher, L.E.; Prahalad, V.; Niedernhofer, L.J.; Robbins, P.D. Targeting Cellular Senescence with Senotherapeutics: Senolytics and Senomorphics. *FEBS J.* **2023**, *290*, 1362–1383. [CrossRef]
62. Anderson, R.; Lagnado, A.; Maggiorani, D.; Walaszczyk, A.; Dookun, E.; Chapman, J.; Birch, J.; Salmonowicz, H.; Ogrodnik, M.; Jurk, D.; et al. Length-Independent Telomere Damage Drives Post-Mitotic Cardiomyocyte Senescence. *EMBO J.* **2019**, *38*, e100492. [CrossRef]
63. Childs, B.G.; Baker, D.J.; Wijshake, T.; Conover, C.A.; Campisi, J.; van Deursen, J.M. Senescent Intimal Foam Cells Are Deleterious at All Stages of Atherosclerosis. *Science* **2016**, *354*, 472–477. [CrossRef] [PubMed]
64. Xu, M.; Palmer, A.K.; Ding, H.; Weivoda, M.M.; Pirtskhalava, T.; White, T.A.; Sepe, A.; Johnson, K.O.; Stout, M.B.; Giorgadze, N.; et al. Targeting Senescent Cells Enhances Adipogenesis and Metabolic Function in Old Age. *eLife* **2015**, *4*, e12997. [CrossRef]
65. Thompson, P.J.; Shah, A.; Ntranos, V.; Van Gool, F.; Atkinson, M.; Bhushan, A. Targeted Elimination of Senescent Beta Cells Prevents Type 1 Diabetes. *Cell Metab.* **2019**, *29*, 1045–1060.e10. [CrossRef] [PubMed]
66. Ogrodnik, M.; Zhu, Y.; Langhi, L.G.P.; Tchkonia, T.; Krüger, P.; Fielder, E.; Victorelli, S.; Ruswhandi, R.A.; Giorgadze, N.; Pirtskhalava, T.; et al. Obesity-Induced Cellular Senescence Drives Anxiety and Impairs Neurogenesis. *Cell Metab.* **2019**, *29*, 1061–1077.e8. [CrossRef] [PubMed]
67. Foretz, M.; Guigas, B.; Bertrand, L.; Pollak, M.; Viollet, B. Metformin: From Mechanisms of Action to Therapies. *Cell Metab.* **2014**, *20*, 953–966. [CrossRef] [PubMed]
68. Martin-Montalvo, A.; Mercken, E.M.; Mitchell, S.J.; Palacios, H.H.; Mote, P.L.; Scheibye-Knudsen, M.; Gomes, A.P.; Ward, T.M.; Minor, R.K.; Blouin, M.-J.; et al. Metformin Improves Healthspan and Lifespan in Mice. *Nat. Commun.* **2013**, *4*, 2192. [CrossRef]
69. Moiseeva, O.; Deschênes-Simard, X.; St-Germain, E.; Igelmann, S.; Huot, G.; Cadar, A.E.; Bourdeau, V.; Pollak, M.N.; Ferbeyre, G. Metformin Inhibits the Senescence-Associated Secretory Phenotype by Interfering with IKK/NF-KB Activation. *Aging Cell* **2013**, *12*, 489–498. [CrossRef]
70. Zhang, C.; Chen, M.; Zhou, N.; Qi, Y. Metformin Prevents H_2O_2-Induced Senescence in Human Lens Epithelial B3 Cells. *Med. Sci. Monit. Basic Res.* **2020**, *26*, e923391. [CrossRef]
71. Farr, J.N.; Xu, M.; Weivoda, M.M.; Monroe, D.G.; Fraser, D.G.; Onken, J.L.; Negley, B.A.; Sfeir, J.G.; Ogrodnik, M.B.; Hachfeld, C.M.; et al. Targeting Cellular Senescence Prevents Age-Related Bone Loss in Mice. *Nat. Med.* **2017**, *23*, 1072–1079. [CrossRef]
72. Jeon, O.H.; Kim, C.; Laberge, R.-M.; Demaria, M.; Rathod, S.; Vasserot, A.P.; Chung, J.W.; Kim, D.H.; Poon, Y.; David, N.; et al. Local Clearance of Senescent Cells Attenuates the Development of Post-Traumatic Osteoarthritis and Creates a pro-Regenerative Environment. *Nat. Med.* **2017**, *23*, 775–781. [CrossRef]
73. Feng, M.; Kim, J.; Field, K.; Reid, C.; Chatzistamou, I.; Shim, M. Aspirin Ameliorates the Long-Term Adverse Effects of Doxorubicin through Suppression of Cellular Senescence. *FASEB Bioadv.* **2019**, *1*, 579–590. [CrossRef] [PubMed]
74. Hybiak, J.; Broniarek, I.; Kiryczyński, G.; Los, L.D.; Rosik, J.; Machaj, F.; Sławiński, H.; Jankowska, K.; Urasińska, E. Aspirin and Its Pleiotropic Application. *Eur. J. Pharmacol* **2020**, *866*, 172762. [CrossRef] [PubMed]
75. Zhang, L.; Zhao, J.; Mu, X.; McGowan, S.J.; Angelini, L.; O'Kelly, R.D.; Yousefzadeh, M.J.; Sakamoto, A.; Aversa, Z.; LeBrasseur, N.K.; et al. Novel Small Molecule Inhibition of IKK/NF-KB Activation Reduces Markers of Senescence and Improves Healthspan in Mouse Models of Aging. *Aging Cell* **2021**, *20*, e13486. [CrossRef] [PubMed]
76. Ogrodnik, M.; Miwa, S.; Tchkonia, T.; Tiniakos, D.; Wilson, C.L.; Lahat, A.; Day, C.P.; Burt, A.; Palmer, A.; Anstee, Q.M.; et al. Cellular Senescence Drives Age-Dependent Hepatic Steatosis. *Nat. Commun.* **2017**, *8*, 15691. [CrossRef]
77. Moncsek, A.; Al-Suraih, M.S.; Trussoni, C.E.; O'Hara, S.P.; Splinter, P.L.; Zuber, C.; Patsenker, E.; Valli, P.V.; Fingas, C.D.; Weber, A.; et al. Targeting Senescent Cholangiocytes and Activated Fibroblasts with B-Cell Lymphoma-Extra Large Inhibitors Ameliorates Fibrosis in Multidrug Resistance 2 Gene Knockout (Mdr2$^{-/-}$) Mice. *Hepatology* **2018**, *67*, 247–259. [CrossRef]
78. Currais, A.; Farrokhi, C.; Dargusch, R.; Armando, A.; Quehenberger, O.; Schubert, D.; Maher, P. Fisetin Reduces the Impact of Aging on Behavior and Physiology in the Rapidly Aging SAMP8 Mouse. *J. Gerontol. Ser. A* **2018**, *73*, 299–307. [CrossRef]
79. Musi, N.; Valentine, J.M.; Sickora, K.R.; Baeuerle, E.; Thompson, C.S.; Shen, Q.; Orr, M.E. Tau Protein Aggregation Is Associated with Cellular Senescence in the Brain. *Aging Cell* **2018**, *17*, e12840. [CrossRef]

80. Bussian, T.J.; Aziz, A.; Meyer, C.F.; Swenson, B.L.; van Deursen, J.M.; Baker, D.J. Clearance of Senescent Glial Cells Prevents Tau-Dependent Pathology and Cognitive Decline. *Nature* **2018**, *562*, 578–582. [CrossRef]
81. Schafer, M.J.; White, T.A.; Iijima, K.; Haak, A.J.; Ligresti, G.; Atkinson, E.J.; Oberg, A.L.; Birch, J.; Salmonowicz, H.; Zhu, Y.; et al. Cellular Senescence Mediates Fibrotic Pulmonary Disease. *Nat. Commun.* **2017**, *8*, 14532. [CrossRef]
82. Pan, J.; Li, D.; Xu, Y.; Zhang, J.; Wang, Y.; Chen, M.; Lin, S.; Huang, L.; Chung, E.J.; Citrin, D.E.; et al. Inhibition of Bcl-2/Xl With ABT-263 Selectively Kills Senescent Type II Pneumocytes and Reverses Persistent Pulmonary Fibrosis Induced by Ionizing Radiation in Mice. *Int. J. Radiat. Oncol. Biol. Phys.* **2017**, *99*, 353–361. [CrossRef]
83. Majewska, J.; Krizhanovsky, V. Breathe It in—Spotlight on Senescence and Regeneration in the Lung. *Mech. Ageing Dev.* **2021**, *199*, 111550. [CrossRef] [PubMed]
84. Hopp, W.J.; Li, J.; Wang, G. *Toward Precision Medicine: Building a Knowledge Network for Biomedical Research and a New Taxonomy of Disease*; National Academies Press: Washington, DC, USA, 2011; Volume 27, ISBN 978-0-309-22222-8.

 *biomolecules*

Review

The Janus-Faced Role of Lipid Droplets in Aging: Insights from the Cellular Perspective

Nikolaus Bresgen [1], Melanie Kovacs [1], Angelika Lahnsteiner [1], Thomas Klaus Felder [2,*] and Mark Rinnerthaler [1,*]

[1] Department of Biosciences and Medical Biology, Paris-Lodron University Salzburg, 5020 Salzburg, Austria; nikolaus.bresgen@plus.ac.at (N.B.)

[2] Department of Laboratory Medicine, Paracelsus Medical University, 5020 Salzburg, Austria

* Correspondence: t.felder@salk.at (T.K.F.); mark.rinnerthaler@plus.ac.at (M.R.)

Abstract: It is widely accepted that nine hallmarks—including mitochondrial dysfunction, epigenetic alterations, and loss of proteostasis—exist that describe the cellular aging process. Adding to this, a well-described cell organelle in the metabolic context, namely, lipid droplets, also accumulates with increasing age, which can be regarded as a further aging-associated process. Independently of their essential role as fat stores, lipid droplets are also able to control cell integrity by mitigating lipotoxic and proteotoxic insults. As we will show in this review, numerous longevity interventions (such as mTOR inhibition) also lead to strong accumulation of lipid droplets in *Saccharomyces cerevisiae*, *Caenorhabditis elegans*, *Drosophila melanogaster*, and mammalian cells, just to name a few examples. In mammals, due to the variety of different cell types and tissues, the role of lipid droplets during the aging process is much more complex. Using selected diseases associated with aging, such as Alzheimer's disease, Parkinson's disease, type II diabetes, and cardiovascular disease, we show that lipid droplets are "Janus"-faced. In an early phase of the disease, lipid droplets mitigate the toxicity of lipid peroxidation and protein aggregates, but in a later phase of the disease, a strong accumulation of lipid droplets can cause problems for cells and tissues.

Keywords: LDs; autophagy; mitochondria; protein aggregates; lipid peroxides; misfolded proteins; mTOR; IIS; lifespan; aging

check for
updates

Citation: Bresgen, N.; Kovacs, M.; Lahnsteiner, A.; Felder, T.K.; Rinnerthaler, M. The Janus-Faced Role of Lipid Droplets in Aging: Insights from the Cellular Perspective. *Biomolecules* **2023**, *13*, 912. https://doi.org/10.3390/biom13060912

Academic Editor: Gabriella D'Orazi

Received: 13 March 2023
Revised: 22 May 2023
Accepted: 29 May 2023
Published: 30 May 2023

1. Introduction

Lipid droplets (LDs) are evolutionary conserved structures that were mentioned for the first time by Van Leeuwenhoek in 1674, but their reassessment as autonomous organelles with important key roles in lipid and energy metabolism occurred many years later [1,2]. LDs originate from the endoplasmic reticulum (ER). In the first step, neutral lipids are synthesized at the ER and are redirected into the bilayer, leading to an aggregation of the highly motile lipids. Morphologically, the accumulation of neutral lipids in the ER bilayer resembles a lens-like structure. Growth in this lens initiates bilayer deformation and the budding-off of LDs to the cytoplasm [3]. Due to this special mode of formation, LDs are surrounded by a lipid monolayer and are filled with neutral lipids, especially triacyclglycerols (TAGs) and sterols. Therefore, LDs are mainly considered fat-storage organelles with high relevance to lipid-metabolism homeostasis. However, in recent years, evidence has accumulated that LDs are also capable of mediating cytoprotective properties by either acting as a "buffer" for toxic lipids [4–6] or serving the cellular clearance of damaged and misfolded proteins [7–11] (Figure 1). There is also growing evidence that LDs are involved in the binding and detoxification of xenobiotics; however, this will not be discussed in detail within this review, which focuses on the aspect of aging. By studying protein composition in LDs in different organisms such as bacteria, plants, insects, yeast, and mammals, hundreds of different LD-surface-associated proteins have been identified.

Although the LD proteome shows qualitative and quantitative variations among different cell types, a typical mammalian LD contains 100–150 different proteins [12]. Surface proteins are important for regulating LD homeostasis and enable the specific contract with other cell organelles. Major LD-associated proteins in mammals belong to the PAT protein family, also known as perilipin 1–5 (PLIN1-5) [13,14], adipocyte differentiation-related protein (ADRP) [15], and tail-interacting protein of 47 kDa (TIP47) [14,15]. Several different LD-resident proteins contributing to lipid biogenesis and degradation, as well as membrane trafficking and signaling, have been well reviewed [16]. Furthermore, it is well established that LDs form defined contacts with several other cellular organelles such as the ER, peroxisomes, lysosomes, and mitochondria (reviewed in [17]). Intriguingly, LDs may also sequester proteins involved in genetic control such as histones [18].

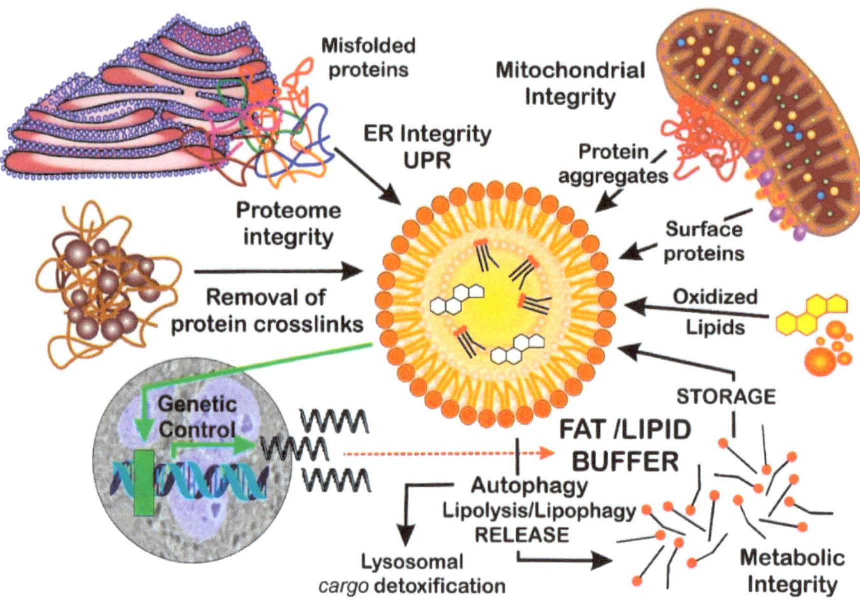

Figure 1. LDs as a cellular "buffer organelle". LDs serve as an intermediate cytosolic lipid buffer and assist the cell in detoxifying lipids, misfolded proteins, and protein aggregates present in the cytosol, ER, and mitochondria. Furthermore, LDs are also involved in adaption to cellular stress by modulating transcriptional control.

Here, we review the existing evidence for a distinct role of LDs in eukaryotic aging as explicitly reflected by the accumulation of LDs at terminal life periods [19–21]. Focusing on the *"physiologic triad"*—metabolic regulation, stress response, and aging—but also covering the evolutionary context, we decided to provide an up-to date, detailed review of a multitude of aging-related aspects of LD biology investigated in the classical biological model system (*Saccharomyces cerevisiae*, *Caenorhabditis elegans*, and *Drosophila melanogaster*) and also associated with age-related human disease. This attempt, by its nature, is complex, and discussing the fascinating, multifaceted role of lipid droplets in the multiple contexts of aging deserves an extended approach. As will be outlined, there is evidence for a Janus-faced role of LDs, their cellular accumulation counteracting stress-associated, disease-provoking forces. Thus, this is beneficial to aging, but conversely accelerates disease progression at advanced stages, which promotes the aging process. As we show in the course of this review, there is a close interplay between cellular pathways that regulate aging processes on the one hand, and on the other hand also affect the biogenesis of LDs and run like a thread through evolution (see Figure 2).

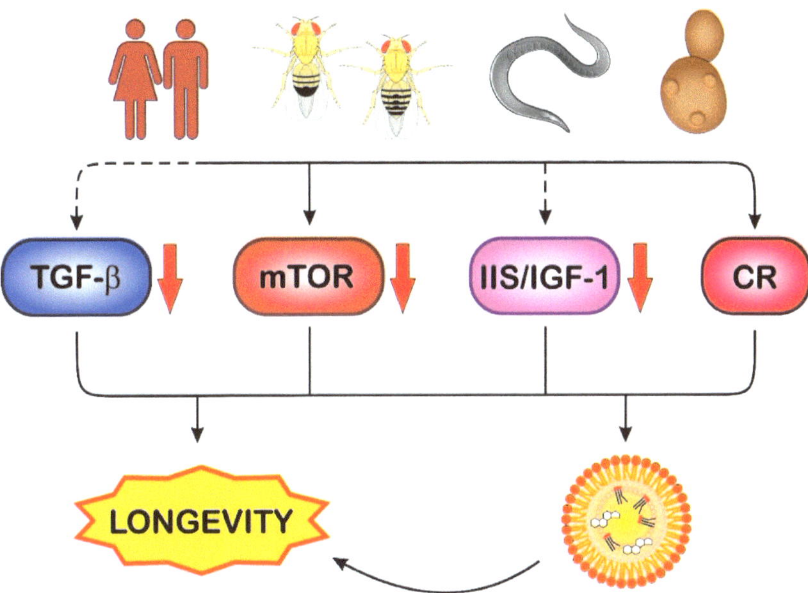

Figure 2. The longevity–LD interaction network. In most model organisms relevant for human aging research, inhibition of (1) TGF-β signaling, (2) mTOR signaling, (3) insulin/IGF-1 signaling and caloric restriction (CR) promotes both longevity and LD formation. Dotted lines indicate the absence of the respective pathways in yeast cells.

2. Lipid Droplets in *Saccharomyces cerevisiae*

The baker's yeast *Saccharomyces cerevisiae* is a valuable tool for aging research, as many aging- and disease-associated pathways such as DNA repair mechanisms, lipostasis, proteostasis, oxidative stress responses, regulated cell death, nutrient signaling, autophagy, and regulation of the cell cycle are evolutionarily conserved to a high degree [22]. Based on sequence similarity, about 30% of the yeast genome is conserved in the human genome [23]. Important for aging research is the fact that when adequate and sufficient nutrients are provided, *S. cerevisiae* cells grow exponentially via asymmetric budding of daughter cells from bigger mother cells [24].

2.1. Replicative and Chronologic Lifespan

In general, in yeast cells, two forms of aging mechanisms can be distinguished, namely, replicative and chronological aging. For both, cell death terminates the lifespan, caused by the intrinsic, mitochondrial outer membrane permeabilization (MOMP)-based activation of programmed cell death (PCD)/apoptosis emerging from increased reactive oxygen species (ROS) production and genomic instability, which provokes damage to the cellular proteome, lipidome, and organelles such as mitochondria [25]. Upon nutritional stress (i.e., exhaustion of nutrients), yeast cells stop dividing and enter a stationary phase which allows survival up to several weeks depending on strain type and culture conditions [22]. This survival period in the stationary phase is termed the chronological lifespan [26] and has to be distinguished from the yeast replicative lifespan, which is measured by the number of daughter cells that can be formed from a mother cell before it stops dividing [27].

The average lifespan in the yeast background BY4741 (the most used genetic background, generally considered as the wild type) lasts 25 generations. Aged (mother) cells are larger, and reveal a slowing down of the cell cycle and a declined protein synthesis. Each daughter cell that is formed leaves a bud scar on the mother cell surface that can be observed microscopically by calcofluor-white staining [28]. It is believed that damaged proteins and organelles (e.g., mitochondria) are specifically retained by the mother cells, which explains the "rejuvenation" of daughter cells resulting from asymmetric segregation [25,29]. Representing a mitosis-based lifespan definition, replicative aging in yeast cells mimics the limited mitotic capacity of non-transformed proliferating mammalian cells types, including undifferentiated stem cells as defined first by the *Hayflick limit* [30]. On the other hand, many phenotypical characteristic described for the chronological aging of yeast cells residing in the stationary phase share similarity with the phenotype of aged, post-mitotic cells in higher eukaryotes mainly comprising the class of terminally differentiated cell types such as cells of the central nervous system [31].

In some respects, several properties of *S. cerevisiae* render this fungal cell system a preferred aging model advantageous to human in vitro cell culture models. For instance, large numbers of cells can be monitored in comparatively short time periods under in vivo conditions in yeast. Of note, contrasting the well-conserved intracellular aging mechanisms common to both, yeast cells fail to display intercellular effects seen in multicellular organisms, such as inflammatory or systemic responses (e.g., regulated by hormones and/or the immune system), as well as other mechanisms involved in cell–cell communication [22]. However, it should not be overseen that, in vitro cell cultures, as for instance derived from mammalian tissues, are devoid of systemic, physiologic "cross-talks and feedback loops" if used as primary cell lines, and co-culturing with other cell types will reflect only part of the systemic complexity directing individual cell fate in vivo, in particular under the aspect of aging. Moreover, interpretation of experimental findings based on immortalized eukaryotic cell lines, self-evidently, is complicated due to the fundamentally altered growth control.

Both replicative and chronological lifespan in yeast can be extended by caloric restriction, which can be obtained by lowering glucose availability in the culture media (e.g., from 2% to 0.5%) [32]. In the absence of caloric restriction, chronologically aged yeast cells accumulate ethanol produced by glucose fermentation [32]. It is speculated that this counteracts the expression of β-oxidation regulatory enzymes Fox1p, Fox2p, and Fox3p (peroxisomal fatty acid β-oxidation core enzymes) leading to a decline in peroxisomal oxidation of LD-derived non-esterified "free" fatty acids that are synthesized in the ER and are stored in LDs [33,34]. In turn, non-oxidized free fatty acids will accumulate in LDs under normal nutritional conditions (i.e., 2% glucose) which promotes an inhibitory feedback loop on the ER-based synthesis of triacyclglycerols (TAG) [33]. It is hypothesized that lipid dynamic remodeling of this kind can shorten lifespan in chronologically aged yeast cells grown without caloric restriction (i.e., in the presence of 2% glucose) by three different mechanisms: (i) via necrotic cell death ensuing from the peroxisomal failure to oxidize free fatty acids, (ii) apoptosis stimulated by the accumulation of diacylglycerol and free fatty acids in the ER (*"lipoapoptosis"*), or (iii) diacylglycerol initiated protein kinase C-dependent signaling [33].

This accounts for a pivotal role of lipid dynamics in yeast aging, which is further supported by the finding that LD biogenesis in yeast is elevated in the course of replicative and chronological aging as well as under stress conditions [19,35,36]. Of special relevance, Beas et al. reported that overexpression of the *BNA2* gene encoding indoleamine 2,3-dioxygenase (*BNA2* is the yeast homolog of mammalian IDO1) leads to a 40% reduction in LD accumulation during replicative aging, which identifies *BNA2* as an important regulator of LD abundance [22]. Bna2p catalyzes the first step of NAD^+ synthesis converting tryptophan to formyl-kynurenine; hence, this finding reveals a connection between the NAD^+/kynurenine pathway and LD formation in the course of aging. It is proposed that the glycolytic flux in aging yeast cells is directed towards neutral lipid synthesis and LD generation, but Bna2p overexpression diverts the glycolytic flux from pyruvate and acetyl-CoA to the shikimate pathway (responsible for the synthesis of the amino acids phenylalanine, tyrosine, and tryptophan) and as a result lowers LD accumulation in the aged cells. Importantly, this investigation reveals that this kind of Bna2p-mediated "metabolic rewiring" in aged yeast cells is not directly associated with longevity. Moreover, the findings indicate that LD accumulation does not cause lifespan shortening, but, conversely, exerts protection of aged cells under stress conditions, which might provide a selective growth advantage under variable environmental conditions [36].

2.2. Lipid Droplets and Stress Adaptation

This concept is supported by another study that substantiates the role of LDs as key players in cellular stress adaption. The yeast cell growth rate declines when phosphatidylcholine biosynthesis is deficient, which changes the cellular phospholipid content and causes ER stress, alterations in ER morphology, and enhanced LD formation. In this case, an excess of phospholipids is converted to TAG by the acyltransferases Lro1p and Dga1p, which is immediately sequestered by LDs. This LD-generating process allows yeast cells to rebalance the pool of freely available phospholipids as an indispensable prerequisite for organelle morphology retrieval and cell growth [9]. Besides this pathway of ER-based regulation of lipid homeostasis yielding LD formation in yeast, ER stress arising from lipid imbalance is also at risk of activating the unfolded protein response (UPR). In most model organisms, it is shown that the UPR protects cells from the detrimental effects of proteotoxicity and is of great importance for the aging process [37]. Therefore, it is not surprising that all interventions that increase the activity of the UPR clearly extend the replicative lifespan of yeast cells [38].

The UPR provides cellular maintenance by specific handling of accumulated misfolded protein as well as facing lipid bilayer stress in the ER. Besides ER expansion, UPR signaling comprises the activity of a number of UPR-related gene products which direct the response either towards re-established homeostasis or, if not adequately facing a prolonged stress condition, participate in apoptosis onset (for a review see [39]). Essential to a successful outcome is the proper elimination of the ER stressor. A misfolded protein that initially accumulates inside the ER is translocated to the cytosol, where it is polyubiquitinylated by ubiquitin-conjugating enzymes residing at the cytosolic ER surface, the polyubiquitination serving as tag for proteasomal degradation [40]. However, lipid bilayer stress may also stimulate UPR in the ER (UPR^{ER}) [41,42] which converges with the UPR triggered by the misfolded protein at the central UPR effector Ire1p (inositol-requiring enzyme 1) [43]. Interestingly, in mouse hepatocytes, ER stress stimulates Ire-1 and downstream targets such as DGAT2 (diacylglycerol-acyltransferase 2) [44], with DGAT2 (as well as DGAT1) being essential to LD biosynthesis [45]. Referring to this and findings demonstrating ROS-triggered LD biogenesis and antioxidant properties of LDs in *Drosophila* [46], Walther et al. suggested that the Ire1p/DGAT2-stimulated LD formation could antagonize phospholipid oxidation via LD-mediated ROS scavenging [47]. This also underlines the importance of LDs for the aging process as the accumulation of ROS is one of the most prominent features at the terminal lifespan [48].

Moreover, linking LD formation to UPR-dependent responses in yeast, it was shown that ER-derived LDs can be associated with polyubiquitinylated proteins and also can be enriched in Kar2p, an ER chaperone involved in protein folding [9]. This led to the conclusion that un-/misfolded proteins accumulating in the ER are cleared from this compartment via LD formation, the released LDs being degraded terminally in the yeast vacuole by a process resembling microautophagy, termed *microlipophagy*. It has to be emphasized that this process differs from starvation-induced macroautophagy, since it does not involve the ATG-dependent initiation of (macro)autophagosomes, but instead requires ESCRT components (endosomal sorting complexes required for transport) and the ER-stress response factor Esm1 (ER stress-induced microlipophagy protein 1) [9,10]. Both stimulation of autophagy and ESCRT components extend the chronological lifespan of yeast cells [49]. A further study also clearly links LDs with the removal of aggregates consisting of misfolded proteins. Moldavski et al. showed that so-called inclusion bodies (IBs) are functionally and spatially linked to LDs [8]. Upon stress induction, unfolded or misfolded proteins, which cannot be cleared by the quality control machinery (e.g., due to quality control system overload or failure) aggregate and form inclusion bodies. In an extensive screening approach, Moldavski and co-workers identified thirteen proteins that are crucial for an efficient and rapid IB clearance. Interestingly one of these proteins, namely, Iml2p, strongly associates with LDs via interaction with the LD-resident proteins Pet10p and Pdr16p. It should be noted that Pet10p is the yeast perilipin, which is the only perilipin discovered so far in *S. cerevisiae* [50]. This interaction especially happens during cell stress, when Iml2 is exclusively located in inclusion bodies. Under such stress conditions, a physical tethering between LDs and IBs can be monitored, the physical binding of LDs to IBs allowing aggregate clearance. Iml2 is essential to this clearance process, which is considered to be mediated by a soluble sterol derivate effusing from LDs via interaction with Iml2 [8]. These findings highlight the role of LD-dependent protein aggregate clearance during aging, which is still poorly studied considering the substantial influence of cellular aging on both protein misfolding and protein toxicity [51]. Besides Pet10p and Pdr16p, another LD-resident protein, Ubx2p, could be involved in protein homeostasis [52,53]. This UBX-domain-containing protein resides in the ER but relocates to LDs upon their formation. UBX2 deletion leads to abnormal cellular numbers of LDs of reduced size and TAG content [54]. At the same time, this protein is also involved in protein homeostasis, in that Ubx2p recruits Cdc48p and both interact to support ER-associated protein degradation [55].

2.3. Lipid Droplets: Guardians of Mitochondrial Integrity

In line with these findings, our research also indicates a linkage between LD formation and the removal of un-/misfolded, potentially harmful proteins in yeast and mammalian cells. Moreover, we demonstrated that several, proteins including yeast Mmi1p and Erg6p, as well as mammalian BAX, BCL-X_L, and TCTP, can be transferred from mitochondria to LDs via a V-shaped domain consisting of two alpha helices [35]. The V-domain shows a higher binding affinity to the LD membrane than to the outer membrane of mitochondria, which explains the directed transfer [7,35]. Among different possible contexts, this directed protein shuttling is of special relevance to the control of PCD/apoptosis onset mediated by the pro-apoptotic bcl-2 family members BAX and BAK. It has to be clearly stated that apoptosis and aging are deeply interconnected in yeast as well as in mammalian cells [56–58], and LDs seem to be involved in both processes. In most cells, apoptosis is increased with the dysregulation of the apoptotic program, enhancing the risk of cancer and cellular senescence [58]. Induced by a plethora of potential intrinsic cell death stimuli, BAK and BAX translocate to the mitochondrial outer membrane where they form the mitochondrial-apoptosis-induced channel (MAC), resulting in MOMP. As a consequence, the release of cytochrome C from the mitochondrial intermembrane space to the cytosol promotes apoptosome formation, caspase 9 activation, and the terminal progression of intrinsic apoptotic signaling [59,60]. Particularly under cellular stress conditions, the anti-apoptotic mammalian bcl-2 family member BCL-X_L, as well as TCTP, also translocate to mitochondria but suppress MOMP by antagonizing BAX/BAK oligomerization [61]. In a similar way, Mmi1p, the yeast homolog of TCTP, also participates in the apoptotic machinery, with the deletion of Mmi1p leading to an extended replicative lifespan [62,63]. From this, it can be speculated that under a given stress condition both pro-and anti-apoptotic proteins locate to the outer mitochondrial membrane, continuously challenging MOMP onset. Such potentially harmful mitochondria may be specifically removed by mitophagy, a selective mode of macroautophagy [64].

Emphasizing its specificity for mitochondria, mitophagy in yeast depends on the activity of Uth1p which localizes to the outer mitochondrial membrane and is required for mitophagy, but not for starvation-induced bulk macroautophagy [65]. As previously stated, mitophagy is crucial to cellular maintenance under stress conditions by eliminating dysfunctional mitochondria, which is complicated by the fact that stress-induced macroautophagy/mitophagy may confer cell protection in one stress context, but conversely can contribute to cell death (i.e., autophagic cell death) under different stress conditions [66,67]. Besides BAX/BAK-mediated MAC, excessive ROS generation can lead to the formation of another mitochondrial permeability transition pore (mPT). The mPT pore complex is composed of VDAC (voltage-dependent anion channel) in the outer membrane, cyclophilin D in the matrix, and ANT (adenine-nucleotide translocator) in the inner membrane, and opening of the mPT, leading to mitochondrial swelling in many cases followed by necrotic cell death [59]. However, mPT opening may also initiate BAX/BAK-mediated MAC/MOMP; to a large degree, the outcome of this depends on cellular ATP availability comprising cell death by either necrosis or apoptosis, which also may involve enhanced autophagy/mitophagy [68]. Reminiscent of this, for yeast mutants lacking Mdm38p, a K^+/H^+ exchange-regulator residing in the inner mitochondrial membrane has been reported, which develops a drop of the mitochondrial membrane potential that is accompanied by mitochondrial swelling, deterioration in mitochondrial morphology, and vacuolar changes indicative of mitophagy [69].

LDs and mitochondrial homeostasis. It has to be emphasized that mitophagy does not necessarily need to be associated with conditions of enhanced stress, but represents an important physiological regulator of mitochondrial homeostasis. In postmitotic mammalian cells, mitophagy is crucial to the control of mitochondria numbers under normal physiologic conditions, as well as the removal of dysfunctional mitochondria in starving cells [70]. In this context, the age-dependent decline in autophagic activity seen in mammalian cells [71] deserves particular attention since it may weaken the cellular clearance from dysfunctional mitochondria. Hence, it is conceivable that additional mechanisms may support cellular maintenance in aged cells by protecting them from the onset of premature cell death via apoptosis caused by "stressed" mitochondria. The above-mentioned V-domain-based shuttling of Mmi1p, BAX, and other MOMP agonists to LDs could fulfill this task considering that LDs closely locating to mitochondria are capable of sequestering pro-apoptotic proteins, and as a result antagonize the onset of MOMP-dependent apoptosis [35]. Terminally, such potentially harmful BAX-enriched LDs will be degraded in the yeast vacuole. Indeed, in yeast cells, we demonstrated the V-domain/LD based protection from apoptosis, but, conversely, human HepG2 hepatoma cells treated with the apoptosis inducer staurosporine revealed a substantially elevated susceptibility for apoptosis upon the V-domain-mediated translocation of BAX and Bcl-X_L from mitochondria to LDs [35]. Explaining this, we observed the translocation of pro-apoptotic Bcl-X_S to the mitochondria in staurosporine-treated HepG2 cells. Opposing anti-apoptotic Bcl-X_L (i.e., the long isoform), Bcl-X_S (the short isoform) is a pro-apoptotic splice variant of Bcl-X, the Bcl-X_L/Bcl-X_S ratio being defined by the cell type and cell differentiation, which are dependent (e.g., non-transformed versus tumor cells) by numerous determinants including transcription factors and cytokine signaling [72]. Importantly, we found Bcl-X_S to be devoid of a V-domain [23], which may explain the enhanced onset of apoptosis in staurosporine-treated HepG2 cells. Taken together, this emphasizes the dependence of V-domain/LD-based MOMP inhibition on additional regulatory elements, in particular in mammalian cells, rendering the mechanisms cell-type-specific. Ongoing research demonstrates that the V-domain-based mitochondria to LD shuttling is not restricted to the MOMP/apoptotic settings as presented above, but seem to play a more general role in cellular stress responses, as indicated by the marked protein accumulation by LDs seen during replicative aging and in the initiation of proteotoxic stress [7]. In good correspondence with this, Garcia et al. reported a substantial remodeling of the LD proteome in the presence of ER stress [10].

LDs and DNA repair. Moreover, certain yeast haploid *rad*Δ (radiation damage) deletion strains also show altered lipid storage patterns and a reduced lifespan [73]. *RAD* genes are involved in DNA repair (e.g., nucleotide/base excision repair) which is evolutionary highly conserved. In yeast, repair of double-strand breaks via homologous recombination is accomplished by the MRX complex composed of the RAD gene products Mre11p, Rad50p, and Xrs2p [73]. Deletion of one of these three genes leads to higher levels of TAGs and steryl esters, as well as characteristic changes in lipid-metabolism-associated gene expression. The down-regulated expression of lipolysis-associated genes (e.g., *TGL3*) at an augmented expression of genes involved in lipid synthesis (*LPP1, SLC1*), together with high TAG levels, may readily explain the observed increase in LD numbers in *rad*Δ mutants. This is accompanied by chronological lifespan shortening and pronounced mitochondrial fragmentation indicative of premature aging. However, as normally aged cells also displayed higher LD numbers, it is not clear whether the increased LD abundance simply reflects the premature aging process of *rad*Δ mutants or, vice versa, LD accumulation is causal to chronological lifespan shortening [73]. Concerning the considerations made above regarding a cytoprotective role of LD accumulation in stress adaptation, it would be interesting to study the extent to which the severity of the phenotype is altered in *rad*Δ mutants devoid of LDs.

These findings account for a functional triad between LD abundance, mitochondrial integrity, and lifespan in yeast, which is addressed by stress conditions as well as the general aging process. Following the common view of mitochondrial dysfunction as a hallmark of aging [74], the causal relationship between mitochondrial fragmentation and chronological lifespan shortening, as seen in yeast exposed to high glucose levels [75], represents a reliable means of monitoring the aging process already at early stages [29]. Extending this, and in line with the functional triad envisaged above, LD accumulation in the same way may be considered a complementary biomarker for both premature and normal aging, as suggested by Kanagavijayan et al. [73]. With respect to this, determinants of LD synthesis such as the cellular levels of TAG and sterols are of prevalent meaning to the whole context. In yeast, two enzymes are regarded as the main actors in TAG production, Lro1p (lecithin cholesterol acyl transferase related open reading frame) and Dga1p (diacylglycerol acyltransferase 1) [76]. For sterols, the acyl-CoA:sterol acyltransferase Are1p and its paralog Are2p are the main sterol esterification tools in yeast [77]. Together these enzymes regulate the TAG:sterol balance to a ratio of 1:1 in yeast LDs [78]. We showed that the simultaneous overexpression of all Lro1p and Dga1p enzymes, as well as Are1p and Are2p (single overexpression of each enzyme), yields an extension of both the chronological and replicative lifespan of *S. cerevisiae* [19].

This stimulation of LD synthesis resulted in less mitochondrial fragmentation and reduced production of ROS, which normally increase during aging. Contrarily, a mutant strain devoid of LDs (*lro1*Δ, *dga1*Δ, *are1*Δ, *are2*Δ) suffers from a significantly shortened chronological lifespan and experiences a burst of ROS production already within one day of cultivation, suggesting severe mitochondrial defects [19]. According to the assumptions made above, mitochondrial functionality is an essential target for age-related cellular decline, and it seems plausible that "fitter" mitochondria with maintained integrity will be beneficial to a prolonged lifespan.

Furthermore, mitochondria have been identified recently to assist the cytosolic proteasome in protein degradation, especially during stress conditions. Underlying this is a process termed MAGIC (<u>m</u>itochondria <u>as</u> guardian <u>in</u> <u>c</u>ytosol), which mediates the import of misfolded proteins into mitochondria where protein degradation is performed by the matrix-resident protease Pim1p [79]. Yeast Pim1p is an ATP-dependent mitochondrial Lon protease required for the degradation of misfolded mitochondrial proteins, which is essential to mitochondrial function and maintenance [80]. With aging, the activity of Pim1p ceases, and *pim1*Δ yeast mutants lacking Pim1p are marked by a shortened replicative lifespan and show reduced proteasomal activity connected with an increased accumulation of oxidized and aggregated proteins in the cytosol [80]. In line with this, we also observed a significant shortening of both the replicative and chronological lifespan in *pim1*Δ cells [19]. In addition, the mitochondria of *pim1*Δ cells showed an abnormal morphology accompanied by enhanced ROS production, enlarged LDs, and a delay in the cell cycle. This premature aging phenotype of *pim1*Δ cells could be reversed partially by overexpressing Lro1p. [19] This suggests an important role of LDs in the detoxification/sequestration of the non-degraded, oxidized protein, which underlines the beneficial role of LDs in cell integrity by assisting cellular clearance from protein aggregates.

It is noteworthy that the advantageous effects of LDs cannot be seen solely as a function of LD abundance, but also as a function of LD size and morphology. This is indicated by the observation that *pim1Δ* cells treated with oleate and olive oil showed a reduced lifespan, revealing a drop in the LD number, with the LDs themselves becoming massively enlarged. In contrast, overexpression of Lro1p/Dga1p on the *pim1Δ* restored the strains' normal replicative lifespan but led to numerous but smaller LDs [19]. Furthermore, cells of the mutant strain *sei1Δ* (SEI, yeast seipin controls LD size, number, and morphology) show a reduced replicative lifespan but no significant differences to wild-type cells in overall neutral lipid levels. Different from the wild type, the LDs of *sei1Δ* cells are smaller and show LD clustering. Hence, LD size and distribution also obviously play an important role in the effect of LDs on lifespan in yeast [19]. In this context, it is worth mentioning that, in yeast cells, life-prolonging interventions such as caloric restriction [81], rapamycin treatment (blockage of the TOR kinase; for details see the following chapters) [82], and sirtuin inhibition [83] induce the formation of LDs [84–86]. In fact, in our own experiments we observed a modest 1.15–1.20-fold increase in the LD content upon treatment of BY4741 cells with 10 µM resveratrol (unpublished data).

Similar research was performed in the filamentous ascomycete *Podospora anserina* [33]. Here, deletion of the gene *PaATG24*, encoding a sorting nexin, resulted in impaired autophagy, a reduced vacuolar size, lowered growth rate, and lifespan shortening. Addition of oleic acid stimulates LD production and gives rise to an extended lifespan in wild-type as well as *PaATG24Δ* cells, revealing a restored autophagic flux and normal vacuolar phenotype. Interestingly, oleic acid treatment also diminishes ROS production in *Podospora* as result of a bypass of complex I and II of the mitochondrial electron transport chain [87].

Taken together, the research on LDs in yeast provides substantial evidence that LDs, apart from their well-defined role in lipid metabolism, can also serve as hitherto underrated "detoxification organelles", which in orchestration with other processes involved in cellular maintenance, in particular the autophagic flux, serve as lifespan determinants. Such protective roles (both for proteotoxic and lipotoxic intervention) were clearly demonstrated for the model organisms discussed below.

3. Lipid Droplets in *Caenorhabditis elegans*

Caenorhabditis elegans has proven to be one of the most important model organisms in aging research. Several milestones in this specific scientific field were achieved in this nematode. It was shown for the very first time in this worm that a mutation in a single gene (*age-1*) can extend the lifespan of a whole organism [88]. Further aging pathways that were unraveled in *C. elegans* or were studied in great detail include the insulin/IGF-1 signaling (IIS) pathway [89], TOR signaling pathway [90,91], caloric-restriction-induced signaling [92], TGF-β-signaling [93], AMPK signaling [91,94] and the HIF-1-dependent hypoxic response [95].

3.1. The C. elegans "Dauer-Larva"

To gain a better understanding of the role of LDs in the aging process of *C. elegans*, it is appropriate to provide a short overview of its lifecycle, in particular with respect to the diapause stage of the "dauer larva" resembling a suitable aging model. In *C. elegans*, two sexes can be distinguished, self-fertilizing hermaphrodites and males, each composed of an exactly defined number of somatic cells. Upon fertilization, eggs are laid. After embryonic development and hatching from these eggs, the nematodes have to pass four larval stages, each of which ends with a molt, before adulthood is reached [96]. Spectacular in this life cycle and important for aging research is the formation of a so-called dauer larva. Environmental cues including starvation-, heat-, or population-density-dependent pheromone secretion at the L2 molt phase are potential inducers of the dauer larva. In this specific phase, the worm stops eating and ceases muscular activity in the pharynx but retains full mobility. The dauer larva has a reduced intestinal lumen and specialized cuticle. As soon as the harmful environmental influences end, the larva exits the dauer stage and, after the third and fourth molts, forms an adult worm. Strikingly, this dauer stage can extend the lifespan up to 70 days, which is close to four-fold the average lifespan of an adult nematode (about 18–20 days at 20 °C) [97,98]. In the development of many longevity concepts, the C. elegans dauer larva played a significant role since all of the above-mentioned pathways (e.g., TOR, IIS, TGF-β signaling) that affect worm longevity also modulate entry of the worm into the dauer larva stage. Critically, many of the mechanisms contributing to lifespan extension have to be considered dauer-related, but also dauer-independent [99]. As we discuss in the following section, this may also apply to LDs; the dauer stage phenotype of *C. elegans* shows a close linkage to the detoxifying effects of LDs, a LD function that may also play a role in cellular maintenance in the adult worm as well as higher organisms.

3.2. Detoxifying Role of Lipid Droplets

The primary sites for fat storage in *C. elegans* are cells of the intestine and the hypodermis. In these cells, three fat deposits were identified, namely LDs, lysosome-related organelles (LROs), and undefined vesicles [100]. These sites for fat storage differ in their abundance and lipid composition. Inspection of intestinal cells by Raman scattering microscopy revealed that 18% of the cellular area is covered by LROs and 4% by LDs [101]. In contrast to yeast cells in which TAG and sterol esters are stored in LDs, there is a clear separation between LROs and LDs in *C. elegans* fat storage. In the worm, LDs are enriched in TAG, whereas cholesterol is mainly deposited in LROs [101]. Connecting LDs with LROs, it is speculated that LROs mediate the flux of fatty acids from LDs to either mitochondria or peroxisomes [102]. Until recently, the discrimination between LROs and LDs in the lipid management in *C. elegans* was widely neglected, and some phenotypes attributed to LDs more likely may be associated with LROs. Today, however, several specific approaches and staining protocols are available, which allow a clear distinction between these two fat-storing cell organelles. Among these, three methods should be mentioned here briefly: (1) In transmission electron microscopy, LDs appear to be electron lucent, whereas the more dense LROs appear to be electron-dense and opaque [103]. (2) Both Nile Red and BODIPY are established as vital dyes for monitoring LDs in a broad variety of organisms. In *C. elegans*, both dyes show a high affinity for LROs, whereas Nile Red fails to stain LDs in living nematodes [100]. (3) Some bona fide LD-resident proteins have been identified in *C. elegans*. One of these proteins is the triacylglycerol lipase ATGL1 which, upon fusion with GFP, specifically stains LDs but not LROs [103].

3.3. Lipid Droplets, Insulin Signaling, and Autophagy

As already addressed in the preceding chapter, caloric restriction represents one of the best-known and most reproducible interventions to prolong eukaryotic lifespan. This phenomenon was first observed in rodents [104], and among others was confirmed in yeast cells [81], *C. elegans* [92], *Drosophila melanogaster* [105], and primates [106]. In *C. elegans*, the intestine as a central organ of the worm is tightly linked to the aging process [107]. Therefore, nutrition is of eminent importance to *C. elegans* and all life-prolonging processes relate directly or indirectly to caloric restriction. In *C. elegans*, nutritional supply is covered by the ingestion of bacteria, and reducing the number of bacteria that are experimentally fed allows extension of the lifespan of up to 70%. This effect was observed in all phases of the worm's life cycle (either growth, reproduction, or post-reproduction phase) [108,109]. Caloric restriction leads to obvious changes in the *C. elegans* phenotype, foremost the reduction in body size, and leads to characteristic changes in lipid metabolism as reflected by an increased TAG: protein ratio observed in L4 larvae as well as in the adult worm. As consequence, this increase in TAG levels also manifests in LD size and abundance. For the wild type, depending on body region and developmental stage, an up to 15-fold increase in the number of enlarged LDs upon caloric restriction has been reported. The very same enlargement in LDs was seen in *eat-2* mutant worms (suffering from a feeding defect) that serve as a genetic model for caloric restriction [108]. Moreover, in L2 larva, starvation can induce development to the dauer larva state. This transition is marked by fat accumulation serving as an internal energy reserve, which occurs in conjunction with a substantial increase in LD number and density in the dauer larvae [110]. The exact mechanics underlying the outcome of caloric restriction are not completely clear, but have to be considered multifactorial. Involved processes may, inter alia, comprise (i) the down-regulation of insulin/insulin-like growth factor 1 (IGF-1) signaling (IIS), (ii) a decline in TOR signaling yielding elevated autophagy, (iii) increased activation of sirtuins resulting in gene silencing, and (iv) a more complex regulated decline of the metabolic rate [111–113].

IIS-pathway/FOXO. The evolutionary highly conserved IIS pathway plays an important role in nutrient sensing and maintenance of glucose homeostasis. Central to this is the IIS-regulated expression of a set of genes involved in stress response, energy generation, drug metabolism, and chaperone activity [114]. Concerning the heavily discussed life-prolonging effect of caloric restriction, certain arguments account for a contribution of IIS, while others are conflicting, such as the additive effect of caloric restriction and IIS repression on life extension [111,113]. In fact, entry of *C. elegans* L2 larva into the dauer larva stage is blocked by activated IIS. Screens searching for mutations that promote the L2/dauer larva transition led to the identification of several IIS-pathway elements. In toto, the associated genes were given the name *daf*, as an abbreviation for "dauer formation" variant [99]. The first, upstream component of the IIS pathway is a receptor tyrosine kinase (DAF-2). Upon binding of insulin-like molecules, DAF-2 activates the PI3P pathway that comprises sequential signaling via phosphoinositide-3-kinase (AGE-1), the 3-phosphoinositide-dependent kinase 1 (PDK-1), and the serine-threonine kinases AKT-1/2. The final targets of this kinase cascade are the transcription factors DAF-16 (a FOXO transcription factor; FOXO, forkhead box O) and SKN-1 (a Nrf1,2,3 transcription factor) which, upon phosphorylation, are blocked from entering the nucleus [99,115]. Hence, reduced IIS upon caloric restriction will allow DAF16 (FOXO) shuttling into the nucleus and promotion of its activity as a transcriptional regulator. With DAF-16, a central pleiotropic mediator of cellular stress responses was identified in *C. elegans* that increases resistance against stressors such as heat or pro-oxidant regimens, but also promotes fat storage [116]. Strikingly, most of the mutations that affect genetic control of IIS in a way that terminally promotes shuttling of non-phosphorylated DAF-16 into the nucleus prolong the lifespan of *C. elegans* in a drastic way; *daf-2* [117], *age-1* [88], and *pdk-1* [118] are such examples. On the contrary, mutations in *daf-16* itself suppress the increased longevity [119]. Furthermore, either mutation or knockdown of some of these IIS-associated genes resulted in a clear increase in the cellular LD content [20,120]. This suggests an association of IIS-controlled LD biogenesis with longevity in *C. elegans*. Supportive of this, Suriyalaksh et al. showed that long-lived worms reveal a strong tendency for an increased LD content [20]. However, it is also reported that an extreme excess of LDs upon passing a certain cut-off is negatively associated with lifespan [20]. These findings perfectly match with our observations in yeast. These demonstrate that a moderate increase in LDs (achieved by overexpression of Dga1p and Lro1p) results in the prolongation of both replicative and chronological lifespan, while overloading yeast cells with oleate (i.e. monounsaturated fatty acids) resulted in super-sized LDs and a clear trend to lifespan shortening (unpublished data and [19]).

TOR pathway and nutrient sensing. Another important rheostat of caloric restriction responses is the nutrient-sensing TOR (target of rapamycin) complex that exists in all eukaryotes, from yeast to humans. As the name indicates, the central component of the TOR pathway is the serine/threonine protein kinase TOR, and it is best described in mammals (termed mTORC, mammalian TOR complex). This kinase is either associated with the binding protein *raptor* (Regulatory Associated protein of mTOR), forming the TOR Complex 1 (TORC1) or *rictor* (Rapamycin-Insensitive Companion of mTOR) forming the TOR Complex 2 (TORC2) [121]. The regulation of TOR activity is highly complex, combining several input signals, such as availability of nutrients (e.g., glucose), growth factors, amino acids, and oxygen. Of importance, active TOR/mTORC inhibits autophagy, the depletion of nutrients such as those seen under starvation conditions, but also stress-derived signals, as well as the pharmacological inducer rapamycin leading to TORC decomposition, in which the deactivation of TORC results in the de-repression of autophagy [122]. Relevant to the role of IIS in caloric restriction, mTOR signaling shares a certain cross-talk with the IIS-pathway. Underlying this is the IIS-related activation of AKT-1/2, which leads to phosphorylation and inactivation of the tuberous sclerosis complex (TSC) consisting of TSC1 and TSC2. As part of an active TSC, TSC2 serves as a GTPase-activating protein (GAP) for Rheb, a small GTPase acting as a positive regulator of mTORC1. Upon IIS/AKT-dependent phosphorylation, TSC2 becomes destabilized, rendering the TSC inactive, which results in mTORC1 activation [123]. In the same way, TORC1 activity is also regulated by TORC2, which also leads to AKT-1/2 phosphorylation [124]. In addition, TORC1 is associated with further GTPases such as the Rag GTPases RAGA and RAGC, which are controlled by glucose- as well as amino-acid-pool-dependent signaling [125]. Among the manifold downstream targets of TOR, 4E-BP (eIF4E-binding protein) and S6K1 (S6 kinase 1) are the best known. 4E-BP is an inhibitor of the eIF4E translation initiation factor 4E and, by forming a complex with eIF4E, blocks translation. Upon TOR-mediated phosphorylation, 4E-BP is released from the eIF4E/4E-BP complex and translation is initiated [126]. S6K1 is another target of TOR belonging to the AGC family of protein kinases. TOR together with PDK1 phosphorylates S6K1 and, in a progression of this kinase cascade, leads to phosphorylation of the ribosomal protein S6, resulting in the translation of specific mRNAs [127]. Most (but not all) components of these highly conserved pathways are existent in *C. elegans* and higher eukaryotes (the *C. elegans* homologues are given in brackets): TOR (LET-363); Raptor (DAF-15); Rictor (RICT-1); RHEB (RHEB-1); RAGA (RAGA-1); RAGC (RAGC-1); 4E-BP (IFET-1); S6K1 (RSKS-1) [121]. In *C. elegans*, it was shown that deletion of the TOR homolog LET-363, and as a result deletion, of the central element of the TOR pathway, leads to an arrest of the larva in the L3 stage [128]. On the other hand, the RNAi-mediated knockdown of let-363 resulted in a dramatically increased mean lifespan and elevated lipid accumulation that is seen most obviously in intestinal cells [128]. Moreover, as revealed either by gene mutation or knockdown (RNAi) experiments, the reduced expression of TOR-pathway-associated compounds RICT-1 [129], DAF-15 [90], RHEB-1 [130], RSKS-1 [131], RAGA-1 [129], and RAGC-1 [129] extends the lifespan of *C. elegans*, thus mimicking the effect of nutrient depletion/caloric restriction (for a detailed review see [121]). In addition, mutations in rct-1 [132,133] and rsks-1 [134], and a RNAi-mediated knockdown of DAF-15 [90], resulted in the formation of numerous enlarged LDs, especially in intestinal cells, with the phenotype seen in daf-15 RNAi experiments marked by the increased presence of autofluorescent granules, most probably representing LROs [90]. In this context, it appears noteworthy that the aging-dependent sequestration of protein aggregates by LDs has been reported to occur in mouse intestinal tissue, where this process may serve the removal of protein aggregates for subsequent autophagic digest via lipophagy [135]. Hence, it cannot be excluded that the appearance of enlarged LDs/LROs in intestinal cells of *C. elegans* is also connected with a similar LD/autophagy-associated process of protein aggregate clearance. The intrinsic linkage between TOR inhibition and LD synthesis seen in *C. elegans* represents an evolutionary recurring motif, as discussed below for *Drosophila melanogaster* and *Homo sapiens*, and also holds true for unicellular eukaryotes (Figure 2 and Section 2). In *S. cerevisiae*, treatment with substances such as rapamycin or methionine sulfoximine inhibits TOR signaling and promotes chronological lifespan of the yeast cells [82]. Another consequence of rapamycin exposure is an increased TAG synthesis that is accompanied by increased LD numbers [84]. However, both investigations did not show whether the observed life-prolonging effect of rapamycin in yeast cells is due to an interplay between TOR inhibition and LD biogenesis, or whether these outcomes represent independent effects of rapamycin.

3.4. Lipid Droplets and TGF-β Signaling

As a further pathway contributing to lifespan expansion upon caloric restriction, we discuss the influence of TGF-β signaling on longevity in *C. elegans* [136]. Five members of the TGF-β superfamily have been identified in *C. elegans*, and with respect to its implication in aging, dauer larva formation, and fat storage, we focus on the TGF-β homolog DAF-7 [137]. Produced under favorable conditions by sensory, amphid ASI neurons, DAF-7/TGF-β stimulates TGF-β receptor (TGF-βR)/Smad-based signaling. The final downstream target affected upon DAF-7 ligation in the TGF-β/Smad pathway is the nuclear factor co-SMAD DAF-3, which binds to the Sno/Ski transcriptional co-factor DAF-5 and promotes by this expression the genes responsible for dauer larva formation. Homodimeric DAF-7 binds to a heterotetramer consisting of two molecules, DAF-1 and DAF-4, resembling the TGF-βR homologous receptor localizing to the plasma membrane in *C. elegans*. In a canonical mode, DAF-7 ligation leads to activation of DAF-4, a Type II TGF-βR which phosphorylates and activates the type I TGF-βR DAF-1 that itself is a serine/threonine kinase. Downstream to this, activated DAF-1 phosphorylates the R-Smad homologs DAF-8 (Smad2) and DAF-14 (Smad8) which, upon heterodimerization, translocate to the nucleus. In the nucleus, heterodimeric DAF-8/DAF-14 inhibits the Co-factor/Co-Smad DAF-3 (Smad4) and the Sno/Ski homologous transcription factor DAF-5, and, due to this, blocks the transcription of dauer-specific genes [138–140]. Each intervention that blocks TGF-β/Smad signaling (i.e., mutations in daf-7, daf-4, daf-8, daf-1, and daf-14) prolongs *C. elegans* lifespan, whereas each opposite intervention boosting TGF-β signaling (mutations in daf-3 and daf-5) shortens the lifespan [141]. In animals with mutated daf-7, daf-1, and daf-4, the improved lifespan was paralleled by a 2.5-fold increased fat accumulation, most probably in LDs compared to wild-type worms. It is noteworthy that this increased fat storage was independent of a reduced food intake, but was a specific outcome of defective (inhibited) TGF-β/Smad signaling [142].

3.5. Significance of Lipid Droplet Accumulation to C. elegans Lifespan

These findings together demonstrate the strong interference of aging-associated pathways with nutrient sensing (IIS, TOR signaling) and developmental growth regulation (TGF-β signaling) in controlling/blocking dauer larva transition [143]. Accordingly, inactivation of each of these pathways will support longevity by promoting the exit from normal development to the dauer larva state. Of high relevance to this, all of the "anti-aging interventions" in *C. elegans* addressed in this review were accompanied by a strong accumulation of LDs. Since the formation of dauer larvae is also inextricably linked to a shifted LD abundance, it may be questioned whether this reflects an evolutionary developed mechanism of intrinsic energy supply under poor environmental conditions and/or the degree to which LD accumulation is an active, driving force in the aging process. The following section aims at addressing this question by discussing two findings, which suggest an active role of LDs in coping with cellular stress conditions.

The Lapierre group showed that overexpression of the autophagy receptor sequestosome (SQST-1) resulted in a decreased lifespan of *C. elegans* at 25 °C [144]. Upon applying a genome-wide RNAi screen, they identified candidates that were able to alter the protein content of a SQST-1-GFP fusion protein, and the candidate list revealed a huge overlap with proteins that were found to be part of the LD proteome of nematodes. In order to boost the cellular LD content, the atgl-1 lipase was silenced, which shifted LDs in numbers and size, and, as a result, replicated observations frequently made in long-lived worms. This increase in lipid storage was accompanied by a strong lifespan extension and accumulation of SQST-1 at LDs. Interestingly, the SQST-1 relocalization to LDs was not restricted to this autophagic receptor, but was also observed for misfolded as well as ubiquitinylated proteins, suggesting a general role of LDs in protein homeostasis [144]. This matches perfectly with our observations made in yeast cells, showing that stimulating LDs can prolong both the replicative and chronological lifespan, most likely by detoxifying harmful proteins [19,35].

Independent of caloric restriction, but the same as in *S. cerevisiae*, there appears to be an interplay between lipid droplets and the ER-associated degradation machinery (ERAD). It was shown that an oleate-rich diet stimulates LD levels, ERAD activity, and longevity in *C. elegans*. The life-prolonging effect of oleate was also strongly dependent on LD-associated proteins such as Plin-1 and Fitm-2 (fat-storage-inducing transmembrane 2) [145]. Plin-1 is the only known perilipin in *C. elegans* [146,147], whereas Fitm-2 is essential for the budding of LDs from the ER [148]. Another publication also confirmed the life-prolonging effect of monounsaturated fatty acids such as oleate [149].

As seen in most eukaryotes, a fraction of LDs can reside in the nucleoplasm; in *C. elegans*, such nuclear LDs are found in the nuclei of intestinal cells, especially under stress conditions [89]. Some of these LDs were shown to be covered by heterochromatin, which is translocated apart from the nuclear lamina. Mosquera et al. speculate that this inward movement of heterochromatin could result in relieved gene silencing and, as a result, promote the aging process. Furthermore, the authors also reported that giant nuclear LDs may come in close contact with the nuclear lamina, especially in areas devoid of lamina. According to the authors' assumptions, these giant LD/nuclear lamina-contact zones could be responsible for ruptures of the nuclear lamina, which are seen frequently in *C. elegans* intestinal cells [150]. Although this may hold true, it is tempting to speculate in a different direction by considering the lamina/nuclear LD/chromatin association areas as sites specialized for chromatin/DNA repair.

In the model organism *Drosophila melanogaster*, with the development of specialized cells and tissues, the evaluation of the role of LDs in the aging process is much more difficult, but here, too, a clear interconnection between lifespan extension, LDs, mTOR, and IIS is beginning to emerge, as the following section shows.

4. Lipid Droplets in *Drosophila melanogaster*

The fruit fly *Drosophila melanogaster* is a well-described animal model organism in genetics, developmental biology, and cell and molecular biological research on the mechanics of senescence and aging [151,152]. Throughout evolution, the "hallmarks" defined for mammalian aging [74] are highly conserved and can also be investigated in *Drosophila*. Consequently, studies in flies identified evolutionary conserved gene mutations, endocrine and cellular signaling mechanisms, and tissue- and environment-specific factors including their interactions with the genetic background, that affect lifespan [153]. Focusing on the metabolic aspect, several lipid-metabolism-associated contexts of substantial physiological and pathophysiological relevance have been addressed in *Drosophila*. These comprise research on TAG storage and mobilization from LDs that have been addressed in *Drosophila*, which also reflect lipid metabolism in humans, including age- and lipid-associated diseases. As will be outlined in this section, evidence is increasing that LDs are deeply involved in the interconnection of nutritional, metabolic, and stress-associated signaling, emphasizing their role as cell organelles with multifaceted implications in lifespan control.

4.1. Lipid Droplets and Drosophila Development

Development of the fruit fly proceeds in an indirect mode, with each developmental stage (egg/embryo–larva–pupa/metamorphosis–imago) differing under nutritional aspects. Lipid homeostasis is regulated in a food-dependent mode during the larval-hood and in the adult fly (imago). In contrast, "nutritional supply" for embryogenesis depends on the maternal deposition of LDs during oocyte maturation, and the energy needed for metamorphosis (pupa, imaginal disc development) is supplied by the LD-rich fat body established during the larval stage, which shares functional equivalence with the mammalian liver and adipose tissue. Reflecting the adverse effects of diet-associated obesity in higher organisms predisposing for a number of pathological conditions, excessive consumption of a high-fat diet decreases lifespan in *Drosophila* [154]. In line with this, *Drosophila* mutants devoid of adipokinetic hormone *Akh* (a functional analog of glucagon), serving as a genetic model of obesity, suffer from lifespan shortening accompanied by characteristic, age-dependent changes in the lipid profile that especially affects the TAG signature [155]. This accounts for a selective TAG degradation from LDs in moribund flies yielding a senescence-specific lipid signature.

As stated above, LDs play crucial roles during all stages of *Drosophila* ontogenesis. For instance, oocytes are loaded with TAG-rich LDs to comply with the metabolic demands of embryogenesis [156–159]. This maternally driven process is regulated by the *Drosophila* perilipin 2 (PLIN2) homologue LSD-2 (lipid storage droplet 2) [160–162]. Beyond this, LSD-2 and LSD-1, the homologue of human perilipin 1(PLIN1) [163], act as central regulators of LD growth and fat storage over the whole lifespan of *Drosophila* [164]. During the larval stage, LDs are indispensable to fat body growth, which is controlled via metabolic signaling involving the IIS/FOXO (dFOXO in *Drosophila*) pathway as well as endocrine signaling [165–167]. Moreover, LDs connect fat body growth with molting, since synthesis of ecdysone (a precursor of the molting regulatory steroid hormone 20-hydroxyecdysoen) requires cholesterol trafficking from LDs to autophagosomes, cholesterol-rich LDs accumulating in the larva if autophagy is inhibited by the accumulation of fat [168]. Larval development is influenced in a nutrition-dependent mode by the LD-associated protein CG9186/Sturkopf, which regulates larval growth by connecting LD biogenesis to nutritional supply via interaction with the IIS/dFOXO pathway and hormone (juvenile hormone) signaling [169]. In the absence of LDs, the CG9186/Sturkopf protein localizes to the ER, but translocates to LDs upon induction of lipid storage [170]. Interestingly, in *Drosophila* CG9186/Sturkopf null mutants, TAG storage is neither affected in embryos (laid down by mutant mothers) nor in mutant larva containing normal LDs, an effect that is not completely understood at present, but could be based on distinct dFOXO targets such as the *Drosophila* lipase *brummer*, an antagonist of LD-regulatory LSD-2 (see the next section) [169]. In contrast, CG9186/Sturkopf null mutations become manifest only in adult flies, which show a markedly reduced TAG storage. In addition, adult CG9186/Sturkopf mutants also reveal a markedly increased protection from desiccation stress (supposedly due to a changed hydrocarbon composition of the cuticula) and reduced locomotor activity [169]. Moreover, as we discuss below, the reduced LD content seen in adult CG9186/Sturkopf mutants negatively affects stress resistance as well as lifespan of the adult flies.

4.2. Control of the Lipid Droplet Pool in Drosophila Adipocytes

Interestingly, *Drosophila* adipocytes bear spatially and functionally distinct LD pools that access distinct lipid pools for their individual maintenance [171]. Larger LDs, residing in the central cell body, require supply by fatty acid synthetase FASN-1 de novo lipogenesis, whereas smaller LDs locating to the cell periphery require gut-derived lipophorin shuttle (Lpp)-dependent lipid supply. The population of small peripheral LDs stays in direct contact with the plasma membrane and its organization changes during fasting periods. This starvation-associated effect is regulated by the protein *Snazarus*, which binds LDs at the ER-plasma membrane contact sites via a C-Nexin domain. TAG storage is enhanced upon Snazarus overexpression, which confers resistance to starvation conditions and yields lifespan prolongation [171]. Conversely, starvation also leads to the up-regulation of the *Drosophila* lipase *brummer* (bmm), an orthologue of mammalian adipose triglyceride lipase, which binds to LDs via a so-called brummer domain and elevates TAG mobilization from LDs [157]. Under normal feeding conditions, loss of bmm activity causes a moderate lifespan reduction, with the mutant flies developing an obese phenotype with adipocytes containing markedly enlarged LDs. In contrast, starving bmm mutants show a marked lifespan extension [157], which is considered to be due to the decelerated TAG mobilization from LDs [172], albeit the involvement of other lipases such as *doppelgänger von brummer* (dob) or other putative starvation-induced lipases may also play a role [157].

As mentioned before, the LD pool of *Drosophila* adipocytes is controlled by LSD-1 (PLIN-1) and LSD-2 (PLIN-2), which affect bmm lipase activity in opposite directions. While LSD-1 supports lipolysis by recruiting bmm to LDs [173], LSD-2 antagonizes bmm access and protects LDs from TGA mobilization [173]. Conversely, binding of bmm to LDs under starvation conditions antagonizes the anti-lipolytic activity of LSD-2 [157]. Concerning the LD phenotype, LSD-2 mutants show no peculiar alterations [164]. In contrast, LSD-1 deficiency promotes LD accumulation [168], with the adipocytes developing a reduced number of markedly enlarged, "giant" LDs during the larval hood and in the adult fly [163]. Moreover, it is clearly shown that the lifespan of LSD-1 mutants is reduced under starvation conditions [173]. Bi and co-workers proposed that LSD-1 and LSD-2 regulate lipolysis in an opponent, LD-size-dependent mode, with LSD-1 promoting TAG mobilization from large LDs, but LSD-2 protecting small LDs from bmm-induced lipolysis [173]. The same study also demonstrated that LSD-1 is able to adopt the anti-lipolytic property of LSD-2 under certain conditions, which points to a functional redundancy between these perilipin homologs in *Drosophila*.

LSD-2 affects the *Drosophila* LD pool already at the earliest developmental stages, with the oocytes of LSD-2 mutant females showing a substantially reduced TAG content, causing impaired embryogenesis [162]. Similarly, the fat body of LSD-2$^{-/-}$ homozygous larva contains less than 75% TAG as seen in normal flies. Pointing to additional functions not directly related to lipid storage, LSD-2 is also required for the endoreplication of cells in the larval salivary gland, with the loss of LSD-2 activity resulting in enhanced ROS generation and stimulation of JNK (c-jun amino-terminal kinase)-dependent apoptotic cell death [174]. Similarly, LSD-2 provides lipid storage in imaginal discs during metamorphosis, but also plays a distinct role in imaginal disc cell growth and differentiation. For instance, LSD-2 expression is stimulated in the wing imaginal disc upon overexpression of the gene vestigial (vg), which encodes a transcription factor controlling wing cell proliferation and differentiation [175]. Correspondingly, knockdown of LSD-2 promotes cell death in the wing imaginal disc, which involves the dFOXO-dependent up-regulation of pro-apoptotic *reaper* [176]. Notably, *reaper* is an orthologue of human autophagy/mTOR-regulatory TSC1 (see Section 3.3) and stimulates apoptosis in *Drosophila* by suppressing anti-apoptotic Diap1 [177]. Diap1, in turn, represents the orthologue of mammalian caspase-activation-inhibiting XIAP, which connects developmental (TGF-β1/BMP) and (oxidative) stress-associated (NF-κB and Nrf2) signaling with apoptosis in mammalian cells [178–180]. These findings point to a distinct role of LDs and LD-associated perilipins (LSD-1, LSD-2) in cellular growth control, interconnecting metabolic regulation with cell cycle/cell death checkpoints, which also affects *Drosophila* lifespan. As already stated, this was shown for LSD-1 under starvation conditions [173] and the involvement of LSD-2 in lifespan control under a high-fat diet has also been discussed [181].

4.3. Lipid Droplets and Lifespan Extension in Drosophila

Complementary to studies on *Drosophila* lipid/LD biology, further investigations addressed the effects of caloric restriction, pharmacological intervention, and the combination of both on the genotype–environment interaction and the underlying mechanisms in *Drosophila* [182–184]. In agreement with the above-mentioned lifespan-shortening effect seen for the Akh-deficiency-based genetic model of obesity [155], these studies revealed a positive effect of dietary restriction and nutritional balance on longevity in *Drosophila*, for which nutrient sensing and the associated downstream signaling is of pivotal relevance [184,185]. Similar to the role of IIS/FOXO in *C. elegans* dauer larva transition and, as stated, for the involvement of CG9186/Sturkopf in *Drosophila* development, the IIS-dependent regulation of dFOXO is also involved in *Drosophila* lifespan control. This is indicated by the finding that reduced IIS (as seen upon caloric/dietary restriction) results in enhanced dFOXO activity and lifespan extension in the fly [186,187]. The IIS-dependent effects are largely mediated by the JNK/dFOXO stress response pathway [187–190] and other transcription factors acting downstream of dFOXO such as AOP (Anterior open, an E-twenty-six (ETS)-family transcriptional repressor), which, in a coordinated manner, mediate the lifespan extension in *Drosophila* [191]. It is noteworthy that dietary inputs address organ-specific interactions. For instance, overexpression of dFOXO in the fat body of the head leads to an extended lifespan only under high protein conditions [192].

Interestingly, flies bearing lifespan-extending deteriorations in IIS, such as those caused by the deletion of insulin-like peptide 2, show an increased body fat storage and are more resistant to lipophilic toxins and oxidative stress [193]. In sum, the regulatory network mediated by IIS that controls LD size and number is highly complex. In *Drosophila* nurse cells (which dump LDs to the oocyte), strong IIS stimulation, e.g., due to the loss of the IIS antagonist PTEN (phosphatase and tensin homologue), leads to the accumulation of abnormally enlarged LDs [194,195]. This is counterbalanced by dFOXO, which induces lipases such as *brummer* [196], which is antagonized by LSD-2, protecting LDs from bmm lipase access as explained above. In this context, it is worthwhile to mention again that LSD-2 can only be found on small LDs, whereas LSD-1 can anchor to LDs of different sizes [173]. Hence, IIS/dFOXO-dependent responses are specifically controlled at the LD level by LSD-1 and LSD-2.

Moreover, excessive intestinal stem cell (ISC) proliferation causing intestinal dysplasia in aged flies is suppressed by reduced IIS/JNK signaling, which restores proliferative homeostasis and extends *Drosophila* lifespan [197]. Connected with this, ER stress-associated UPR (UPR^{ER}, see Section 2) has been shown to also play an important regulatory role in ISC proliferation, with the chronic UPR^{ER}-dependent hyper-stimulation of ISC proliferation being causal to age-related intestinal dysplasia [198]. Central to this UPR^{ER} response in ISC proliferation in *Drosophila* is an orchestrated interplay involving (i) the activity of PKR-like ER kinase (PERK) which is controlled by the JAK/Stat pathway, (ii) transcriptional control via Ire1 (endoribonuclease 1)-mediated splicing of the transcription factor Xbp1 (X-Box binding protein 1), and (iii) the activation of another transcriptional regulator ATF6 [198,199]. With respect to lipid metabolism, UPR^{ER}-dependent Ire1/Xbp1 signaling is of special relevance since it connects ER stress to triacylglycerol synthesis and lifespan extension in *Drosophila* [200]. Upon caloric restriction, Ire1/Xbp1 signaling promotes lipogenesis and TAG accumulation in intestinal enterocytes and prolongs lifespan of the fly; this effect also involves activity of the transcription factor *sugarbabe* (a Gli-like zinc finger transcription factor involved in the carbohydrate metabolism). With respect to this, the concept that an Ire1/DGAT2-based shift of LD biogenesis, such as that seen under conditions of ER stress in the mouse liver, could improve LD-mediated protection from phospholipid oxidation [47] (see Section 5), is of particular interest. Indeed, LDs have been shown to confer antioxidant properties in *Drosophila* by incorporating TAG redistributed from PUFAs and, as a result, protect PUFAs from lipid peroxidation in neural glial cells [201]. Hence, it would be interesting to investigate whether aging- and/or starvation-induced ER stress leading to the stimulation of Ire-1 and Xbp1-dependent LD biogenesis also confers lifespan extension in *Drosophila* by improving the anti-oxidative capacity. If so, however, this protection will demand tight control of LD abundance since the ROS-stimulated accumulation of LDs is at risk of promoting neurodegeneration in *Drosophila* [46]. According to this study, ROS generation emerging from mitochondrial dysfunction is causal to glial LD accumulation, a finding that is considered to indicate an evolutionary conserved process of pathogenic relevance to neurodegeneration. It is tempting to speculate whether harmful mitochondrial proteins can also be shuttled to LDs in *Drosophila*, as shown in yeast [7]. Pointing in this direction, analysis of the LD proteome in embryonic [202] and adult [203] tissue in *Drosophila* revealed the presence of mitochondrial proteins in LDs. In addition, similar to the binding of protein-aggregate-enriched IBs by LDs in yeast [8], protein aggregates formed upon ER stress/oxidative stress could also be sequestered by specific LD binding and, as a result, confer protection in the neuroglia. In fact, LD binding of protein aggregates has been demonstrated recently for mouse intestinal tissue which may be followed by lipophagic digest [135]. Thus, a similar process could contribute to the degradation of protein aggregates via lipophagy in neuroglia, but also other tissues, such as the intestine in *Drosophila*.

Autophagy–TOR pathway. Similar to *C. elegans*, the autophagy–regulatory TOR pathway is also intimately involved in metabolic homeostasis and lifespan control in *Drosophila* [204]. In general, the inhibition of mTOR signaling results in lowered translational activity and elevated autophagy which improves proteostasis [205], a critical hallmark of aging [74]. It should already be mentioned at this point that, in *Drosophila*, both nutrient deficiency and TOR inhibition also lead to an increase in LD size [206]. The necessity of enhanced, functional autophagy for lifespan extension in *Drosophila* was demonstrated by feeding experiments in adult flies using the TOR inhibitor rapamycin, and it was shown that the stimulation of autophagy resulted in prolonged survival of starving wild-type animals but also enhanced the lifespan of *Drosophila* IIS mutants [207]. Moreover, this investigation also demonstrated that rapamycin-induced autophagy also improves the resistance towards paraquat (1,1′-Dimethyl-4,4′-bipyridin)-induced oxidative stress. Paraquat serves as an insecticide via the cytochrome P450/Fenton-reaction-dependent formation of free hydroxyl radicals [208], which impair intestinal regeneration and cause substantial cell damage in the aging fly [170]. Revealing the involvement of LDs, flies with a reduced LD content such as that found in adult CG9186/Sturkopf mutants show reduced survival and a decreased lifespan upon paraquat treatment [169]. Pointing further to the role of autophagy, *Drosophila* mutants lacking a proper autophagic flux due to mutation of the autophagy–regulatory gene Atg8a (autophagy-related 8a) show a shortened lifespan accompanied by enhanced protein oxidation and ubiquitination, effects which are aggravated by pro-oxidant conditions [209]. Contrarily, the age-dependent decline in autophagy seen in *Drosophila*, especially in the nervous tissue as a result of reduced Atg-expression, is counteracted by Atg8 overexpression, which improves oxidative stress tolerance and longevity in aged flies. As stated above, the transfer of cholesterol from LDs to Atg8-rich vesicles (autophagosomes) is essential to larval hormone synthesis, with the up-regulation of TOR limiting Atg8 expression and autophagosome formation, which leads to LD accumulation [168]. With respect to this, it is tempting to hypothesize about a similar Atg8-related mechanism shifting LD numbers in the adult fly in support of cell surveillance if autophagy declines. In addition to Atg8, a connection with increased longevity and enhanced stress responses was reported also for other compounds, enhancing autophagy especially in the nervous system of *Drosophila*. These include AUTEN-67 and 99 (autophagy enhancer-67 and -99), acting downstream of TOR at the level of autophagosome membrane formation [210,211] and spermidine, which elevates autophagy upon interference with epigenetic control and yields lifespan expansion in yeast, *C. elegans*, *Drosophila*, and human cell lines [210–212].

Furthermore, autophagy is also stimulated by nutritional factors such as flavonoids, a class of plant polyphenolic compound with well-known antioxidant properties [213]. Among these, isoquercetin and xanthohumol have been proven to boost LD formation in *Drosophila*, especially in cells of the nervous system [214]. In addition, xanthohumol was shown to increase the resistance of adult flies to several stressors (e.g., hydrogen peroxide, paraquat, starvation, and heat) and prolong lifespan in *Drosophila* [215]. As discussed below, flavonoids also exert similar protective, antioxidant effects in the human system [216]; for instance, quercetin counteracts liver steatosis by inhibiting lipid peroxidation [217].

As mentioned above, TOR (mTOR) signaling directly affects the translational activator S6 kinase [204,207] and the translational repressor 4EBP [207,218,219]. In *Drosophila*, additional nutritional sensors have been described that modulate lifespan, including the transcription factor ATF4 [220], amino acid deprivation–activated kinase GCN2 acting upstream of ATF4 [221], GCN2 deficiency leading to a massive loss of TAGs and thus LDs [222], and AMP-activated protein kinase (AMPK). From the mechanistic point of view, AMPK acts as a central, positive regulator of autophagy by inhibiting mTORC1 (via phosphorylation of TSC-associated TSC2). AMPK extends the lifespan of adult flies in an autophagy-dependent mode, specifically affecting the central nervous system and intestinal tissue [223,224]. In contrast, reduced AMPK activity is associated with an enhanced susceptibility to starvation-induced lethality (especially in the larva; AMPK null mutants are larval lethal) and abnormal lipid accumulation marked by the accumulation of enlarged LDs in larval tissue [225]. Of note, this LD-related phenotype is seen in normally fed, mutant larva lacking AMPK activity and resembles the phenotype observed in oenocytes of starving wild-type larva [226]. Oenocytes are the primary site of LD accumulation seen in starving *Drosophila* larva and, analogous to the role of human hepatocytes in systemic lipid homeostasis, are sites of lipid release from the larval fat body during starvation. Strengthening the crucial role of autophagy on lifespan control, rapamycin partially improved the survival of adult flies with reduced AMPK activity under starvation conditions. Interestingly, the type-2 diabetes therapeutic compound metformin, which antagonizes IIS but stimulates AMPK signaling, and, as a result, also shifts autophagy (recently reviewed in [227]), extended lifespan in *C. elegans* and healthy mice [228,229]. However, Slack et al. reported that metformin failed to prolong lifespan in adult *Drosophila*; at the same time, the metformin-mediated activation of AMPK resulted in a drop in TAG levels [230]. With respect to this, it is noteworthy that metformin inhibits TOR in *Drosophila* independently of AMPK stimulation [231] and Slack et al. suspected a potential negative interference of metformin effects, not related to AMPK, with positive effects of AMPK activation on *Drosophila* lifespan [230]. Support to this is provided by the finding that overexpression of the serine/threonine kinase LKB1, another positive regulator of AMPK, indeed extends lifespan in Drosophila [232]. LKB1-null mutant flies with reduced AMPK activation show decreased TAG levels, a phenotype that can be compensated by transgenic expression of wild-type AMPK [233]. Accordingly, this suggests that AMPK activation at reduced LD numbers does not affect lifespan, whereas AMPK activation at an increased LD abundance extends the lifespan in *Drosophila*.

4.4. Lipid Droplets, Transsulfuration, and Cellular Antioxidant Defenses

Together, these findings emphasize the pivotal role of metabolic regulation and LD homeostasis in lifespan determination. At its most basic, this applies to the maintenance of metabolite pools, especially under dietary restriction [134], and in particular to the amino acid balance, which is controlled on the anabolic (translation) and catabolic (protein degradation, autophagy) levels as a fundament of cellular proteostasis, with its dysregulation representing another hallmark of aging [74]. This was demonstrated in a study by Grandsion et al., which shows that dietary-restriction-based lifespan extension in *Drosophila* is abolished by feeding the flies a mix containing all essential amino acids (EAA feed), while using an EEA-feed omitting methionine does not affect starvation-induced lifespan prolongation [234]. Further research revealed that trans-sulfuration (i.e., the production of cysteine from methionine-derived homocysteine or cystathionine) plays an important role in extending survival under starvation conditions in *Drosophila* [235]. Upon dietary restriction, enhanced trans-sulfuration preserves lipid storage in LDs (to levels seen in fully fed flies) and, due to the excessive consumption of methionine, lowers overall protein synthesis. With respect to this, the positive effect of limited methionine availability on lifespan seems paradoxical. However, the beneficial effect of methionine restriction on longevity is conserved from yeast to mammals as demonstrated by the improved longevity seen in yeast, *Drosophila*, and rodents upon sulfur-amino acid starvation (SAAR, also referring to cysteine) [236–239]. Of particular relevance to this, in yeast, methionine starvation contributes to longevity by stimulating autophagy and an improved vacuolar acidification [239]. This puts emphasis on the specific role of sulfur-containing methionine and cysteine pools in cellular responses to nutritional stress and aging, primarily acting on the autophagic flux. Similar to starving flies in which inhibition of trans-sulfuration (resembling SAAR) lowers TAG levels [235], dietary SAAR reduces fat deposition and the TAG content in the rodent liver [240,241]. Stressing the aspect of autophagy, SAAR acts on the same metabolic regulators in the mammalian system—GCN2, ATF4, and AMPK (for a review see [236])—that link nutritional sensing to the onset of autophagy, and act as lifespan modulators in *Drosophila* as outlined above.

Pointing to a further, indirectly nutrition-associated aspect of profound relevance to lifespan control, the extended survival seen in rodents upon SAAR is accompanied by increased systemic levels of the cysteine-containing antioxidant glutathione (GSH) [237]. This agrees well with the inverse correlation existing between aging and systemic GSH levels [242]. In addition, GSH levels are also elevated in long-lived flies upon dietary restriction [235], an effect that is attributable to the trans-sulfuration-based refueling of the cellular cysteine pool. Explaining this, dietary restriction up-regulates the expression of cystathionine β-synthase (CBS), a key enzyme catalyzing the conversion of homocysteine to the cysteine precursor cystathionine, and elevation of CBS synthesis is essential to lifespan extension in the starving fly [235]. In good accordance with this, blocking the final step of trans-sulfuration-based cysteine synthesis, which is catalyzed by cystathionine-γ-lyase, leads to a marked drop in GSH levels and abolishes the starvation-based lifespan–extension in *Drosophila* [235]. Connecting the trans-sulfuration pathway to LD biogenesis, inhibition of cystathionine-γ-lyase also lowers the overall LD content in *Drosophila* [235]. Of note, this effect of trans-sulfuration inhibition on LD abundance seems to represent an evolutionarily conserved motive, since in human ovarian cancer cells the knockdown of CBS also results in reduced LD numbers [243]. Taking into consideration the crucial role of the GSH/GSSG (the oxidized GS=SG disulfide) redox balance in cellular antioxidant capacity, the elevation of trans-sulfuration has to be considered pivotal to cellular stress adaption by connecting metabolic competence to pro-/antioxidant balance. Accordingly, it makes sense that an increase in GSH is associated with an increase in LDs under oxidative stress. Considering the pivotal role of GSH in cellular detoxification of peroxides (H_2O_2, but also lipoperoxides) and lipid-peroxidation-derived metabolites such as 4-hydroxy-2-nonenal (HNE), limitations of GSH availability, ensuing from enhanced GSH consumption and/or inadequate GSH synthesis, are at considerable risk of promoting oxidative damage to lipids, DNA, and proteins [244,245]. Hence, LDs could support cell survival under conditions of GHS depletion by eliminating oxidized lipids and misfolded proteins as discussed above. Serving a similar, cytoprotective task, LDs have been shown to adopt the role of a cellular antioxidant in *Drosophila* larva by sequestering polyunsaturated fatty acids from the cell membranes, which yields protection of these lipids from peroxidation [201].

Moreover, as a sulfhydryl group donor, methionine also undertakes the synthesis of iron–sulfur clusters (Fe-S) via the mitochondrial ISCU (iron–sulfur cluster forming unit) [246]. Fe-S clusters are indispensable for electron transfer in the respiratory chain and energy charge in all aerobic organisms. In addition, Fe-S clusters are essential to the citrate dehydrogenase activity of aconitase and are co-factors of DNA repair enzymes [247,248]. Interestingly, ablation of ISCU-mediated Fe-S biogenesis leads to increased citrate concentrations, generated from glucose-derived acetyl CoA, an elevated fatty acid synthesis, and the accumulation of LDs in human HEK293 embryonic kidney cells [249]. Furthermore, in a mouse model of Friedreich's Ataxia (FRDA), an autosomal recessive disease marked by substantially reduced levels of the mitochondrial ISCU regulatory protein frataxin (Ftx) [250,251], the absence of Ftx function resulted in Fe-S protein deficiency, mitochondrial iron accumulation, and increased LD abundance in cardiac muscle cells [252]. Similar effects were seen in a *Drosophila* model for FRDA, where the Ftx deficiency stimulated both fatty acid synthesis and lipid peroxidation, as well as LD accumulation in glial cells [253]. Both findings suggest that hampered Fe-S cluster synthesis caused by Ftx deficiency leads to a disturbance of lipid homeostasis. However, it should not be overlooked that the FRDA/Ftx deficiency also shifts the iron content of mitochondria as demonstrated by the FRDA mouse model. It is well known that labile iron (i.e., free Fe^{2+}) acts as a central cellular source for the Fenton reaction-based generation of hydroxyl radicals. In turn, these hydroxyl radicals readily react with polyunsaturated fatty acids (PUFAs) and, as a result, initiate the lipid peroxidation chain reaction (LPO), causing potentially lethal cellular damage, which is aggravated by the geno- and cytotoxic properties of LPO metabolites such as malondialdehyde and HNE [254–256]. Therefore, the LD accumulation seen in FRDA/Ftx-deficiency could indicate the specific up-regulation of LD biogenesis as cytoprotective response to iron-mediated oxidative stress. Support to this come from the recent finding that LDs participate in cellular responses to the pro-oxidant effects of paraquat (also a source for Fenton chemistry-based hydroxyl radical formation) in *Drosophila* [257]. It was shown that the RNA binding protein *Spen* (Split ends; the *Drosophila* orthologue of SPEN/SHARP, a regulator of NOTCH signaling) modulates the LD content in adult glial cells and provides protection from paraquat cytotoxicity. Conversely, LD biogenesis can also be stimulated by iron deficiency, which was demonstrated in human ARPE19 retinal pigment epithelium cells treated with the iron chelator deferiprone (DFP) [258]. DFP induces marked changes in lipid metabolism, including enhanced TAG synthesis, and leads to the accumulation of LDs in proximity to mitochondria followed by mitophagy. Diacylglycerol O-acyltransferase 1 has been identified as a stimulus of LD biogenesis under conditions of iron depletion, enabling the re-esterification of fatty acids regenerated upon macroautophagy [258].

In summary, these findings link metabolic homeostasis to stress tolerance and shed light on a particular role of LDs in cellular antioxidant defense and cytoprotection also affecting longevity. Evidence exists that the LD-mediated clearance of oxidized compounds is essential to this task, as suggested by the sequestration of LPO-products by LDs in *Drosophila* protecting larval tissue, and especially neuroblasts in imaginal discs, from hypoxia-induced oxidative stress [201]. Therefore, LDs could participate in cellular maintenance under pro-oxidant conditions by acting as a sink for lipid peroxidation products and other oxidized cellular compounds. However, considering the transient, dynamic nature of LDs, such "sinks" do not necessarily need to resemble long-term deposits for the potentially harmful "oxidized waste". Findings in glial cells of the *Drosophila* eye point in this direction. In these cells, loss of the metalloproteinase ADAM17, a trigger of tumor necrosis factor (TNF)-based signaling, as well as lack of TNF and the *Drosophila* TNF receptor homologue *Grindelwald* [259], causes an age-related degeneration of neuronal and glial cells [260]. In this case, accumulation of LDs in glial cells prior to the degradative process confers an initial protection from glia- and neuron-derived ROS, while the subsequent metabolic decomposition of LDs leads to the release of toxic lipid peroxides causing cell damage and neurodegeneration. Hence, LDs may exert opposing effects in oxidative stress/LPO-dependent contexts, with the outcome being strongly dependent on additional factors such as nutritional sensing and lipid turnover, as well as different stress qualities (also in terms of stress duration: short, intermittent, or chronic), linking the protective capacity of LDs to the aging process.

4.5. Lipid Droplets and (Epi)Genetic Control

Addressing a further aspect of LD biology that may also be connected to cytoprotection, LDs may locate to the cytosol as well as to the cell nucleus. As already stated for *C. elegans*, nuclear LDs (nLDs) exist in many organisms, including *Drosophila*. Analysis of the nLD-proteome isolated from rat liver identified a number of proteins, including histones, cytoskeletal elements (e.g., cytokeratins), proteins involved in transcriptional and translational control, protein folding, and post-translational modification, and lipid-metabolism-associated carboxylesterase 1d (Ces1d; cholesteryl-ester hydrolase) [261]. Among different functions in lipid metabolism, mammalian Ces1d contributes to cellular detoxification by hydrolyzing lipid esters, either derived from xenobiotic or endogenous sources, especially in the liver and intestine [262]. With respect to LD biogenesis, Ces1d deficiency has no effect on cytosolic (ER-based) LD formation itself, but yields increased numbers of small-sized cytosolic LDs, an effect that is attributable to a lower lipid transfer to LDs [263,264]. Referring to the presence of Ces1d in nLDs in *Drosophila*, carboxylesterases may also participate in nLD biogenesis. In mammalian cells, nLDs are formed either de novo at the inner nuclear membrane or upon translocation of cytosolic LDs, originating from ER resident lipoprotein precursors, to the nucleoplasmic reticulum terminally moving to the nucleoplasm (recently reviewed by Fujimoto [265]). In hepatocytes, ER stress promotes LD shuttling to the nucleus and requires the activity of promyelocytic leukemia protein (PML) locating to the inner nuclear membrane. PML, which is critical to nuclear signaling, can be retained in nLDs (then termed lipid-associated PML structures; LAPS) and due to the PML binding properties, nLDs/LAPS may regulate the PML-mediated control of the gene-expression, for instance, as part of lipid stress responses [265,266]. The above-mentioned association between stress responses and nLDs in *C. elegans* (see Section 3) may point in a similar direction and it cannot be excluded that nLDs assist cytosolic LDs in directing the expression of lipid-metabolism-linked genes. Moreover, the association between nLDs and heterochromatin seen in *C. elegans* under stress conditions could also indicate a functional role of nLDs in aging-related gene silencing [118]. Indeed, aging in *Drosophila* is associated with changes in heterochromatin structure yielding a repression of gene silencing, which also activates the expression of transposable elements (TEs) residing in heterochromatic areas of adipocyte nuclei in the fat body (as mentioned, the *Drosophila* equivalent to the human liver) and brain tissue [267,268]. Antagonistic to this, dietary restriction counteracts TE activation and extends lifespan. Similarly, mutation of the *Drosophila* gene *Argonaute 2* (Ago2), a regulator of TE silencing [269], results in enhanced TE expression, impaired neuronal function, and reduced longevity [270]. Of note, the enhanced TE expression seen in the aged fat body of old flies is accompanied by increased DNA damage and declined levels of the nuclear-lamina protein lamin-B, with the depletion of lamin-B yielding a similar phenotype (enhanced TE expression and DNA damage) in the fat tissue of larva and young adults [267]. These findings shed light on the critical role of the nuclear lamina in gene silencing, including TF expression and genome/DNA integrity surveillance in lifespan control.

Interestingly, in early *Drosophila* development, binding of extranuclear histones to cytosolic LDs enables the storage of histones required for chromatin organization, a task that is conferred by the protein *Jabba* [18,271]. Extranuclear histone stores are specific for very early, syncytial blastoderm stages of *Drosophila* embryogenesis that are marked by a rapid series of consecutive nuclear divisions (without accompanying cell divisions, thus forming a syncytium), with the Jabba-based recruitment of histones to LDs providing an adequate extranuclear histone supply. Although this mode of *Jabba*-aided histone-to-LD binding may be specific for early *Drosophila* embryogenesis, it is tempting to speculate that similar histone–LD interactions play a role in aging-associated changes in chromatin organization in adult flies as well as higher organisms. These may comprise chromatin remodeling and histone methylation, which are both considered further hallmarks of aging [74]. In *Drosophila*, repressive histone methylations contribute to heterochromatin stability and their disruption results in lifespan shortening [272,273]. Histones found on LDs in *Drosophila* include histones H2A and H2B, which both show an age-specific ubiquitination [18]. Reduction of the ubiquitinated form of H2A prolongs the flies' lifespan [274]. Notably, binding of H2A by LDs would naturally serve the same purpose. In H2B, the ubiquitination is a prerequisite for the trimethylation of the histone H3K4 [18], which upon trimethylation (H3K4me3) promotes the aging process in *Drosophila* [275]. Accordingly, the possible binding of H2B by LDs would lead to a reduced amount of H3K4me3 and thus prolong the lifespan of the flies.

Moreover, the aging-associated loss of histones (originally identified as a lifespan-restricting determinant in yeast by Feser et al. [276]) and changes in the eu-/heterochromatin ratio are linked to changed histone methylation patterns that affect gene expression and may promote aging-associated DNA damage [277,278]. This fits well with the afore-mentioned *Argonaute 2* mutant phenotype in *Drosophila* marked by lamin depletion, enhanced TE expression, and DNA damage. Similarly, down-regulation of lamin-B1 accelerates the senescence of proliferating human fibroblast cells and has a profound impact on chromatin structure and gene expression [279]. Finally, it has been demonstrated that recruitment of the chromatin remodeler SNF2h (enhancing DNA accessibility in DNA repair) by *sirtuin* (SIRT6), which deacetylates histones, protects human and mouse cells from genotoxic damage [280]. On the other hand, SIRT6 also serves as a negative regulator of lipid metabolism [281], with the EGF-dependent down-regulation of SIRT6 (FOXO3/SIRT6) resulting in enhanced LD biogenesis in human colon cancer cells [282]. Taking into account the inverse nature of SIRT6 regulation on LD formation and the protective role of SIRT6 on DNA/chromatin integrity, sirtuins (histone deacetylases) could play a critical role in balancing LD abundance. However, this could also limit the cellular "lipid mass", including LDs, serving as substrate for lipid-peroxidation-derived genotoxic effects [283] under pro-oxidant conditions. Summarizing, these findings account for a distinct role of cytosolic and nuclear LDs in chromatin and genome surveillance with a particular impact on the aging process, for which the underlying mechanisms remain to be elucidated by further investigation.

4.6. Intracellular Lipid Droplet Trafficking

Finally, additional findings point to intracellular LD trafficking as an important further aspect in LD biology. In *Drosophila*, LD transport is mediated by the interaction of the motor proteins kinesin and dynein with microtubules [284–287]. Involved in the coordination of LD movement conferred by cytoskeletal interaction are proteins such as *Bicaudal D* (interacting with dynein) [288] and perilipin-homolog LSD-2 [289], which physically interacts with the gene product *klarsicht* (Klar is identical to the *Drosophila* gene *marbles*) [286,290]. *Klarsicht* mutants develop more or less normally, but due to disturbed LD transport show a markedly reduced lipid/LD deposition in the blastoderm, yielding enhanced transparency of the embryo (hence the name of the mutant). In addition, absence of Klar also leads to mispositioning of photoreceptor nuclei in the developing eye and affects trafficking of secretory vesicles in the salivary gland [291]. Three Klar isoforms (Klar α, β, γ) have been described [292]. While no function is known for Klar γ, the isoform Klar α is involved in linking cytosolic and nucleoplasmic proteins and, as a result, affects positioning of photoreceptor nuclei. Although not shown, it cannot be excluded that Klar α and LD-associated Klar β together with LD-resident Jabba participate in the recruitment of extranuclear histones to LDs during early *Drosophila* embryogenesis, as discussed above. Klar β, in a Klar α-analogous fashion, targets LDs for cytoskeletal interaction, which is mediated by a distinct, C-terminal LD domain [292]. It is noteworthy that Klar via its LD domain serves intracellular, microtubule-based transport of LDs, not only in embryonic but also in adult *Drosophila* tissue, as well as in cultured insect S2 cells [293]. Summarizing, this indicates that the association between LDs and the cytoskeletal/microtubule network may enable distinct intracellular LD positioning mechanics in developmental, physiological, and pathological contexts.

Taken together, *Drosophila* is a model organism with outstanding findings for LD biology that foreshadow the integrative functionality of LDs in higher organisms connecting environmental, dietary, and hormonal inputs, and assisting in their translation to metabolism and signaling cascades. The interaction with other cellular organelles, primarily mitochondria and the ER, make LDs a useful and highly dynamic organelle apart from lipid storage. However, it has to be admitted that the complexity of the LD interaction network increases with the number of different cell types and tissues, although common central regulatory pathways are conserved from yeast to the mammalian system (Figure 2). In the following chapter dedicated to mammals, we show that LDs, as a kind of "Janus-faced organelle", essentially fulfill a cytoprotective function, but contribute to age-related diseases if LD accumulation becomes inadequately excessive.

5. LDs in Human Disease

5.1. Caloric Restriction, Lifespan Control, and Age-Related Disease

It is a common thread in evolution that caloric restriction promotes health and prolongs lifespan, which has been documented for a variety of organisms such as *S. cerevisiae*, protozoans, rotifers, crustaceans, *C. elegans* and nematodes in general, *Drosophila melanogaster*, and fish (*Lebistes reticulates* and *Danio rerio*) [32,105,108,294–299]. The underlying molecular mechanisms seem to be similar or identical in all the organisms studied and converge to common pathways—TGF-β signaling, IIS/IGF-1 signaling, mTOR, and stimulation of autophagy—as illustrated in Figure 2. As outlined in this figure, these processes also share the common motive that they stimulate or are otherwise associated with LD biosynthesis. Aging research has identified several compounds that preserve health and extend lifespan in different model organisms: resveratrol [83], rapamycin [82], spermidine [212], 2-deoxy-D-glucose [300], curcumin [301], quercetin [83], metformin [302], and NAD$^+$ precursors [303].

Many if not all of these substances have been shown to be caloric-restriction mimetics [304] and, as discussed in the previous sections, many of these substances also stimulate LD biogenesis. This provokes the central question of whether the life-prolonging effects of caloric-restriction mimetics observed in non-mammalian animal model systems also apply to mammals and humans, and whether these processes are also LD-driven.

In rodents, the life-prolonging effects of caloric restriction have been known for a long time [104,305,306]. Over the last few years, data on the life-prolonging effect of caloric restriction also became available for non-human primates, which now allows conclusions to be drawn for humans. Since the 1980s, the effects of caloric restriction have been investigated in the rhesus monkey *Macaca mulatta* by three different organizations (University of Maryland, University of Wisconsin Madison, and the National Institute on Aging). After settling some controversy over the study design, the participating organizations agreed in concluding that caloric restriction has a positive effect on survival and aging-associated diseases [106]. Since these studies did not investigate the involvement of LDs in the life-prolonging effect seen in the rhesus monkeys, a distinct role of LDs in lifespan control can only be conjectured from observations made in other model organisms. Indeed, various findings in mammals, including humans, suggest a positive effect of caloric restriction on LD biology. In humans, a distinction can be made between white, brown, and beige adipose tissue. For a long time it was considered that brown adipose tissue is present only in the newborn, contributing to the regulation of body temperature, and that it is rapidly lost after birth. More recently, however, brown adipose tissue was also found in adults [307], being involved in lipid and glucose oxidation as well as insulin-independent glucose uptake [308]. Based on this, it could be shown that activity of brown adipose tissue increases during adolescence and rapidly ceases at higher age [309]. The implication of brown adipose tissue in the aging process is also supported by the finding that the activity of brown adipose tissue is significantly higher in long-lived than in short-lived animals [308]. Importantly, the influence of caloric restriction leads to the "browning" of white adipose tissue, meaning that adipocytes, instead of forming one large lipid droplet (as in white adipose tissue), constitute many small LDs in brown adipose tissue [310]. This is accompanied by distinct changes in the LD proteome [311]. Due to the long life of primates (in rhesus monkeys, between 30–40 years), the study on caloric restriction referenced above is the only one of its kind hitherto that proves the influence of an anti-aging strategy on primates/humans. Since no data are available at present referring to LDs at the organismic level in aged individuals, we set the focus of this chapter to diseases whose prevalence and severity generally increases with age, and are hence widely accepted as being age-associated [312], and for which, in many cases, a contribution of LDs to the disease pattern have been reported.

In contrast to the previously discussed models such as *C. elegans* and *D. melanogaster*, the role of LDs in human and mammalian aging is poorly defined. Representing a general conceptual flaw for most pathologies that show elevated cellular LD levels, it is not clear at present whether this LD accumulation is causal to the diseased state or rather is a consequence of disease-related changes in lipid metabolism. In particular, evidence for a regulatory, cytoprotective effect of increased LD levels, such as that indicated by the lifespan-extending effects seen in other, less-complex model systems, is widely missing in the mammalian/human system. Nevertheless, the research listed in Table 1 provides clear evidence that LD levels increase in several, if not all, age-related diseases (ARDs). In ARDs, the irreversible cessation of cell proliferation that demarcates the progression to cellular senescence is discussed as a major pathological criterion [313]. Essential to this, the deregulation of cell-cycle-regulating genes is considered a hallmark of cellular senescence which, surprisingly, also applies to lipid-related pathways [314]. Evidence is accumulating that the deregulation of nutrient-sensing pathways, such as growth hormone and IIS pathway [315–317], autophagy–regulatory mTOR and AMPK signaling, and the histone deacetylase activity of sirtuins, play key roles in ARD development [318]. As discussed above, exactly the same pathways have been shown to be important stimuli of LD biogenesis in the other animal models of aging. Insulin signaling is also intrinsically tied to trafficking and storage of lipids in lipid droplets. With respect to the wide range of ADR-associated processes, this review focuses on aging-associated aspects of LD biology in several selected tissues.

Table 1. Age-related diseases associated with an increase in cellular LD numbers.

Disease	Main Affected Cell Type/Tissue	References
Alzheimer's disease	neurons, glia, myeloid cells, ependymal cells, astrocytes	[319–323]
Parkinson's disease	neurons, microglia	[319,324,325]
Age-related macular degeneration	retinal pigment epithelium	[326]
Stroke	microglia	[327]
Atherosclerosis	Foam cells	[328,329]
Cardiovascular disease	myocardium	[330,331]
Sarcopenia	muscle cells	[332,333]
Rheumatoid arthritis	T-cells	[334]
Chronic obstructive pulmonary disease (COPD)	macrophages	[327]
Periodontitis	monocytes, macrophages	[335,336]
Osteopenia	osteoblasts, osteocytes	[337,338]
Osteoarthritis	chondrocytes, cartilage	[339–341]
Diabetes	β-cells	[342,343]
Liver disease (NAFLD) [1]	parenchymal hepatocytes	[344] [2]
Cancer	several	[343,345,346]
Senescence	several	[347–349]

[1] Non-alcoholic fatty liver disease per se is characterized by progressive steatosis involving LD accumulation.
[2] Review of the aging aspect of NAFLD to HCC progression.

5.2. Bone Marrow Aging–Epigenetic Mechanisms

In both children and adults, bone mineral density is inversely correlated with bone marrow fat abundance [350,351] and the aging process in bone marrow is characterized by an expansion in marrow adipose tissue (MAT), which impairs bone stability, thereby yielding an enhanced bone fracture risk [352]. Essential to this, non-differentiated bone marrow stromal cells (BMSCs), amongst others, serve as progenitors for osteoblast and adipocyte differentiation. BMSCs are marked by a low LD content but LD abundance increases upon osteogenesis due to enhanced energy demands, and the blockade of LD formation by Triacsin C (an inhibitor of fatty acyl CoA synthase) results in impaired osteogenic differentiation [353]. In line with this, LD-associated PLIN2 is generally expressed in bone tissue with the highest levels seen in osteo-progenitor cells [353]. Increased serum levels of fatty acids stimulate the adipocytic differentiation of bone marrow progenitor cells [354], the MAT-resident adipocytes containing large LDs that serve as fatty acid and adipokine reservoirs [355]. Of special relevance, MAT expansion may promote free-fatty-acid-based lipotoxicity and negatively affect bone marrow osteoblast proliferation [356], and the inhibition of adipocyte-derived fatty acid synthesis may protect osteoblasts from the lipotoxic effect [357].

Epigenetic mechanisms may play a crucial role in bone marrow aging. Underlying this, histone deacetylases (HDACs) such as sirtuins remove acetyl groups from lysine residues of histone tails, which leads to chromatin condensation and thereby alters gene expression. For example, activation of the murine class III HDAC sirtuin 1 (SIRT1), a nutritional sensor responding to NAD^+/NADH changes, directs mouse mesenchymal C3H10T1/2 cell lines and primary rat bone marrow stromal cells towards enhanced osteoblastic and reduced adipocytic differentiation, while SIRT1 inhibition shows exactly the opposite effect [358]. It is also worth mentioning that LD-derived mono-unsaturated fatty acids are strong activators of SIRT1 [359]. Moreover, a direct interaction has been shown between the peroxisome proliferator-activated receptor γ2 (PPARγ2), a key transcription factor for the differentiation of progenitors into adipocytes, and Sirt1 through its catalytic core domain, forming a stable transcription-inhibiting complex. Binding of this complex to the SIRT1 promoter represses SIRT1 transcription via a self-regulatory feedback loop [360]. This report shows a decline in SIRT1 mRNA and protein levels in older compared to younger human lung, heart, and fat tissues, thus indicating that the SIRT1/PPARγ interaction is a senescence-associated (epi)genetic mechanism [360]. In accordance with this, elevated PPARγ2 levels were also detected in aged compared to young bone marrow stromal cells [361] and it is proposed that the SIRT1/PPARγ2 negative feedback loop lowers SIRT1 expression in bone marrow, and, as a result, promotes MAT expansion and the aging process.

In addition, studies in mice have shown that deletion of another HDAC—Hdac3—leads to lipid accumulation in osteochondrocyte progenitor cells and promotes MAT expansion in young mice [337]. Compared to the wild type, the Hdac3 knockout also leads to a substantial shift in LD/lipid storage-associated Plin1 and Fsp27/Cidec (Fspe27, with fat-specific protein 27 belonging to the family of death-inducing DFF45-like effectors (CIDE) [362]), and a minor but still significant increase in lipolysis-associated lipases (Pnpla2 and Lipe). Further transcriptome analysis in Hdac3 knockout mice revealed a highly abundant expression of 11b-hydroxysteroid dehydrogenase type 1 (Hsd11b1), a gene encoding an enzyme involved in the activation of intracellular glucocorticoids participating in glucocorticoid receptor-based signaling. Inhibition of Hsd11b1 by carbenoxolone resulted in reduced expression of Plin1 and Cidec in Hdac3-deficient cells, which identifies Hdac3 as a crucial regulator of glucocorticoid-induced LD formation in osteoprogenitor cells [337].

5.3. Lipid Droplets in Neurodegeneration

Neurodegenerative diseases such as Morbus Alzheimer and Morbus Parkinson represent another highly prevalent complex of age-related pathologies. Aging of the central nervous system is associated with progressive myelin degeneration at a reduced myelin renewal [363], overall loss of total brain volume [364], glia activation (microglia and astrocytes) and cilia loss in ependymal cells [365], and changes in neuron morphology leading to neuron dysfunction [366]. Lipid homeostasis is of central importance to the functionality of the nervous system since neuron function is considerably impaired by the accumulation of fatty acids that promotes ER stress, lipotoxicity, and mitochondrial damage (recently reviewed in [319]). Notably, the lipid composition of the normal human brain is about 60% by dry weight, which ranks directly after adipose tissue in terms of the tissues with the highest fat fraction, and the brain fat content varies markedly between different brain areas, being highest in myelin (78–81% of the dry weight) and lowest in grey matter (36–40%) [367]. Triacylglycerol levels are low in neurons, probably due to the constant lipid turnover generating the phospholipid mass required for cell membrane maintenance [368]. Although only sparse evidence exists for the in vivo LD formation in neurons, the LD content of neurons increases under "lipid stress" arising from fatty acid treatment [369,370]. In addition, LD numbers are raised upon expression of mutant huntingtin protein [371]. In the aged brain, LD accumulation has been shown in neurons, microglia, astrocytes, and ependymal cells [372,373]. ROS-induced LPO, considered a hallmark of neurodegeneration, serves as a driving force since, similar to the case in *Drosophila*, LPO caused by ROS originating from mitochondrial dysfunction stimulates LD biogenesis by JNK-mediated responses, which precedes the neurodegenerative process in the mammalian system [46]. Astrocytes are important regulators of oxidative stress adaptation in the brain by providing homeostatic, antioxidant support for neurons. This macroglial cell type shows an enhanced resistance towards oxidative stress regimens [374], which might be attributable to its well-equipped antioxidant defenses such as the glutathione redox system, manganese superoxide dismutase, and catalase [375]. Intriguingly, it has been shown that peroxidized lipids (Lox) formed by ROS/LPO in neurons bind to apolipoproteins (especially ApoE) and these ApoE/Lox complexes can be shuttled from neurons to astrocytes, where they are decomposed by lysosomal processing and the liberated fatty acids are stored transiently in LDs [370]. Terminally, the "imported", LD-bound Lox are oxidized in mitochondria via β-oxidation, which stimulates ROS formation; however, due to the enhanced antioxidant defense, astrocytes may successfully prevent substantial ROS build-up. In good accordance with the findings of Schroeter et al. [375], Ioannu et al. [370] identified the up-regulated expression of several genes associated with antioxidant defenses (Gpx8, glutathione peroxidase 8; superoxide dismutase Sod1 and Sod3; catalase and fatty acid transporters FabP5 and FabP7) in cultured, LD-rich astrocytes. This emphasizes the integration of LDs in enhanced astrocytic antioxidant defenses that confer particular antioxidant robustness needed in these macroglial cells to serve as a "detoxifying recipient" for neuron-derived LPO products. Hence, it appears likely that disturbances of this neuron-to-astrocyte lipid transfer and LD storage detoxification mechanism will enhance the risk of neuron damage and the development of Alzheimer's disease (AD). In humans, three APO-E alleles have been described: APOE2, APOE3, and APOE4. Compared to APOE2 homozygotes, individuals bearing the APOE4 allele either in a heterozygous or homozygous genotype show a 9–15-fold increased risk of acquiring AD, rendering the APOE4 allele as one of the main risk factors for AD [376,377]. In fact, it was shown that the APOE4 allele dampens the "neuron to glia" lipid transfer, thus promoting neurodegeneration [378]. With respect to this, it is tempting to speculate that, downstream of ApoE4/Lox-shuttling, the inability of astrocytes to store Lox/lipid peroxides in LDs also promotes the risk of developing AD.

LDs and α-synuclein. In Parkinson's disease (PD), the second most common neurodegenerative disease after AD, LDs also move into the focus of research. Central to PD pathogenesis is the loss of dopaminergic neurons in the substantia nigra pars compacta accompanied by the appearance of Lewy bodies [379]. Lewy bodies are insoluble aggregates of misfolded proteins, of which α-synuclein comprises the main constituent [379,380]. Oligomers of α-synuclein are neurotoxic to themselves and are considered as the main drivers of neurodegeneration [381], with mutations of α-synuclein dramatically increasing the prevalence of PD [382]. It was shown that α-synuclein forms di- and trimers that accumulate at the surface of LDs in human embryonic kidney cells as well as hippocampal neurons treated with high concentrations of oleic acid [325]. Taking into account the observations made in *S. cerevisiae*, *C. elegans*, and *Drosophila*, this suggests the existence of LD-based detoxification mechanisms for α-synuclein aggregates in higher organisms, which hypothetically may provide neuroprotection. Whether this holds true and the extent to which it may play a role in PD pathogenesis remains to be addressed by further research.

LD accumulation in Huntington's disease. Apart from AD and PD, accumulation of LDs is seen in several other neurological and neurodegenerative disorders, including Huntington's disease and amyotrophic lateral sclerosis (ALS), as reviewed recently by Islyme et al. [383]. However, the authors also leave open the question of whether LD accumulation mitigates or promotes the progression of neurodegenerative disorders. This emphasizes the ambiguity of LD biogenesis in the pathogenic context. LD accumulation may enable dynamic lipid storage, and, as a result, act as a potential substrate for LPO, in one pathological condition, but opposite to this, in another diseased state, resemble a sink for the "safe" sequestration of LPO-derived compounds including peroxidized fatty acids as well as aggregates of potentially harmful proteins such as α-synuclein oligomers. It is reasonable to consider that the direction of the LD-based response is dependent on the influence of additional factors. For instance, in Huntington's disease, the mutation-based excessive N-terminal poly-glutamylation of the protein huntingtin (polyG-Htt) is causative of neuronal cell death, which to a substantial degree is due to the disturbed interaction between poly-glutamylated huntingtin and the cytoskeleton [384]. Huntingtin is involved in several cellular transport processes and, in concert with the microtubule network in particular, participates in the axonal transport of organelles and neurotransmitters in neurons [385]. Interestingly, in a yeast model of poly-glutamylation, polyG-Htt aggregation leads to the formation of inclusion bodies and cell death, the degree of which correlates with an aberrant LD morphology that is indicative of a disturbed TAG storage function [386]. Taking into consideration the role of cytoskeletal alterations in neurodegenerative disease [387], an inadequate binding of protein aggregates to LDs could hypothetically also result in a disturbance of cytoskeleton-based intracellular LD/lipid trafficking, for instance, by impairing the afore-mentioned interaction of LDs with motor proteins (see Section 4), and, as a result, aggravate disease progression. Therefore, although the clearance of potentially dangerous protein aggregates by LDs may be considered beneficial to cell integrity, evaluating the impact of this LD-based mechanism on disease outcome deserves a rather holistic approach integrating a multiplicity of interlinked accessory factors.

5.4. Lipid Droplets in Metabolic Disease

Intimately associated with the essential, lipid-metabolism-linked function of LDs, the metabolic process itself represents a critical element of aging. This is indicated by the fact that the risk for metabolic diseases such as the metabolic syndrome, type II diabetes, non-alcoholic fatty liver disease/steatohepatitis (NASH), and cardiovascular diseases and atherosclerosis, is increasing with age [388]. For example, in the pathogenesis of type II diabetes, the development of insulin resistance represents an early, critical issue leading to the disturbance of glucose homeostasis. As a pathophysiological response, insulin production by the β-cells of the pancreatic Langerhans islets becomes elevated, which partially compensates for the incremental insulin resistance. However, with progression of the diabetic condition, the β-cell mass declines and, with this, insulin production ceases. Nutrition-derived lipid stress arising from a high-fat diet is under discussion as a main driver of type II diabetes development, with LDs playing a central role in β-cell lipid management [342]. In fact, it was shown that knockdown of perilipin 2 (PLIN2), an essential LD scaffold protein, resulted in reduced insulin production by β-cells, whereas PLIN2 overexpression boosted insulin secretion. Moreover, β-cells devoid of LDs are prone to ER stress, which results in an impairment of β-cell functionality [343]. This accounts for a critical role of LDs in protecting β-cells from the toxic effects of lipids, and, as a result, acting as an antagonist of type II diabetes progression.

LD accumulation in NASH. Insulin resistance is closely associated with obesity marked by the enhanced accumulation of LDs in epithelial cells and other non-adipose tissues, which establishes a pro-inflammatory microenvironment in the affected tissues as commonly found in metabolic diseases. This also applies to the development of NASH, the inflammatory type of non-alcoholic fatty liver disease (NAFLD) [389]. NASH, which is closely associated with insulin resistance, represents a well-studied liver pathology emerging from chronic fat-rich alimentation. NASH is characterized by the marked accumulation of triglycerides in liver epithelial cells (i.e., parenchymal hepatocytes) causing liver steatosis and the accompanying, chronic inflammation that promotes fibrotic/cirrhotic remodeling of liver tissue (recently reviewed in [390]). Of particular relevance, the development of NASH is accompanied by several processes including mitochondrial dysfunction, ER stress, and enhanced ROS formation, as well as tissue-specific changes, primarily the activation of hepatic stellate cells (HSCs), which promotes the inflammatory process and liver fibrosis upon trans-differentiation of activated HSCs into the extracellular matrix (ECM) producing myofibroblast-like cells [344]. In the *"multiple hit pathogenesis"* of NASH, toxic lipids play an essential role, affecting different liver cell populations, in particular parenchymal hepatocytes, HSCs/myofibroblast-like cells, and *Kupffer* cells (i.e., liver macrophages) in different ways [391]. Contrasting with the well-defined general understanding of NASH development and its clinical manifestation, little is known about age-associated alterations in lipid metabolism and LD biogenesis in NAFLD/NASH progression. Nevertheless, ROS and the ROS-mediated senescence of hepatic cells may also play a pivotal role in NAFLD/NASH pathogenesis. For instance, it was shown that liver steatosis is promoted by a decline in mitochondrial fatty acid metabolism in senescent hepatocytes, an effect that was abolished by the antioxidant, lipid peroxidation inhibitory flavonoid quercetin [217,392].

Role of ROS. Several investigations account for a distinct role of the ECM-associated matrix protein CCN1 (central communication network factor 1; formerly termed Cyr61, cysteine-rich protein 61 [393]) in NASH progression. CCN1 plays an important role in wound-repair-associated ECM remodeling via binding to integrin $\alpha_V\beta3$ and $\alpha_V\beta5$ of epithelial cells and myofibroblast integrin $\alpha6\beta1$ [394]. In line with this and accounting for a role in liver fibrosis, an increased expression of CCN1 has been shown in the liver of NASH patients [395], in hepatocytes of the human cirrhotic liver, and as a reaction to liver injury [396]. Furthermore, it has been shown that CCN1 stimulates hepatic steatosis in obese mice and promotes LD accumulation in hepatocytes treated with free fatty acids [395]. Of considerable relevance, CCN1 expression is enhanced by ROS. This was demonstrated in skin fibroblasts exposed to hydrogen peroxide, which resulted in the c-jun/AP1-dependent up-regulation of CCN1 expression yielding a repression of collagen synthesis and fibroblast senescence [397,398]. Similarly, the overexpression of CCN1 also lowers the production of collagen type1α1 (col1α1) in HSCs [399]. Moreover, CCN1, by acting via integrin $\alpha6\beta1$, shifts ROS/RAC1-dependent NOX1 (NADPH oxidase 1) activity and promotes senescence of HSCs as well as myofibroblasts, as a result conferring an anti-fibrotic response, and in addition stimulates liver-regeneration-associated signaling via IL-6 (interleukin-6) and CXCR2 (chemokine receptor 2) ligands [396,400].

These observations suggest a seemingly ambiguous involvement of CCN1 in NASH: promoting steatosis by LD accumulation in parenchymal hepatocytes, but counteracting NASH-associated fibrosis by mediating senescence, in non-parenchymal HSCs and myofibroblasts. It is noteworthy that in HSCs, the overexpression of CCN1 is also able to trigger ER stress and UPR due to the high abundance of CCN1 protein, which renders these cells susceptible to apoptotic cell death [399]. Hence, elevated CCN1 levels may exert a particular challenge for cell integrity in HSCs. Importantly, parenchymal hepatocytes, and not HSCs or Kupffer cells, represent the major hepatic source of CCN1. This was shown by employing CCl_4, a hepatotoxic compound that generates free radicals upon cytochrome P450-based metabolization and shifts CCN1 expression only in parenchymal hepatocytes [396]. It has to be mentioned that CCN1 null mutant mice reveal a normal hepatic function in the absence of CCl_4, which suggests a specific association of hepatocytic CCN1 expression with stress/ROS-mediated conditions, such as existing in liver injury and inflammation. This notion is supported by the further observation that CCN1 expression in parenchymal hepatocytes is also up-regulated by the pro-inflammatory cytokine TNFα in a ROS-dependent mode [401]. Conversely, the before-mentioned stimulation of LD biogenesis upon CCN1 overexpression in primary mouse hepatocytes is accompanied by an enhanced expression of TNFα [42]. TNFα, acting via the TNFα receptor (TNFR), stimulates mitochondrial ROS generation via JNK signaling and downstream ER stress, which leads to the activation of ATF6 (activating transcription factor 6) and elF2α (eukaryotic initiation factor 2α). In turn, ATF6 and elF2α transduce the signal to nuclear transcriptional control via C/EBP (CCAAT/enhancer-binding protein α) homologous protein (CHOP), which may stimulate ER-stress-dependent apoptosis in hepatocytes [402].

Taken together, these findings account for the existence of an autocrine amplification loop in hepatocytes, established by CCN1 and TNFα/TNFR under pro-oxidant conditions that enables the secretion of CCN1 to the extracellular space and drives activated HSCs and myofibroblasts towards senescence, eventually counteracting the HSC/myofibroblast-driven fibrotic process. Indeed, an inflammation-associated feed-forward loop of cytokine secretion including TNFα has been discussed recently [402]. In extension to this, CCN1-stimulated LD biogenesis could play a central, albeit ambiguous, role in this regulatory network. With respect to the causative role of increased CCN1 and TNFα levels in ER stress, UPR, and apoptosis in HSCs, as well as hepatocytes, it is appropriate to consider that hepatocytes, as the main hepatic source for CCN1 under stress conditions, need to be protected from the cytotoxic potential established by the self-amplifying CCN1–TNFα

circuit. With respect to this, the finding of Ju et al. revealed that overexpression of CCN1 in hepatocytes leads to both (i) the up-regulation of lipid-metabolism-associated genes (ii) and the up-regulation of sirtuins (Sirt 1, 2 and 3), Nrf1, BMP2 (a member of the TGF-β superfamily), and AMP kinases [395]. This connects the CCN1–TNFα circuit to hepatic LD biogenesis, and by this steatosis to most of the LD-associated signaling pathways discussed above, which are capable of exerting a cytoprotective function in cells under oxidative stress. Hence, it is tempting to speculate that enhanced LD biogenesis allows hepatocytes to synthesize CCN1 in potentially cytotoxic amounts, which are needed for counteracting inflammation-driven fibrosis by the paracrine induction of HCS/myofibroblast senescence. Obviously, if this holds true, such a mechanism would assign a yet ambiguous context to LD biogenesis in NAFLD/NASH (as well as alcoholic AFLD): slowing disease progression in a paracrine mode via the CCN1-mediated deceleration of inflammation-associated fibrosis at the cost of promoting steatosis progression via an autocrine amplification loop.

The stimulation of NOX1-based ROS (superoxide) production following CCN1 signaling via integrinα6β1 in HSCs represents a further critical element of such auto- and paracrine regulatory networks. Concerning this, Kim et al. proposed that HSC senescence is caused by CCN1/NOX1-mediated ROS leading to genotoxic damage, and p53 and p16/pRb–dependent, senescence-related responses [396]. Moreover, similar to LD accumulation mediated by CCN1 in hepatocytes [395], Long et al. showed that the up-regulation of NOX1 also stimulates lipid-metabolism-related gene expression and LD accumulation in mouse hepatocytes. and also leads to the up-regulation of ER-stress-associated genes ATF6 and eIF2, effects which were antagonized by the antioxidant N-acetylcysteine (NAC) [403]. This strengthens the assumptions made above regarding a protective role of LD biogenesis counteracting the pro-oxidant effects on the CCN1–TNFα circuit, as well as altered NOX1 activity in hepatocytes under conditions of inflammation. It is noteworthy to emphasize that in the experiments conducted by Long et al., NOX1 overexpression was accomplished in hepatocytes via knockdown of the transcription factor hepatocyte nuclear factor 1β(HNF1β), which identifies HNF1β as a negative regulator of NOX1, and, as a result, a suppressor of both NOX1-mediated superoxide formation and LD biogenesis in hepatocytes [403]. Interestingly, steatotic livers of obese mice show a reduced expression of HNF1β, and treatment of mouse hepatocytes with palmitic acid also lowers HNF1β expression, an effect that is suppressed by NAC [403]. In addition, the HNF1β knockdown also stimulated insulin resistance in hepatocytes, which was also ameliorated by NAC. These observations account for a further feedback mechanism in hepatocytes, leading to an elevated LD accumulation that is driven by the lipid/ROS-based down-regulation of HNF1β, and resulting in NOX1-mediated LD biogenesis and causing increased insulin resistance. With respect to this, HNF1β connects NAFLD with diabetes, which is underlined by the finding that mutated HNF1β alleles are associated with diabetes type MODY5 (maturity-onset diabetes of the young) [404] as well as type-II diabetes [405], suggesting that HNF1β activity mitigates insulin resistance, at least in these pathologies.

Lipid droplets, lipophagy, and hepatic lipid homeostasis. The hepatic lipid flux is marked by diurnal oscillations of fed and fasted states reflected by LD catabolism (fed state) and LD storage (fasted state), with the hepatocyte LD lifecycle playing a pivotal role in systemic lipid homeostasis. Regarding this, the blood insulin concentration is critical by linking systemic fed/fasted states to oscillating high (fed) and low (fasted) insulin signaling. Among several physiologic contexts, this nutrition-dependent systemic insulin dynamics also affects cellular LD dynamics, in particular intracellular LD trafficking. Similar to the case of *Drosophila* (see Section 4), LDs may also become connected to the cytoskeleton in primary hepatocytes where LD binding of the motor protein kinesin-1 mediates their transport along microtubules, a process that is regulated by insulin signaling [406] and serves the delivery of LDs to the smooth ER (sER) for VLDL (very-low-density lipoprotein) production [407]. Essential to this, insulin signaling enhances the GDP-dependent binding of GTPase ADP-ribosylation factor 1 (ARF1) to LDs, rendering them "reactive" [408], the bound ARF1 in turn recruiting phospholipase-D1 (PLD1), which generates phosphatic acid

(PA), and, as a result, shifts the PA content of reactive LDs [409]. As a consequence, via binding to PA, these LDs recruit kinesin-1, which terminally mediates LD shuttling to the sER. Therefore, in the fed state, high systemic insulin levels will promote LD–sER shuttling and fuel VLDL production as well as secretion by hepatocytes, while the postprandial, low-blood insulin levels characteristic of the fasted state will antagonize the shuttling process and thus limit VLDL synthesis and secretion. Notably, starvation conditions markedly enhance the hepatic clearance of adipose-tissue-derived lipids from the circulation, which causes a substantial shift in the hepatocyte LD content. Hence, the down-regulation of LD trafficking entailed by low insulin signaling serves as a "bottleneck" for VLDL production in LD-rich hepatocytes, limiting VLDL under fasted conditions. This puts emphasis on the role of LDs participating in the physiological regulation of systemic lipid homeostasis serving as a hormonally controlled, dynamic lipid buffer in the liver.

At the cellular level, LDs can be selectively degraded by autophagy, a process termed (macro)lipophagy, which sequesters cytosolic LDs for autophagolysosomal digest and is pivotal to lipid metabolism [410,411]. In addition to its role in mobilizing fatty acids from cellular LD-based lipid stores, lipophagy represents a pivotal "guardian" of cellular LD abundance. This holds particularly true when LD biogenesis is substantially stimulated in hepatocytes in response to an enhanced clearance of lipids from the blood stream when facing the risk of systemic lipotoxicity arising from an excess of circulating free fatty acids. Emphasis on this is provided by experiments showing that lipid treatment of cultured hepatocytes stimulates lipophagy while lipophagy inhibition by the macroautophagy inhibitor 3-methyladenine shifts the number of LDs in hepatocytes even under normal culture conditions [410]. Moreover, lipophagy is also enhanced in the mouse liver under starvation conditions that stimulate LD biogenesis. Importantly, these experiments revealed further that enhanced exogenous lipid supply such as that provoked by a fat-rich diet negatively affects LD breakdown by lipophagy. In good agreement with this, it was reported recently for a mouse model of obesity that lipophagy declines upon feeding a high-fat diet, resulting in liver steatosis [412]. Of special relevance to lipotoxic side effects, the observed drop in autophagic/lipophagic efficiency was accompanied by the enhanced accumulation of HNE-modified proteins. This connects lipophagy with reparative autophagy (i.e., detoxification of the aggregated modified proteins), and, as a result, LD abundance, with the critical interference existing between LPO and proteostasis. Hence, it is not surprising that a reduced lipophagic flux plays an important pathogenic role in NAFLD and other lipid-metabolism-associated diseases such as atherosclerosis [413], and may affect the aging-associated transition from NAFLD into primary hepatocellular carcinoma (HCC) [344].

LD accumulation, oxidative stress, and cell death. Finally, the excessive cytosolic accumulation of fatty acids seen under "hyperlipidemic" states such as obesity, NAFLD/NASH, and diabetes may lead to an elevated susceptibility to lipotoxicity-induced cell death via apoptosis. The specific term lipoapoptosis was coined for apoptotic cell death stimulated by fatty acid derivatives such as ceramide [414], and as a terminal issue in lipotoxic settings, lipoapoptosis is of particular relevance to several lipid-associated pathologies including NAFLD [415], as well as others such as vascular and cardio-metabolic diseases [416], which are discussed below. In addition, ROS-stimulated LPO yields further metabolites such as HNE, which interferes with anti-apoptotic and cell-proliferation-associated intracellular signaling [417], but is also capable of inducing apoptosis per se [418]. Moreover, the concept of ferroptosis, which has gained substantial interest over the last decade, also represents, in principle, an LPO-dependent mode of cell death. In ferroptosis, LPO is initiated by iron-derived •OH radicals at inadequate antioxidant defenses (i.e., weakening of the GSH/GSSG redox system due to GSH-peroxidase 4 deficiency) leading to a non-apoptotic, necrotic mode of cell death [419]. Taking into consideration that hepatocytes represent an iron-rich cell type serving systemic iron buffering, ferroptosis represents a considerable issue in NAFLD/NASH that also occurs with high levels of PUFAs [420]. In addition, it should not be overlooked that ferroptosis may not only lead to hepatocyte loss, but as a necrotic mode of cell death will also aggravate the pro-inflammatory condition. Therefore, lipid stress arising from an inappropriate lipid accumulation will affect many cellular targets and pathways, among which LDs are of central relevance, either fueling lipid (per)oxidation or attenuating lipotoxicity by aiding cellular lipid detoxification and interfering with cytotoxic responses such as lipoapoptosis and possibly also other lethal outcomes such as ferroptosis.

5.5. Lipid Droplets in Vascular Disease

In hepatic Kupffer cells residing in the liver sinusoids, the augmented uptake of cholesterol and free fatty acids enhances the development of a lipid-rich macrophage phenotype [421], characterized as foam cells, and the aggregation of such lipid-rich Kupffer cells contributes to NAFLD/NASH-associated lipogranulomas that are built from inflammatory cells, ECM (collagen), and LDs [422,423]. Of particular pathological relevance, the conversion of macrophages into foam cells represents a major issue in atherosclerosis development (atherogenesis), since the "foamy" macrophages build up atherosclerotic plaques in arterial walls, leading to inflammation and progressive damage to the vessel wall (reviewed in [328]). Underlying this is the oxidation of low-density lipoprotein (LDL) bound to proteoglycans of the extracellular matrix and the vessel endothelium, which yields oxidized LDL (oxLDL), the oxLDL in turn triggering the release of monocyte chemoattractant protein (MCP-1) by vascular endothelial cells and vascular smooth muscle cells (VSMCs). The attracted monocytes migrate to the arterial wall and differentiate into macrophages that endocytose the oxLDL particles via scavenger receptor A (SR-A) and CD36 [424], although alternative uptake mechanisms may exist [425]. Upon lysosomal processing of the oxLDL particles, the free fatty acids and cholesterol molecules are released to the cytoplasm, where they are either stored in LDs [426,427] or are released via high-density lipoprotein (HDL) [428]. The enhanced uptake of oxLDL by macrophages will nourish the accumulation of cholesterol-rich LDs, and, as a result, stimulate foam cell and plaque formation [328]. It should not be overlooked that LDs are also essential to reverse cholesterol transport (RCT), a process by which cholesterol sequestered from the circulation is stored transiently in LDs, from which it can be liberated via lipophagy and be re-released from the cell, for instance, to the bile for fecal excretion. Although RCT can be accomplished by several cell types, macrophage-associated RCT is considered causal to the atherosclerotic process [429].

LDs and cholesterol homeostasis. From this, it becomes clear that the ability of LDs to store cholesterol is essential to cellular cholesterol homeostasis and the protection from cholesterol lipotoxicity. Essential to this is the esterification of free, unesterified cholesterol by Acetyl-coenzyme A cholesterol *O*-acyltransferase-1 (ACAT1) as a prerequisite for the incorporation of the cholesteryl-esters into LDs [430]. In line with the protective role of LDs, stimulation of ACAT1 is essential for proper cellular cholesterol management and, by facilitating LD-based cholesterol clearance, counteracts cholesterol toxicity [431]. In addition, oxysterol-binding protein-related proteins ORP2, ORP5, and ORP8 can stimulate LD biogenesis and can bind oxysterols such as 25-hydroxycholesterol and 7-ketocholesterol, as well as cholesterol itself, to LDs [432,433]. Considering the toxic effects of oxysterols, oxysterol binding to LDs clearly represents a cytoprotective function of LDs. This holds particularly true for 7-ketocholesterol, which accumulates in foam cells (reviewed in [434]) and is known as a stimulator of oxiapoptophagy, a distinct mode of oxysterol/oxidative-stress-associated cell death involving apoptosis and autophagy with particular pathogenic relevance to age-related diseases including atherosclerosis [435–439]. The cholesteryl ester-driven biogenesis of LDs in vascular macrophages is considered causal to foam cell development [429], which likewise also holds true for oxysterol binding to LDs, and both processes represent driving forces of atherogenesis. As stated by Lee-Rueckert et al. [440], the development of foam cells may be accompanied by the reduced expression of pro-inflammatory genes (characteristic of the activated M1 macrophage phenotype) converting the phenotype into an anti-inflammatory one (i.e., activated lipid-rich M2 macrophages [441], leading to the concept that foam cell development represents an anti-atherogenic effect [440]. Hence,

LD accumulation in atherosclerosis serves as a further example of the ambiguous role of LDs in pathophysiological settings, as mentioned above for NAFLD: protection from acute lipotoxicity, thus aiding cell survival at the cost of promoting a chronic process such as atherosclerosis and liver fibrosis/cirrhosis. Interestingly, Lathe et al. followed a similar concept in discussing the effects of the pathogen (virus)-induced stimulation of 25-hydroxycholesterol, which via ACAT1 esterification can also bind to LDs, and contributes to both atherosclerosis as well as Alzheimer's disease, postulating that 25-hydroxycholesterol protects from *"infectious agents at the expense of longer-term pathology"* [442]. Emphasizing the ambivalent role of LD formation in foam cell development and the atherogenic context, plaque formation is accompanied by a decline in the lipophagic flux, which limits LD breakdown, and, as a result, excessive liberation of cholesterol [443], but aggravates atherogenesis due to the continuous stimulation of foam cell formation driven by LD accumulation. Finally, it appears noteworthy that VSMCs may translocate to the arterial intima in the course of atherosclerosis progression and transdifferentiate into a macrophage-like foam cell phenotype, revealing an enhanced oxLDL content, although containing fewer LDs and showing a reduced lipophagic flux compared to macrophage-derived foam cells [444,445]. Nevertheless, VSMC-derived foam cells can comprise about 50% of the foam cell content seen in human atherosclerotic plaques [444], and thus represent an atherogenesis-associated cell population of considerable interest.

Finally, LDs may assist macrophage integrity not only by the control of lipid balance and oxLDL/oxysterol sequestration, but also by aiding the "clearance" of other aging and stress-associated compounds, especially protein aggregates. This is indicated by a recent investigation that demonstrates the aging-dependent binding of protein aggregates to LDs in mouse intestinal tissue, supposedly followed by the terminal degradation of the critical matter via lipophagy [135]. It is conceivable that a similar, LD-based cytoprotective mechanism is involved in aging and lifespan control of other organisms such as *C. elegans* and *Drosophila*, and also likely in yeast, considering the binding of IBs to LDs as discussed in this review.

5.6. LD Accumulation in Cardiomyocytes: Role of PPARs

As in other tissues, LDs also play a dual role in the cardiac system. To overcome the enhanced energy demand of cardiomyocytes, long-chain fatty acids (LCFAs) such as palmitate and oleate (due to their higher energy yield per carbon molecule as compared to glucose) are the primary fuel needed for ATP synthesis [330]. Subsequent to esterification by acyl-coenzyme A synthetase (CoA), these CoA-fatty acyls are further esterified to a glycerol backbone and stored as TAGs in LDs. Upon LD lipolysis and lipase-mediated TAG hydrolysis, the liberated fatty acids will fuel mitochondrial β-oxidation or serve as ligands for the nuclear peroxisome proliferator-activated receptor α (PPARα), a transcription factor that is central to the control of intracellular TAG turnover and fatty acid metabolism [446,447]. It is noteworthy that another PPAR species, PPARγ, stimulates lipid uptake and LD biogenesis in cardiac tissue and confers protection of cardiomyocytes from ROS-mediated damage via regulating the expression of the Sod2 gene, encoding manganese superoxide dismutase [448]. Hence, PPARs represent important determinants of cardiac LD turnover, which is central to cardiac lipid management and protects the heart from organ dysfunction caused by lipotoxicity [330]. Interestingly, it has been shown that PPARα together with mTOR also regulate a reciprocal mode of LD biogenesis and mTORC1-containing stress granule formation in lipid-stressed HEK239T and SH-SY5Y cells [449]. It cannot be excluded that this also applies to heart tissue and this interconnects cardiac LD turnover with autophagy via mTOR/PPAR signaling.

Role of PLIN5 in cardiac disease. Of particular pathologic relevance, heart failure in obesity and diabetes mellitus is associated with hyperlipidemia resulting in lipid accumulation and an expansion of the myocardial LD content [450]. In diabetic heart disease, particular attention has been paid to the role of the LD-associated protein perilipin 5 (PLIN5) (recently reviewed in [451]. Reflecting the context-dependent role of LDs in lipid homeostasis, PLIN5, under normal conditions, inhibits lipolysis via binding comparative gene identification-58 (CGI-58), which otherwise binds triglyceride lipase, but under stress conditions (e.g. fasting, exercise) stimulates lipolysis as a result of phosphorylation by PKA (protein kinase A), leading to the release of CGI-58 from PLIN5, which in turn activates triglyceride lipase [452]. Connected with this, cells with a high oxidative capacity show an enhanced expression of PLIN5, as holds especially true for cardiomyocytes, where PLIN5 is responsible for the tethering of LDs to mitochondria [453]. Compared to healthy control donors, expression of PLIN5 is reduced in samples drawn from patients with heart failure, showing a decline in direct LD-mitochondria contacts and reduced fatty acid usage for energy supply [454]. On the other hand, by inhibiting triglyceride lipase, PLIN5 supports the sequestration of TAG by LDs, which will limit fatty acid availability, and, as a result, protect the heart from lipotoxicity; however, dysregulation of this process upon PLIN5 overexpression will promote cardiac steatosis and hypertrophy [455,456]. Conversely, LDs are absent from myocardial tissue in PLIN5 knockout mice and myocytes isolated from these PLIN5$^{-/-}$ mice show an increased fatty acid oxidation in vitro compared to the wild type. Moreover, ROS production is enhanced in the heart tissue of PLIN5$^{-/-}$ mice, which aggravates the age-related cardiomyopathy, but can be antagonized by the glutathione-precursor N-acetylcysteine [457]. In addition, PLIN5 may also exert protection from lipotoxicity by antagonizing ER stress, which has been shown for pancreatic β-cells upon chronic exposure to free fatty acids [458]. Finally, a recent report demonstrated a regulatory role for cardiac PLIN5 in cardiac Ca^{2+} signaling and muscle contractility that is based on the interaction between PLIN5 and sarcoplasmic/endoplasmic reticulum Ca^{2+} ATPAase2 [459]. Summarizing, these findings put emphasis on the proper LD balance and expression of LD-associated PLIN5 on heart integrity maintenance.

5.7. Lipid Droplets and Cancer—A General Outline

Growing evidence suggests a manifold involvement in LDs in cancer (reviewed in [460]), however, it still is not clear whether LD accumulation plays a causative role in carcinogenesis (as, for instance, is discussed above for the transition from NAFLD to HCC), or is a consequence of increased lipid demands of tumor cells; or—most likely—both may even apply. In many aspects, aging and tumorigenesis show opposing phenotypes, which led to the proposal that anti-aging strategies can be developed based on tumor cells [461]. Attributable to the altered energy demands of tumor cell proliferation, elevated LD accumulation is observable in different kinds of tumors, such as colorectal cancer, hepatocellular and pancreatic carcinoma, renal cell carcinoma, prostate and breast cancer, lung cancer, and glioblastoma [462–464]. For several cancers, a direct correlation between tumor cell survival, tumor aggressiveness, and LD numbers has been documented and tendencies exist to consider cancer as a "LD-driven metabolic disease" [465]. A series of excellent reviews addresses the question of how LDs can promote tumorigenesis [460,462,465], which are recapitulated here briefly. As stated above, the most obvious role of LDs in cancer growth is energy supply, with LD-derived FAs serving as fuel for β-oxidation and mitochondrial ATP production [466]. LD-resident PLIN5 is essential to the FA flux between LDs and mitochondria [453], which is essential to cellular lipid supply coping with increased energy demands such as seen in tumor cells, but also in normal cells under stress conditions as discussed above for cardiovascular disease. Of special pathophysiological relevance, the FAs released from LDs are not only used for energy production, but also act as signaling molecules (e.g., lysophosphatidic acid) regulating tumor progression and metastasis [460,467]. Furthermore, LDs are able to modulate cell cycle checkpoints and gene expression in tumor cells (e.g., G_0/G_1 bypass and regulation of FOXO3A activity) [460,468,469]. In addition, LDs also enable the intracellular trafficking of growth-signaling proteins such as PI3K, ERK1, ERK2, p38, and PKC, as well as endo-/transcytosis-regulating caveolin, which are also involved in tumorigenesis [470,471].

Moreover, LDs seem to be especially important for tumor initiation during early carcinogenesis. In the so-called elimination phase, tumor defense by the both the innate and adaptive immune systems is based on the detection of potentially malignant cells and their targeted elimination via apoptosis [472,473]. Besides acting inside the tumor cells, LDs also interfere with the tumor microenvironment [460]. As an illustrative example of the highly complex interactions between LDs, tumor cells, and the tumor cell microenvironment, the role of LDs in cellular eicosanoid production [474] should be mentioned here. Eicosanoids (e.g., prostaglandins, leukotrienes, and lipoxins) are important PUFA-derived (e.g., arachidonic acid) signaling molecules, which are secreted from tumor cells into their microenvironment where they exert autocrine and paracrine activities. For instance, prostaglandin E2 (PGE$_2$) is mainly synthesized from LDs in cancer cells [475] and immune suppression conferred by tumor-derived PGE$_2$ is deeply involved in the tumor escape from immune surveillance [476]. In a complementary mode, PGE$_2$ is also involved in tumor cell proliferation, angiogenesis, and metastasis [477]. Conversely, dendritic cells enriched in LDs containing oxidized TAG show a dysfunctional antigen presentation [478], which will also impair the host tumor defense.

Acting on the central balance of homeostatic growth control, LD accumulation may also affect the onset of apoptosis [479,480], probably by delaying the accumulation of toxic fatty acids inside the affected tissue [481,482]. Several findings account for the direct involvement of LDs in cancer cell apoptosis. Notably, in both tumorous and non-tumorous cell lines, stimulation of apoptosis occurs in conjunction with enhanced LD biogenesis [35,483], and evidence exists that an increased LD content improves the tumor cells' resistance to pro-apoptotic stimuli. This may be due to the enhanced sequestration of a pro-apoptotic stimulus by LDs, as was demonstrated for curcumin. This plant polyphenol stimulates apoptosis via intrinsic, mitochondria-dependent signaling in several cell lines [484–487], but fails to do so in glioblastoma cells [488]. In these cells, curcumin is efficiently sequestered by the high LD content. Lowering LD numbers via inhibition of cytosolic phospholipase A2 restores the sensitivity to curcumin-mediated apoptosis [488]. In a similar way, the enhanced sequestration of chemotherapeutic drugs by LDs may render cancer treatment inefficient and finally promote drug resistance [460,489].

However, further approaches exist to explain the anti-apoptotic role of LDs, addressing intrinsic, mitochondria-dependent (MOMP/apoptosome), and extrinsic, death-receptor-dependent (TNFα receptor superfamily /DISC) apoptotic signaling [490]. For instance, it was shown that alterations of the cholesterol content of lipid rafts blocks the onset of apoptosis induced upon TRAIL (tumor necrosis factor-related apoptosis-inducing ligand) ligation to death receptors DR4 and DR5 in non-small cell lung carcinoma cells [491]. Considering that LDs serve as a reservoir for cholesterol, LDs could hypothetically contribute to the suppression of extrinsic apoptosis. In addition, our own findings suggest a direct involvement of LDs in intrinsic apoptotic signaling. We showed (for details see Section 2.3) that mitochondria-localized apoptotic proteins (pro- as well as anti-apoptotic) contain a V-domain that enables shuttling of these proteins from mitochondria to LDs. The affinity of this V-domain is higher for LDs than for mitochondria and, upon an increase in the cytosolic LD content (as seen in tumor cells), these apoptotic proteins are cleared from the mitochondria, with the relocalization to LDs interrupting the apoptotic program [35]. In fact, both pro- and anti-apoptotic proteins such as BAX [12,35,492], BCL-X$_L$ [35], Bcl-w [12], AIFM1, AIFM2 [12], CCAR2 [12], API5 [492], and TPT1 [35,492] were shown to localize to LDs in tumor cells.

Taken together, it is likely that LDs play a hitherto underestimated role in cancer biology, addressing tumorigenesis at several critical instances. As discussed in this review, LDs may confer protection by exerting antioxidant properties including lipid stress (LPO) under healthy conditions. On the contrary, LDs may promote carcinogenesis in diseased contexts, especially in chronic, inflammation-associated settings such as the *"malignant"* transition from NASH to HCC, and may terminally also contribute to tumor progression and metastasis by interfering with vascularization and proliferation–regulatory cell signaling in the tumor environment.

6. Concluding Remarks

In most model organisms, a clear picture seems to emerge that LDs, despite their rather negative appraisal as a mere "fat-particle", fulfill a cytoprotective role. This is due to the "buffering" function of LDs, which enables them to take up lipid peroxides and other oxidized lipid derivatives (e.g., oxLDL), as well as to detoxify misfolded proteins and protein aggregates in and on various cell organelles. Accordingly, it is not surprising that lifespan extension is positively correlated with LD abundance (at least to some extent). Similar experimental evidence can be found throughout a diversity of biological model systems, suggesting LDs inherit highly conserved functions. This picture is clearest in simple organisms such as *S. cerevisiae* or *C. elegans*, but is also presented by the more complex organism *D. melanogaster*. Central to this LD–aging connection seems to be metabolic pathways such as TOR signaling or IIS (see Figure 2), which upon inhibition lead to both prolonged lifespan and elevated LD synthesis. The situation in mammals and humans is more difficult to interpret due to markedly larger cell numbers, the enhanced diversity of differentiated cell types, and the complex interaction among diverse tissues. It is striking that LDs are concomitant to age-related disease. Concerning this, however, we want to question the still prevailing concept that LDs, by oversimplification understood as "monofunctional" fat-accumulating vesicles, are causative of the pathogenesis of age-related diseases. Taken together, the existing literature advocates a different view, suggesting that LDs represent multifunctional organelles of particular physiological relevance that play a subtle, *Janus-faced* role in disease: LDs essentially fulfilling a protective, retarding function during early pathogenetic stages, but converting to the opposite function in the course of disease progression when an excessive accumulation of LDs amplifies a phenotype characteristic of advanced disease states.

Moreover, disregarding the pathogenic aspect, there is evidence for a physiological role of LDs as important "players" in healthy aging in humans. It is well accepted that the Mediterranean diet has numerous beneficial effects on human health. Many studies have shown that this diet reduces mortality and lowers the risk of developing cancer, neurodegenerative diseases, and cardiovascular diseases [493]. Some of the effects of the Mediterranean diet can be attributed to sirtuins [494], which have been addressed as regulators of LD biogenesis at several instances in this review. This puts emphasis on beneficial nutritional aspects, in particular focusing on two essential pillars of the Mediterranean diet: red wine and olive oil. In fact, it was shown that resveratrol, a polyphenol enriched in red wine, is an activator of sirtuin Sir2p (the yeast homologue of Sirt1) that has the capability to extend the lifespan in a broad variety of organisms [83]. It has to be noted critically that the activation of yeast Sirt1 by resveratrol occurs in an indirect mode via the cAMP-Epac1-AMPK-Sirt1 pathway, with Sirt1 being likely to be activated by increased cellular amounts of NAD^+ [495]. Recently it was shown that monounsaturated fatty acids such as oleic acid, the main component of olive oil, allosterically activate Sirt1 at a magnitude many times higher than that of resveratrol [359]. Considering the connection between Sirt signaling and LD biogenesis, it would be thrilling to see, in the future, if some of the positive effects of the Mediterranean diet can be attributed to the stimulation of LD biogenesis.

In synopsis, it is obvious that lipid metabolism is closely linked to aging and cellular stress responses via highly complex interactions that are not yet fully understood. According to our recent knowledge, it can be concluded that LDs participate in these complex metabolic, aging-associated networks by playing a "Janus-faced" role, as illustrated in Figure 3, and it will be the subject of future investigation to elucidate the exact, underlying contexts in detail.

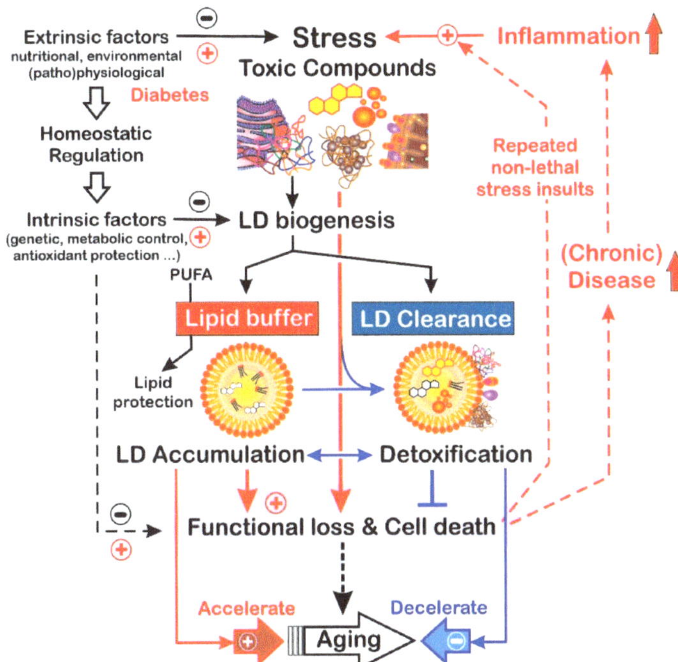

Figure 3. Model for the Janus-faced role of LDs in the aging process. Central to this explanatory approach is the bifunctional involvement of LDs in cellular maintenance, with LDs serving as both (i) a dynamic lipid / fat buffer, and (ii) a "sink" for toxic compounds upon LD clearance (proteins, lipids, and toxic compounds). In addition, LDs may also provide lipid protection by preserving PUFAs from excessive LPO and can quench stress derived from extrinsic factors. Under stress conditions, such as those created by ROS, the ER, or mitochondria, stimulation of LD biogenesis becomes the most important factor, and excessive LD formation depends on a variety of intrinsic factors (TOR, IIS, and TGF-β). The increase in the LD pool establishes a delicate balance between LD accumulation and LD-based detoxification. This balance determines the outcome of the stress response, which protects against cell death but may result in a chronic process (disease / inflammation) based on it. In contrast with the initial beneficial effects, the accumulation of LD accelerates the progression of chronic disease and thus the "aging" process. Hence, in stressed cells, LD biogenesis and LD functionality, both indirectly and directly, intervenes with multifaceted cellular life–death decisions (shaded blue area).

Author Contributions: Conceptualization, M.R.; writing—original draft preparation, N.B., M.K., A.L., T.K.F. and M.R.; writing—review and editing, N.B., M.K., A.L., T.K.F. and M.R.; visualization, N.B. and M.R.; funding acquisition, M.R. All authors have read and agreed to the published version of the manuscript.

Funding: This research was funded by the Austrian Science Fund (FWF) with the grant P33511 to M.R.

Institutional Review Board Statement: Not applicable.

Informed Consent Statement: Not applicable.

Data Availability Statement: Not applicable.

Acknowledgments: This scientific work is dedicated to the great scientist Breitenbach, with whom it was always a pleasure to cooperate and discuss science.

Conflicts of Interest: The authors declare no conflict of interest.

Abbreviations

AD, Alzheimer's disease; AMPK, AMP-activated protein kinase; ANT, adenine-nucleotide translocator; Apo and APO, apolipoproteins; ARDs, age-related diseases; ATF6, activating transcription factor 6; ATG, autophagy related genes; ALS, amyotrophic lateral sclerosis; CBS, cystathionine β-synthase; Ces1d, cholesteryl-ester hydrolase; CCN1, central communication network factor 1; DFP, deferiprone; ECM, extracellular matrix; ER, endoplasmic reticulum; ESCRT, endosomal sorting complexes required for transport; Fe-S, iron-sulfur clusters; FOXO, forkhead box O; FRDA, Friedreich's Ataxia; Ftx, frataxin; GAP, GTPase activating protein; GSH, glutathione (reduced from); GSSG, the oxidized (disulfide) form of GSH; HCC, hepatocellular carcinoma; HDACs, histone deacetylases; HNE, 4-hydroxy-2-nonenal; HNF1β, hepatocyte nuclear factor 1β; HSCs, hepatic stellate cells; IBs, inclusion bodies; IIS, Insulin/Insulin growth factor -1 signaling; ISC, intestinal stem cell; ISCU, iron–sulfur cluster forming unit; JNK, c-jun amino-terminal kinase; LDs, Lipid droplets; LDL, low-density lipoprotein; oxLDL, oxidized LDL; Lox, peroxidized lipids; LPO, lipid peroxidation; LROs, lysosome-related organelles; MAC, mitochondrial-apoptosis-induced channel; MAGIC, mitochondria as guardian in cytosol; MAT, marrow adipose tissue; MOMP, mitochondrial outer membrane permeabilization; MSCs, bone marrow stromal cells; mPT, mitochondrial permeability transition pore; NAFLD, non-alcoholic fatty liver disease; NASH, non-alcoholic steatohepatitis; nLDs, nuclear LDs; NOX1, NADPH oxidase 1; PA, phosphatic acid; PCD, programmed cell death; PD, Parkinson's disease; PERK, PKR-like ER kinase; PLIN, perilipin; PML, promyelocytic leukemia protein; PPAR, peroxisome proliferator-activated receptor; PUFAs, polyunsaturated fatty acids; rad, radiation damage; *raptor*, regulatory associated protein of mTOR; RCT, reverse cholesterol transport; *rictor*, rapamycin-insensitive companion of mTOR; ROS, reactive oxygen species; SIRT, sirtuin; TAG, triacylglycerol(s); TE, transposable element; TNFα, tumor necrosis factor α; TNFR, TNFα-receptor; TOR, target of rapamycin; mTORC, mammalian TOR complex; TSC, trans-sulfuration pathway; UPR, unfolded protein response; UPR^{ER}, ER stress-associated UPR; TSC, tuberous sclerosis complex; VDAC, voltage-dependent anion channel; VLDL, very-low-density lipoprotein; VSMCs, vascular smooth muscle cells; Xbp1, X-Box binding protein 1.

References

1. Murphy, D.J.; Vance, J. Mechanisms of lipid-body formation. *Trends Biochem. Sci.* **1999**, *24*, 109–115. [CrossRef]
2. Martin, S.; Parton, R.G. Lipid droplets: A unified view of a dynamic organelle. *Nat. Rev. Mol. Cell Biol.* **2006**, *7*, 373–378. [CrossRef] [PubMed]
3. Wilfling, F.; Haas, J.T.; Walther, T.C.; Farese, R.V., Jr. Lipid droplet biogenesis. *Curr. Opin. Cell Biol.* **2014**, *29*, 39–45. [CrossRef]
4. Geltinger, F.; Schartel, L.; Wiederstein, M.; Tevini, J.; Aigner, E.; Felder, T.K.; Rinnerthaler, M. Friend or foe: Lipid droplets as organelles for protein and lipid storage in cellular stress response, aging and disease. *Molecules* **2020**, *25*, 5053. [CrossRef] [PubMed]
5. Plotz, T.; Hartmann, M.; Lenzen, S.; Elsner, M. The role of lipid droplet formation in the protection of unsaturated fatty acids against palmitic acid induced lipotoxicity to rat insulin-producing cells. *Nutr. Metab.* **2016**, *13*, 16. [CrossRef]
6. Listenberger, L.L.; Han, X.; Lewis, S.E.; Cases, S.; Farese, R.V., Jr.; Ory, D.S.; Schaffer, J.E. Triglyceride accumulation protects against fatty acid-induced lipotoxicity. *Proc. Natl. Acad. Sci. USA* **2003**, *100*, 3077–3082. [CrossRef] [PubMed]
7. Geltinger, F.; Tevini, J.; Briza, P.; Geiser, A.; Bischof, J.; Richter, K.; Felder, T.; Rinnerthaler, M. The transfer of specific mitochondrial lipids and proteins to lipid droplets contributes to proteostasis upon stress and aging in the eukaryotic model system *Saccharomyces cerevisiae*. *Geroscience* **2020**, *42*, 19–38. [CrossRef]
8. Moldavski, O.; Amen, T.; Levin-Zaidman, S.; Eisenstein, M.; Rogachev, I.; Brandis, A.; Kaganovich, D.; Schuldiner, M. Lipid Droplets Are Essential for Efficient Clearance of Cytosolic Inclusion Bodies. *Dev. Cell* **2015**, *33*, 603–610. [CrossRef]
9. Vevea, J.D.; Garcia, E.J.; Chan, R.B.; Zhou, B.; Schultz, M.; Di Paolo, G.; McCaffery, J.M.; Pon, L.A. Role for Lipid Droplet Biogenesis and Microlipophagy in Adaptation to Lipid Imbalance in Yeast. *Dev. Cell* **2015**, *35*, 584–599. [CrossRef]

10. Garcia, E.J.; Liao, P.C.; Tan, G.; Vevea, J.D.; Sing, C.N.; Tsang, C.A.; McCaffery, J.M.; Boldogh, I.R.; Pon, L.A. Membrane dynamics and protein targets of lipid droplet microautophagy during ER stress-induced proteostasis in the budding yeast, *Saccharomyces cerevisiae*. *Autophagy* **2021**, *17*, 2363–2383. [CrossRef]
11. Kumar, R.; Nawroth, P.P.; Tyedmers, J. Prion Aggregates Are Recruited to the Insoluble Protein Deposit (IPOD) via Myosin 2-Based Vesicular Transport. *PLoS Genet.* **2016**, *12*, e1006324. [CrossRef]
12. Bersuker, K.; Peterson, C.W.H.; To, M.; Sahl, S.J.; Savikhin, V.; Grossman, E.A.; Nomura, D.K.; Olzmann, J.A. A Proximity Labeling Strategy Provides Insights into the Composition and Dynamics of Lipid Droplet Proteomes. *Dev. Cell* **2018**, *44*, 97–112.e117. [CrossRef]
13. Greenberg, A.S.; Egan, J.J.; Wek, S.A.; Garty, N.B.; Blanchette-Mackie, E.J.; Londos, C. Perilipin, a major hormonally regulated adipocyte-specific phosphoprotein associated with the periphery of lipid storage droplets. *J. Biol. Chem.* **1991**, *266*, 11341–11346. [CrossRef] [PubMed]
14. Cho, S.Y.; Shin, E.S.; Park, P.J.; Shin, D.W.; Chang, H.K.; Kim, D.; Lee, H.H.; Lee, J.H.; Kim, S.H.; Song, M.J.; et al. Identification of mouse Prp19p as a lipid droplet-associated protein and its possible involvement in the biogenesis of lipid droplets. *J. Biol. Chem.* **2007**, *282*, 2456–2465. [CrossRef] [PubMed]
15. Turro, S.; Ingelmo-Torres, M.; Estanyol, J.M.; Tebar, F.; Fernandez, M.A.; Albor, C.V.; Gaus, K.; Grewal, T.; Enrich, C.; Pol, A. Identification and characterization of associated with lipid droplet protein 1: A novel membrane-associated protein that resides on hepatic lipid droplets. *Traffic* **2006**, *7*, 1254–1269. [CrossRef]
16. Onal, G.; Kutlu, O.; Gozuacik, D.; Dokmeci Emre, S. Lipid Droplets in Health and Disease. *Lipids Health Dis.* **2017**, *16*, 128. [CrossRef]
17. Renne, M.F.; Hariri, H. Lipid Droplet-Organelle Contact Sites as Hubs for Fatty Acid Metabolism, Trafficking, and Metabolic Channeling. *Front. Cell Dev. Biol.* **2021**, *9*, 726261. [CrossRef] [PubMed]
18. Li, Z.; Thiel, K.; Thul, P.J.; Beller, M.; Kühnlein, R.P.; Welte, M.A. Lipid droplets control the maternal histone supply of *Drosophila* embryos. *Curr. Biol.* **2012**, *22*, 2104–2113. [CrossRef]
19. Kovacs, M.; Geltinger, F.; Verwanger, T.; Weiss, R.; Richter, K.; Rinnerthaler, M. Lipid Droplets Protect Aging Mitochondria and Thus Promote Lifespan in Yeast Cells. *Front. Cell Dev. Biol.* **2021**, *9*, 774985. [CrossRef]
20. Suriyalaksh, M.; Raimondi, C.; Mains, A.; Segonds-Pichon, A.; Mukhtar, S.; Murdoch, S.; Aldunate, R.; Krueger, F.; Guimera, R.; Andrews, S.; et al. Gene regulatory network inference in long-lived *C. elegans* reveals modular properties that are predictive of novel aging genes. *Iscience* **2022**, *25*, 103663. [CrossRef]
21. Zhao, X.; Li, X.; Shi, X.; Karpac, J. Diet-MEF2 interactions shape lipid droplet diversification in muscle to influence *Drosophila* lifespan. *Aging Cell* **2020**, *19*, e13172. [CrossRef]
22. Zimmermann, A.; Hofer, S.; Pendl, T.; Kainz, K.; Madeo, F.; Carmona-Gutierrez, D. Yeast as a tool to identify anti-aging compounds. *Fems Yeast Res.* **2018**, *18*, foy020. [CrossRef]
23. Stefanini, I.; De Filippo, C.; Cavalieri, D. Yeast as a Model in High-Throughput Screening of Small-Molecule Libraries. In *Diversity-Oriented Synthesis*; Wiley Online Library: Hoboken, NJ, USA, 2013; pp. 455–482.
24. Steinkraus, K.A.; Kaeberlein, M.; Kennedy, B.K. Replicative aging in yeast: The means to the end. *Annu. Rev. Cell Dev. Biol.* **2008**, *24*, 29–54. [CrossRef] [PubMed]
25. Rockenfeller, P.; Madeo, F. Apoptotic death of ageing yeast. *Exp. Gerontol.* **2008**, *43*, 876–881. [CrossRef] [PubMed]
26. Fabrizio, P.; Longo, V.D. The chronological life span of *Saccharomyces cerevisiae*. *Methods Mol. Biol.* **2007**, *371*, 89–95. [CrossRef] [PubMed]
27. Bitterman, K.J.; Medvedik, O.; Sinclair, D.A. Longevity regulation in *Saccharomyces cerevisiae*: Linking metabolism, genome stability, and heterochromatin. *Microbiol. Mol. Biol. Rev.* **2003**, *67*, 376–399. [CrossRef]
28. Pringle, J.R. Staining of bud scars and other cell wall chitin with calcofluor. *Methods Enzymol.* **1991**, *194*, 732–735. [CrossRef] [PubMed]
29. Klinger, H.; Rinnerthaler, M.; Lam, Y.T.; Laun, P.; Heeren, G.; Klocker, A.; Simon-Nobbe, B.; Dickinson, J.R.; Dawes, I.W.; Breitenbach, M. Quantitation of (a)symmetric inheritance of functional and of oxidatively damaged mitochondrial aconitase in the cell division of old yeast mother cells. *Exp. Gerontol.* **2010**, *45*, 533–542. [CrossRef]
30. Hayflick, L.; Moorhead, P.S. The serial cultivation of human diploid cell strains. *Exp. Cell Res.* **1961**, *25*, 585–621. [CrossRef]
31. Sinclair, D.A. Studying the Replicative Life Span of Yeast Cells. In *Biological Aging: Methods and Protocols*, 2nd ed.; Springer: Berlin/Heidelberg, Germany, 2013; Volume 1048, pp. 49–63. [CrossRef]
32. Lin, S.J.; Defossez, P.A.; Guarente, L. Requirement of NAD and SIR2 for life-span extension by calorie restriction in *Saccharomyces cerevisiae*. *Science* **2000**, *289*, 2126–2128. [CrossRef]
33. Goldberg, A.A.; Bourque, S.D.; Kyryakov, P.; Boukh-Viner, T.; Gregg, C.; Beach, A.; Burstein, M.T.; Machkalyan, G.; Richard, V.; Rampersad, S.; et al. A novel function of lipid droplets in regulating longevity. *Biochem. Soc. Trans.* **2009**, *37*, 1050–1055. [CrossRef]
34. Hiltunen, J.K.; Mursula, A.M.; Rottensteiner, H.; Wierenga, R.K.; Kastaniotis, A.J.; Gurvitz, A. The biochemistry of peroxisomal beta-oxidation in the yeast *Saccharomyces cerevisiae*. *FEMS Microbiol. Rev.* **2003**, *27*, 35–64. [CrossRef] [PubMed]
35. Bischof, J.; Salzmann, M.; Streubel, M.K.; Hasek, J.; Geltinger, F.; Duschl, J.; Bresgen, N.; Briza, P.; Haskova, D.; Lejskova, R.; et al. Clearing the outer mitochondrial membrane from harmful proteins via lipid droplets. *Cell Death Discov.* **2017**, *3*, 17016. [CrossRef] [PubMed]

36. Beas, A.O.; Gordon, P.B.; Prentiss, C.L.; Olsen, C.P.; Kukurugya, M.A.; Bennett, B.D.; Parkhurst, S.M.; Gottschling, D.E. Independent regulation of age associated fat accumulation and longevity. *Nat. Commun.* **2020**, *11*, 2790. [CrossRef] [PubMed]
37. Martínez, G.; Duran-Aniotz, C.; Cabral-Miranda, F.; Vivar, J.P.; Hetz, C. Endoplasmic reticulum proteostasis impairment in aging. *Aging Cell* **2017**, *16*, 615–623. [CrossRef]
38. Cui, H.-J.; Liu, X.-G.; McCormick, M.; Wasko, B.M.; Zhao, W.; He, X.; Yuan, Y.; Fang, B.-X.; Sun, X.-R.; Kennedy, B.K.; et al. PMT1 deficiency enhances basal UPR activity and extends replicative lifespan of *Saccharomyces cerevisiae*. *Age* **2015**, *37*, 46. [CrossRef]
39. Walter, P.; Ron, D. The unfolded protein response: From stress pathway to homeostatic regulation. *Science* **2011**, *334*, 1081–1086. [CrossRef]
40. Kopito, R.R.; Sitia, R. Aggresomes and Russell bodies. Symptoms of cellular indigestion? *EMBO Rep.* **2000**, *1*, 225–231. [CrossRef]
41. Celik, C.; Lee, S.Y.T.; Yap, W.S.; Thibault, G. Endoplasmic reticulum stress and lipids in health and diseases. *Prog. Lipid Res.* **2022**, *89*, 101198. [CrossRef]
42. Halbleib, K.; Pesek, K.; Covino, R.; Hofbauer, H.F.; Wunnicke, D.; Hanelt, I.; Hummer, G.; Ernst, R. Activation of the Unfolded Protein Response by Lipid Bilayer Stress. *Mol. Cell* **2017**, *67*, 673–684.e678. [CrossRef]
43. Rubio, C.; Pincus, D.; Korennykh, A.; Schuck, S.; El-Samad, H.; Walter, P. Homeostatic adaptation to endoplasmic reticulum stress depends on Ire1 kinase activity. *J. Cell Biol.* **2011**, *193*, 171–184. [CrossRef] [PubMed]
44. Lee, A.H.; Scapa, E.F.; Cohen, D.E.; Glimcher, L.H. Regulation of hepatic lipogenesis by the transcription factor XBP1. *Science* **2008**, *320*, 1492–1496. [CrossRef] [PubMed]
45. Luo, W.; Wang, H.; Ren, L.; Lu, Z.; Zheng, Q.; Ding, L.; Xie, H.; Wang, R.; Yu, C.; Lin, Y.; et al. Adding fuel to the fire: The lipid droplet and its associated proteins in cancer progression. *Int. J. Biol. Sci.* **2022**, *18*, 6020–6034. [CrossRef]
46. Liu, L.; Zhang, K.; Sandoval, H.; Yamamoto, S.; Jaiswal, M.; Sanz, E.; Li, Z.; Hui, J.; Graham, B.H.; Quintana, A.; et al. Glial lipid droplets and ROS induced by mitochondrial defects promote neurodegeneration. *Cell* **2015**, *160*, 177–190. [CrossRef] [PubMed]
47. Walther, T.C.; Chung, J.; Farese, R.V., Jr. Lipid Droplet Biogenesis. *Annu. Rev. Cell Dev. Biol.* **2017**, *33*, 491–510. [CrossRef]
48. Harman, D. Aging: A Theory Based on Free Radical and Radiation Chemistry. *J. Gerontol.* **1956**, *11*, 298–300. [CrossRef] [PubMed]
49. Cohen, A.; Weindling, E.; Rabinovich, E.; Nachman, I.; Fuchs, S.; Chuartzman, S.; Gal, L.; Schuldiner, M.; Bar-Nun, S. Water-Transfer Slows Aging in *Saccharomyces cerevisiae*. *PLoS ONE* **2016**, *11*, e0148650. [CrossRef] [PubMed]
50. Gao, Q.; Binns, D.D.; Kinch, L.N.; Grishin, N.V.; Ortiz, N.; Chen, X.; Goodman, J.M. Pet10p is a yeast perilipin that stabilizes lipid droplets and promotes their assembly. *J. Cell Biol.* **2017**, *216*, 3199–3217. [CrossRef]
51. Di Gregorio, S.E.; Duennwald, M.L. Yeast as a model to study protein misfolding in aged cells. *FEMS Yeast Res.* **2018**, *18*, foy054. [CrossRef]
52. Currie, E.; Guo, X.; Christiano, R.; Chitraju, C.; Kory, N.; Harrison, K.; Haas, J.; Walther, T.C.; Farese, R.V. High confidence proteomic analysis of yeast LDs identifies additional droplet proteins and reveals connections to dolichol synthesis and sterol acetylation. *J. Lipid Res.* **2014**, *55*, 1465–1477. [CrossRef]
53. Grillitsch, K.; Connerth, M.; Köfeler, H.; Arrey, T.N.; Rietschel, B.; Wagner, B.; Karas, M.; Daum, G. Lipid particles/droplets of the yeast *Saccharomyces cerevisiae* revisited: Lipidome meets proteome. *Biochim. Biophys. Acta* **2011**, *1811*, 1165–1176. [CrossRef] [PubMed]
54. Wang, C.-W.; Lee, S.-C. The ubiquitin-like (UBX)-domain-containing protein Ubx2/Ubxd8 regulates lipid droplet homeostasis. *J. Cell Sci.* **2012**, *125*, 2930–2939. [CrossRef] [PubMed]
55. Neuber, O.; Jarosch, E.; Volkwein, C.; Walter, J.; Sommer, T. Ubx2 links the Cdc48 complex to ER-associated protein degradation. *Nat. Cell Biol.* **2005**, *7*, 993–998. [CrossRef] [PubMed]
56. Laun, P.; Büttner, S.; Rinnerthaler, M.; Burhans, W.C.; Breitenbach, M. Yeast Aging and Apoptosis. In *Aging Research in Yeast*; Breitenbach, M., Jazwinski, S.M., Laun, P., Eds.; Springer: Dordrecht, The Netherlands, 2012; pp. 207–232.
57. Côrte-Real, M.; Madeo, F. Yeast Programed Cell Death and Aging. *Front. Oncol.* **2013**, *3*, 283. [CrossRef] [PubMed]
58. Tower, J. Programmed cell death in aging. *Ageing Res. Rev.* **2015**, *23*, 90–100. [CrossRef]
59. Chipuk, J.E.; Bouchier-Hayes, L.; Green, D.R. Mitochondrial outer membrane permeabilization during apoptosis: The innocent bystander scenario. *Cell Death Differ.* **2006**, *13*, 1396–1402. [CrossRef]
60. Dadsena, S.; King, L.E.; García-Sáez, A.J. Apoptosis regulation at the mitochondria membrane level. *Biochim. Biophys. Acta Biomembr.* **2021**, *1863*, 183716. [CrossRef]
61. Subburaj, Y.; Cosentino, K.; Axmann, M.; Pedrueza-Villalmanzo, E.; Hermann, E.; Bleicken, S.; Spatz, J.; García-Sáez, A.J. Bax monomers form dimer units in the membrane that further self-assemble into multiple oligomeric species. *Nat. Commun.* **2015**, *6*, 8042. [CrossRef]
62. Rinnerthaler, M.; Lejskova, R.; Grousl, T.; Stradalova, V.; Heeren, G.; Richter, K.; Breitenbach-Koller, L.; Malinsky, J.; Hasek, J.; Breitenbach, M. Mmi1, the yeast homologue of mammalian TCTP, associates with stress granules in heat-shocked cells and modulates proteasome activity. *PLoS ONE* **2013**, *8*, e77791. [CrossRef]
63. Rinnerthaler, M.; Jarolim, S.; Heeren, G.; Palle, E.; Perju, S.; Klinger, H.; Bogengruber, E.; Madeo, F.; Braun, R.J.; Breitenbach-Koller, L.; et al. MMI1 (YKL056c, TMA19), the yeast orthologue of the translationally controlled tumor protein (TCTP) has apoptotic functions and interacts with both microtubules and mitochondria. *Biochim. Biophys. Acta* **2006**, *1757*, 631–638. [CrossRef]
64. Ding, W.X.; Yin, X.M. Mitophagy: Mechanisms, pathophysiological roles, and analysis. *Biol. Chem.* **2012**, *393*, 547–564. [CrossRef]

65. Kissová, I.; Deffieu, M.; Manon, S.; Camougrand, N. Uth1p is involved in the autophagic degradation of mitochondria. *J. Biol. Chem.* **2004**, *279*, 39068–39074. [CrossRef]

66. Kim, E.H.; Choi, K.S. A critical role of superoxide anion in selenite-induced mitophagic cell death. *Autophagy* **2008**, *4*, 76–78. [CrossRef] [PubMed]

67. Chu, C.T.; Zhu, J.; Dagda, R. Beclin 1-independent pathway of damage-induced mitophagy and autophagic stress: Implications for neurodegeneration and cell death. *Autophagy* **2007**, *3*, 663–666. [CrossRef] [PubMed]

68. Kim, I.; Rodriguez-Enriquez, S.; Lemasters, J.J. Selective degradation of mitochondria by mitophagy. *Arch. Biochem. Biophys.* **2007**, *462*, 245–253. [CrossRef] [PubMed]

69. Nowikovsky, K.; Reipert, S.; Devenish, R.J.; Schweyen, R.J. Mdm38 protein depletion causes loss of mitochondrial $K^+/H+$ exchange activity, osmotic swelling and mitophagy. *Cell Death Differ.* **2007**, *14*, 1647–1656. [CrossRef]

70. Terman, A.; Kurz, T.; Navratil, M.; Arriaga, E.A.; Brunk, U.T. Mitochondrial turnover and aging of long-lived postmitotic cells: The mitochondrial-lysosomal axis theory of aging. *Antioxid. Redox Signal.* **2010**, *12*, 503–535. [CrossRef]

71. Bergamini, E. Autophagy: A cell repair mechanism that retards ageing and age-associated diseases and can be intensified pharmacologically. *Mol. Asp. Med.* **2006**, *27*, 403–410. [CrossRef]

72. Stevens, M.; Oltean, S. Modulation of the Apoptosis Gene Bcl-x Function Through Alternative Splicing. *Front. Genet.* **2019**, *10*, 804. [CrossRef]

73. Kanagavijayan, D.; Rajasekharan, R.; Srinivasan, M. Yeast MRX deletions have short chronological life span and more triacylglycerols. *Fems Yeast Res.* **2016**, *16*, fov109. [CrossRef]

74. López-Otín, C.; Blasco, M.A.; Partridge, L.; Serrano, M.; Kroemer, G. The hallmarks of aging. *Cell* **2013**, *153*, 1194–1217. [CrossRef]

75. Aung-Htut, M.T.; Lam, Y.T.; Lim, Y.L.; Rinnerthaler, M.; Gelling, C.L.; Yang, H.; Breitenbach, M.; Dawes, I.W. Maintenance of mitochondrial morphology by autophagy and its role in high glucose effects on chronological lifespan of *Saccharomyces cerevisiae*. *Oxidative Med. Cell. Longev.* **2013**, *2013*, 636287. [CrossRef]

76. Sorger, D.; Athenstaedt, K.; Hrastnik, C.; Daum, G. A yeast strain lacking lipid particles bears a defect in ergosterol formation. *J. Biol. Chem.* **2004**, *279*, 31190–31196. [CrossRef] [PubMed]

77. Valachovic, M.; Hronska, L.; Hapala, I. Anaerobiosis induces complex changes in sterol esterification pattern in the yeast *Saccharomyces cerevisiae*. *FEMS Microbiol. Lett.* **2001**, *197*, 41–45. [CrossRef] [PubMed]

78. Leber, R.; Zinser, E.; Zellnig, G.; Paltauf, F.; Daum, G. Characterization of lipid particles of the yeast, *Saccharomyces cerevisiae*. *Yeast* **1994**, *10*, 1421–1428. [CrossRef]

79. Ruan, L.H.; Zhou, C.K.; Jin, E.L.; Kucharavy, A.; Zhang, Y.; Wen, Z.H.; Florens, L.; Li, R. Cytosolic proteostasis through importing of misfolded proteins into mitochondria. *Nature* **2017**, *543*, 443–446. [CrossRef] [PubMed]

80. Erjavec, N.; Bayot, A.; Gareil, M.; Camougrand, N.; Nystrom, T.; Friguet, B.; Bulteau, A.L. Deletion of the mitochondrial Pim1/Lon protease in yeast results in accelerated aging and impairment of the proteasome. *Free Radic. Biol. Med.* **2013**, *56*, 9–16. [CrossRef] [PubMed]

81. Lin, S.J.; Kaeberlein, M.; Andalis, A.A.; Sturtz, L.A.; Defossez, P.A.; Culotta, V.C.; Fink, G.R.; Guarente, L. Calorie restriction extends *Saccharomyces cerevisiae* lifespan by increasing respiration. *Nature* **2002**, *418*, 344–348. [CrossRef] [PubMed]

82. Powers, R.W., 3rd; Kaeberlein, M.; Caldwell, S.D.; Kennedy, B.K.; Fields, S. Extension of chronological life span in yeast by decreased TOR pathway signaling. *Genes Dev.* **2006**, *20*, 174–184. [CrossRef]

83. Howitz, K.T.; Bitterman, K.J.; Cohen, H.Y.; Lamming, D.W.; Lavu, S.; Wood, J.G.; Zipkin, R.E.; Chung, P.; Kisielewski, A.; Zhang, L.L.; et al. Small molecule activators of sirtuins extend *Saccharomyces cerevisiae* lifespan. *Nature* **2003**, *425*, 191–196. [CrossRef] [PubMed]

84. Madeira, J.B.; Masuda, C.A.; Maya-Monteiro, C.M.; Matos, G.S.; Montero-Lomeli, M.; Bozaquel-Morais, B.L. TORC1 Inhibition Induces Lipid Droplet Replenishment in Yeast. *Mol. Cell. Biol.* **2015**, *35*, 737–746. [CrossRef] [PubMed]

85. Vall-llaura, N.; Mir, N.; Garrido, L.; Vived, C.; Cabiscol, E. Redox control of yeast Sir2 activity is involved in acetic acid resistance and longevity. *Redox. Biol.* **2019**, *24*, 101229. [CrossRef] [PubMed]

86. Boender, L.G.; Almering, M.J.; Dijk, M.; van Maris, A.J.; de Winde, J.H.; Pronk, J.T.; Daran-Lapujade, P. Extreme calorie restriction and energy source starvation in *Saccharomyces cerevisiae* represent distinct physiological states. *Biochim. Biophys. Acta* **2011**, *1813*, 2133–2144. [CrossRef]

87. Schurmanns, L.; Hamann, A.; Osiewacz, H.D. Lifespan Increase of Podospora anserina by Oleic Acid Is Linked to Alterations in Energy Metabolism, Membrane Trafficking and Autophagy. *Cells* **2022**, *11*, 519. [CrossRef] [PubMed]

88. Friedman, D.B.; Johnson, T.E. A mutation in the age-1 gene in *Caenorhabditis elegans* lengthens life and reduces hermaphrodite fertility. *Genetics* **1988**, *118*, 75–86. [CrossRef]

89. Iser, W.B.; Wolkow, C.A. DAF-2/insulin-like signaling in *C. elegans* modifies effects of dietary restriction and nutrient stress on aging, stress and growth. *PLoS ONE* **2007**, *2*, e1240. [CrossRef]

90. Jia, K.; Chen, D.; Riddle, D.L. The TOR pathway interacts with the insulin signaling pathway to regulate *C. elegans* larval development, metabolism and life span. *Development* **2004**, *131*, 3897–3906. [CrossRef]

91. Johnson, S.C.; Rabinovitch, P.S.; Kaeberlein, M. mTOR is a key modulator of ageing and age-related disease. *Nature* **2013**, *493*, 338–345. [CrossRef]

92. Lakowski, B.; Hekimi, S. The genetics of caloric restriction in *Caenorhabditis elegans*. *Proc. Natl. Acad. Sci. USA* **1998**, *95*, 13091–13096. [CrossRef]

93. Luo, S.; Kleemann, G.A.; Ashraf, J.M.; Shaw, W.M.; Murphy, C.T. TGF-beta and insulin signaling regulate reproductive aging via oocyte and germline quality maintenance. *Cell* **2010**, *143*, 299–312. [CrossRef]

94. Greer, E.L.; Dowlatshahi, D.; Banko, M.R.; Villen, J.; Hoang, K.; Blanchard, D.; Gygi, S.P.; Brunet, A. An AMPK-FOXO pathway mediates longevity induced by a novel method of dietary restriction in *C. elegans*. *Curr. Biol.* **2007**, *17*, 1646–1656. [CrossRef]

95. Leiser, S.F.; Kaeberlein, M. The hypoxia-inducible factor HIF-1 functions as both a positive and negative modulator of aging. *Biol. Chem.* **2010**, *391*, 1131–1137. [CrossRef]

96. Golden, J.W.; Riddle, D.L. The *Caenorhabditis elegans* dauer larva: Developmental effects of pheromone, food, and temperature. *Dev. Biol.* **1984**, *102*, 368–378. [CrossRef]

97. Klass, M.; Hirsh, D. Non-ageing developmental variant of *Caenorhabditis elegans*. *Nature* **1976**, *260*, 523–525. [CrossRef]

98. Zhang, S.; Li, F.; Zhou, T.; Wang, G.; Li, Z. *Caenorhabditis elegans* as a Useful Model for Studying Aging Mutations. *Front. Endocrinol.* **2020**, *11*, 554994. [CrossRef]

99. Ewald, C.Y.; Castillo-Quan, J.I.; Blackwell, T.K. Untangling Longevity, Dauer, and Healthspan in *Caenorhabditis elegans* Insulin/IGF-1-Signalling. *Gerontology* **2018**, *64*, 96–104. [CrossRef]

100. Zhang, S.O.; Trimble, R.; Guo, F.; Mak, H.Y. Lipid droplets as ubiquitous fat storage organelles in *C. elegans*. *BMC Cell Biol.* **2010**, *11*, 96. [CrossRef]

101. Wang, P.; Liu, B.; Zhang, D.; Belew, M.Y.; Tissenbaum, H.A.; Cheng, J.X. Imaging lipid metabolism in live *Caenorhabditis elegans* using fingerprint vibrations. *Angew. Chem. Int. Ed.* **2014**, *53*, 11787–11792. [CrossRef] [PubMed]

102. Srinivasan, S. Regulation of body fat in *Caenorhabditis elegans*. *Annu. Rev. Physiol.* **2015**, *77*, 161–178. [CrossRef] [PubMed]

103. Mak, H.Y. Lipid droplets as fat storage organelles in *Caenorhabditis elegans*: Thematic Review Series: Lipid Droplet Synthesis and Metabolism: From Yeast to Man. *J. Lipid Res.* **2012**, *53*, 28–33. [CrossRef] [PubMed]

104. McCay, C.M.; Crowell, M.F.; Maynard, L.A. The effect of retarded growth upon the length of life span and upon the ultimate body size. *J. Nutr.* **1935**, *10*, 63–79. [CrossRef]

105. Partridge, L.; Piper, M.D.; Mair, W. Dietary restriction in *Drosophila*. *Mech. Ageing Dev.* **2005**, *126*, 938–950. [CrossRef] [PubMed]

106. Mattison, J.A.; Colman, R.J.; Beasley, T.M.; Allison, D.B.; Kemnitz, J.W.; Roth, G.S.; Ingram, D.K.; Weindruch, R.; de Cabo, R.; Anderson, R.M. Caloric restriction improves health and survival of rhesus monkeys. *Nat. Commun.* **2017**, *8*, 14063. [CrossRef] [PubMed]

107. McGhee, J.D. The *C. elegans* intestine. In *WormBook: The Online Review of C. elegans Biology*; WormBook: Pasadena, CA, USA, 2007.

108. Klass, M.R. Aging in Nematode *Caenorhabditis-elegans*—Major Biological and Environmental-Factors Influencing Life-Span. *Mech. Ageing Dev.* **1977**, *6*, 413–429. [CrossRef]

109. Hosono, R.; Nishimoto, S.; Kuno, S. Alterations of life span in the nematode *Caenorhabditis elegans* under monoxenic culture conditions. *Exp. Gerontol.* **1989**, *24*, 251–264. [CrossRef] [PubMed]

110. Mullaney, B.C.; Ashrafi, K. *C. elegans* fat storage and metabolic regulation. *Biochim. Biophys. Acta* **2009**, *1791*, 474–478. [CrossRef]

111. Wu, Z.; Isik, M.; Moroz, N.; Steinbaugh, M.J.; Zhang, P.; Blackwell, T.K. Dietary Restriction Extends Lifespan through Metabolic Regulation of Innate Immunity. *Cell Metab.* **2019**, *29*, 1192–1205.e1198. [CrossRef] [PubMed]

112. Houthoofd, K.; Gems, D.; Johnson, T.E.; Vanfleteren, J.R. Dietary restriction in the nematode *Caenorhabditis elegans*. *Interdiscip. Top. Gerontol.* **2007**, *35*, 98–114. [CrossRef]

113. Walker, G.; Houthoofd, K.; Vanfleteren, J.R.; Gems, D. Dietary restriction in *C. elegans*: From rate-of-living effects to nutrient sensing pathways. *Mech. Ageing Dev.* **2005**, *126*, 929–937. [CrossRef]

114. McElwee, J.; Bubb, K.; Thomas, J.H. Transcriptional outputs of the *Caenorhabditis elegans* forkhead protein DAF-16. *Aging Cell* **2003**, *2*, 111–121. [CrossRef]

115. Sun, X.; Chen, W.D.; Wang, Y.D. DAF-16/FOXO Transcription Factor in Aging and Longevity. *Front. Pharmacol.* **2017**, *8*, 548. [CrossRef] [PubMed]

116. Baumeister, R.; Schaffitzel, E.; Hertweck, M. Endocrine signaling in *Caenorhabditis elegans* controls stress response and longevity. *J. Endocrinol.* **2006**, *190*, 191–202. [CrossRef] [PubMed]

117. Kenyon, C.; Chang, J.; Gensch, E.; Rudner, A.; Tabtiang, R. A *C. elegans* Mutant That Lives Twice as Long as Wild-Type. *Nature* **1993**, *366*, 461–464. [CrossRef]

118. Paradis, S.; Ailion, M.; Toker, A.; Thomas, J.H.; Ruvkun, G. A PDK1 homolog is necessary and sufficient to transduce AGE-1 PI3 kinase signals that regulate diapause in *Caenorhabditis elegans*. *Gene Dev.* **1999**, *13*, 1438–1452. [CrossRef]

119. Ogg, S.; Paradis, S.; Gottlieb, S.; Patterson, G.I.; Lee, L.; Tissenbaum, H.A.; Ruvkun, G. The Fork head transcription factor DAF-16 transduces insulin-like metabolic and longevity signals in *C. elegans*. *Nature* **1997**, *389*, 994–999. [CrossRef]

120. Zecic, A.; Braeckman, B.P. DAF-16/FoxO in *Caenorhabditis elegans* and Its Role in Metabolic Remodeling. *Cells* **2020**, *9*, 109. [CrossRef]

121. Blackwell, T.K.; Sewell, A.K.; Wu, Z.; Han, M. TOR Signaling in *Caenorhabditis elegans* Development, Metabolism, and Aging. *Genetics* **2019**, *213*, 329–360. [CrossRef] [PubMed]

122. Jung, C.H.; Ro, S.H.; Cao, J.; Otto, N.M.; Kim, D.H. mTOR regulation of autophagy. *FEBS Lett.* **2010**, *584*, 1287–1295. [CrossRef]

123. Inoki, K.; Li, Y.; Zhu, T.Q.; Wu, J.; Guan, K.L. TSC2 is phosphorylated and inhibited by Akt and suppresses mTOR signalling. *Nat. Cell Biol.* **2002**, *4*, 648–657. [CrossRef]

124. Dibble, C.C.; Cantley, L.C. Regulation of mTORC1 by PI3K signaling. *Trends Cell Biol.* **2015**, *25*, 545–555. [CrossRef]

125. Lapierre, L.R.; Hansen, M. Lessons from *C. elegans*: Signaling pathways for longevity. *Trends Endocrinol. Metab.* **2012**, *23*, 637–644. [CrossRef] [PubMed]

126. Gingras, A.C.; Gygi, S.P.; Raught, B.; Polakiewicz, R.D.; Abraham, R.T.; Hoekstra, M.F.; Aebersold, R.; Sonenberg, N. Regulation of 4E-BP1 phosphorylation: A novel two-step mechanism. *Genes Dev.* **1999**, *13*, 1422–1437. [CrossRef] [PubMed]

127. Powers, T. TOR signaling and S6 kinase 1: Yeast catches up. *Cell Metab.* **2007**, *6*, 1–2. [CrossRef]

128. Vellai, T.; Takacs-Vellai, K.; Zhang, Y.; Kovacs, A.L.; Orosz, L.; Muller, F. Genetics: Influence of TOR kinase on lifespan in *C. elegans*. *Nature* **2003**, *426*, 620. [CrossRef] [PubMed]

129. Robida-Stubbs, S.; Glover-Cutter, K.; Lamming, D.W.; Mizunuma, M.; Narasimhan, S.D.; Neumann-Haefelin, E.; Sabatini, D.M.; Blackwell, T.K. TOR Signaling and Rapamycin Influence Longevity by Regulating SKN-1/Nrf and DAF-16/FoxO. *Cell Metab.* **2012**, *15*, 713–724. [CrossRef]

130. Honjoh, S.; Yamamoto, T.; Uno, M.; Nishida, E. Signalling through RHEB-1 mediates intermittent fasting-induced longevity in *C. elegans*. *Nature* **2009**, *457*, 726–730. [CrossRef]

131. Hansen, M.; Taubert, S.; Crawford, D.; Libina, N.; Lee, S.J.; Kenyon, C. Lifespan extension by conditions that inhibit translation in *Caenorhabditis elegans*. *Aging Cell* **2007**, *6*, 95–110. [CrossRef]

132. Jones, K.T.; Greer, E.R.; Pearce, D.; Ashrafi, K. Rictor/TORC2 regulates *Caenorhabditis elegans* fat storage, body size, and development through sgk-1. *PLoS Biol.* **2009**, *7*, e60. [CrossRef]

133. Soukas, A.A.; Kane, E.A.; Carr, C.E.; Melo, J.A.; Ruvkun, G. Rictor/TORC2 regulates fat metabolism, feeding, growth, and life span in *Caenorhabditis elegans*. *Genes Dev.* **2009**, *23*, 496–511. [CrossRef]

134. Hara, K.; Maruki, Y.; Long, X.; Yoshino, K.; Oshiro, N.; Hidayat, S.; Tokunaga, C.; Avruch, J.; Yonezawa, K. Raptor, a binding partner of target of rapamycin (TOR), mediates TOR action. *Cell* **2002**, *110*, 177–189. [CrossRef]

135. Long, L.; Liu, W.; Ruan, P.; Yang, X.; Chen, X.; Li, L.; Yuan, F.; He, D.; Huang, P.; Gong, A.; et al. Visualizing the Interplay of Lipid Droplets and Protein Aggregates During Aging via a Dual-Functional Fluorescent Probe. *Anal. Chem.* **2022**, *94*, 2803–2811. [CrossRef] [PubMed]

136. Fletcher, M.; Kim, D.H. Age-Dependent Neuroendocrine Signaling from Sensory Neurons Modulates the Effect of Dietary Restriction on Longevity of *Caenorhabditis elegans*. *PLoS Genet.* **2017**, *13*, e1006544. [CrossRef] [PubMed]

137. Gumienny, T.L.; Savage-Dunn, C. *TGF-Beta Signaling in C. elegans*; WormBook: Pasadena, CA, USA, 2013; pp. 1–34. [CrossRef]

138. Liu, T.; Zimmerman, K.K.; Patterson, G.I. Regulation of signaling genes by TGFbeta during entry into dauer diapause in *C. elegans*. *BMC Dev. Biol.* **2004**, *4*, 11. [CrossRef] [PubMed]

139. Hu, M.H.; Crossman, D.; Prasain, J.K.; Miller, M.A.; Serra, R.A. Transcriptomic Profiling of DAF-7/TGF beta Pathway Mutants in *C. elegans*. *Genes* **2020**, *11*, 288. [CrossRef]

140. Lant, B.; Storey, K.B. An Overview of Stress Response and Hypometabolic Strategies in *Caenorhabditis elegans*: Conserved and Contrasting Signals with the Mammalian System. *Int. J. Biol. Sci.* **2010**, *6*, 9–50. [CrossRef]

141. Shaw, W.M.; Luo, S.; Landis, J.; Ashraf, J.; Murphy, C.T. The *C. elegans* TGF-beta Dauer pathway regulates longevity via insulin signaling. *Curr. Biol.* **2007**, *17*, 1635–1645. [CrossRef]

142. Greer, E.R.; Perez, C.L.; Van Gilst, M.R.; Lee, B.H.; Ashrafi, K. Neural and molecular dissection of a *C. elegans* sensory circuit that regulates fat and feeding. *Cell Metab.* **2008**, *8*, 118–131. [CrossRef]

143. Rashid, S.; Pho, K.B.; Mesbahi, H.; MacNeil, L.T. Nutrient Sensing and Response Drive Developmental Progression in *Caenorhabditis elegans*. *Bioessays* **2020**, *42*, e1900194. [CrossRef]

144. Kumar, A.V.; Mills, J.; Parker, W.M.; Leitão, J.A.; Rodriguez, D.I.; Ng, C.; Patel, R.; Aguilera, J.L.; Johnson, J.R.; Wong, S.Q.; et al. Lipid droplets modulate proteostasis, SQST-1/SQSTM1 dynamics, and lifespan in *C. elegans*. *bioRxiv* **2022**, 2021.04.22.440991. [CrossRef]

145. Castillo-Quan, J.I.; Steinbaugh, M.J.; Fernández-Cárdenas, L.P.; Pohl, N.K.; Wu, Z.; Zhu, F.; Moroz, N.; Teixeira, V.; Bland, M.S.; Lehrbach, N.J.; et al. An anti-steatosis response regulated by oleic acid through lipid droplet-mediated ERAD enhancement. *bioRxiv* **2022**, 2022.06.15.496302. [CrossRef]

146. Na, H.; Zhang, P.; Chen, Y.; Zhu, X.; Liu, Y.; Liu, Y.; Xie, K.; Xu, N.; Yang, F.; Yu, Y.; et al. Identification of lipid droplet structure-like/resident proteins in *Caenorhabditis elegans*. *Biochim. Biophys. Acta (BBA)—Mol. Cell Res.* **2015**, *1853*, 2481–2491. [CrossRef]

147. Chughtai, A.A.; Kaššák, F.; Kostrouchová, M.; Novotný, J.P.; Krause, M.W.; Saudek, V.; Kostrouch, Z.; Kostrouchová, M. Perilipin-related protein regulates lipid metabolism in *C. elegans*. *PeerJ* **2015**, *3*, e1213. [CrossRef] [PubMed]

148. Choudhary, V.; Ojha, N.; Golden, A.; Prinz, W.A. A conserved family of proteins facilitates nascent lipid droplet budding from the ER. *J. Cell Biol.* **2015**, *211*, 261–271. [CrossRef] [PubMed]

149. Papsdorf, K.; Miklas, J.W.; Hosseini, A.; Cabruja, M.; Morrow, C.S.; Savini, M.; Yu, Y.; Silva-García, C.G.; Haseley, N.R.; Murphy, L.M.; et al. Lipid droplets and peroxisomes are co-regulated to drive lifespan extension in response to mono-unsaturated fatty acids. *Nat. Cell Biol.* **2023**, *25*, 672–684. [CrossRef] [PubMed]

150. Mosquera, J.V.; Bacher, M.C.; Priess, J.R. Nuclear lipid droplets and nuclear damage in *Caenorhabditis elegans*. *PLoS Genet.* **2021**, *17*, e1009602. [CrossRef]

151. He, Y.; Jasper, H. Studying aging in *Drosophila*. *Methods* **2014**, *68*, 129–133. [CrossRef]

152. Piper, M.D.W.; Partridge, L. *Drosophila* as a model for ageing. *Biochim. Biophys. Acta Mol. Basis Dis.* **2018**, *1864*, 2707–2717. [CrossRef]

153. Proshkina, E.N.; Shaposhnikov, M.V.; Sadritdinova, A.F.; Kudryavtseva, A.V.; Moskalev, A.A. Basic mechanisms of longevity: A case study of *Drosophila* pro-longevity genes. *Ageing Res. Rev.* **2015**, *24*, 218–231. [CrossRef]

154. Liao, S.; Amcoff, M.; Nässel, D.R. Impact of high-fat diet on lifespan, metabolism, fecundity and behavioral senescence in *Drosophila*. *Insect Biochem. Mol. Biol.* **2021**, *133*, 103495. [CrossRef]

155. Hofbauer, H.F.; Heier, C.; Sen Saji, A.K.; Kühnlein, R.P. Lipidome remodeling in aging normal and genetically obese *Drosophila* males. *Insect Biochem. Mol. Biol.* **2021**, *133*, 103498. [CrossRef]

156. Buszczak, M.; Lu, X.; Segraves, W.A.; Chang, T.Y.; Cooley, L. Mutations in the midway gene disrupt a *Drosophila* acyl coenzyme A: Diacylglycerol acyltransferase. *Genetics* **2002**, *160*, 1511–1518. [CrossRef]

157. Gronke, S.; Mildner, A.; Fellert, S.; Tennagels, N.; Petry, S.; Muller, G.; Jackle, H.; Kuhnlein, R.P. Brummer lipase is an evolutionary conserved fat storage regulator in *Drosophila*. *Cell Metab.* **2005**, *1*, 323–330. [CrossRef] [PubMed]

158. Song, Y.; Park, J.O.; Tanner, L.; Nagano, Y.; Rabinowitz, J.D.; Shvartsman, S.Y. Energy budget of *Drosophila* embryogenesis. *Curr. Biol.* **2019**, *29*, R566–R567. [CrossRef] [PubMed]

159. Tennessen, J.M.; Barry, W.E.; Cox, J.; Thummel, C.S. Methods for studying metabolism in *Drosophila*. *Methods* **2014**, *68*, 105–115. [CrossRef] [PubMed]

160. Bickel, P.E.; Tansey, J.T.; Welte, M.A. PAT proteins, an ancient family of lipid droplet proteins that regulate cellular lipid stores. *Biochim. Biophys. Acta* **2009**, *1791*, 419–440. [CrossRef]

161. Miura, S.; Gan, J.W.; Brzostowski, J.; Parisi, M.J.; Schultz, C.J.; Londos, C.; Oliver, B.; Kimmel, A.R. Functional conservation for lipid storage droplet association among Perilipin, ADRP, and TIP47 (PAT)-related proteins in mammals, *Drosophila*, and Dictyostelium. *J. Biol. Chem.* **2002**, *277*, 32253–32257. [CrossRef]

162. Teixeira, L.; Rabouille, C.; Rørth, P.; Ephrussi, A.; Vanzo, N.F. *Drosophila* Perilipin/ADRP homologue Lsd2 regulates lipid metabolism. *Mech. Dev.* **2003**, *120*, 1071–1081. [CrossRef]

163. Beller, M.; Bulankina, A.V.; Hsiao, H.H.; Urlaub, H.; Jackle, H.; Kuhnlein, R.P. PERILIPIN-dependent control of lipid droplet structure and fat storage in *Drosophila*. *Cell Metab.* **2010**, *12*, 521–532. [CrossRef]

164. Grönke, S.; Beller, M.; Fellert, S.; Ramakrishnan, H.; Jäckle, H.; Kühnlein, R.P. Control of fat storage by a *Drosophila* PAT domain protein. *Curr. Biol.* **2003**, *13*, 603–606. [CrossRef]

165. Kayukawa, T.; Jouraku, A.; Ito, Y.; Shinoda, T. Molecular mechanism underlying juvenile hormone-mediated repression of precocious larval-adult metamorphosis. *Proc. Natl. Acad. Sci. USA* **2017**, *114*, 1057–1062. [CrossRef]

166. Mirth, C.K.; Shingleton, A.W. Integrating body and organ size in *Drosophila*: Recent advances and outstanding problems. *Front. Endocrinol.* **2012**, *3*, 49. [CrossRef]

167. Mirth, C.K.; Tang, H.Y.; Makohon-Moore, S.C.; Salhadar, S.; Gokhale, R.H.; Warner, R.D.; Koyama, T.; Riddiford, L.M.; Shingleton, A.W. Juvenile hormone regulates body size and perturbs insulin signaling in *Drosophila*. *Proc. Natl. Acad. Sci. USA* **2014**, *111*, 7018–7023. [CrossRef]

168. Texada, M.J.; Malita, A.; Christensen, C.F.; Dall, K.B.; Faergeman, N.J.; Nagy, S.; Halberg, K.A.; Rewitz, K. Autophagy-Mediated Cholesterol Trafficking Controls Steroid Production. *Dev. Cell* **2019**, *48*, 659–671.e654. [CrossRef] [PubMed]

169. Werthebach, M.; Stewart, F.A.; Gahlen, A.; Mettler-Altmann, T.; Akhtar, I.; Maas-Enriquez, K.; Droste, A.; Eichmann, T.O.; Poschmann, G.; Stuhler, K.; et al. Control of *Drosophila* Growth and Survival by the Lipid Droplet-Associated Protein CG9186/Sturkopf. *Cell Rep.* **2019**, *26*, 3726–3740.e3727. [CrossRef] [PubMed]

170. Thiel, K.; Heier, C.; Haberl, V.; Thul, P.J.; Oberer, M.; Lass, A.; Jäckle, H.; Beller, M. The evolutionarily conserved protein CG9186 is associated with lipid droplets, required for their positioning and for fat storage. *J. Cell Sci.* **2013**, *126*, 2198–2212. [CrossRef]

171. Ugrankar, R.; Bowerman, J.; Hariri, H.; Chandra, M.; Chen, K.; Bossanyi, M.F.; Datta, S.; Rogers, S.; Eckert, K.M.; Vale, G.; et al. *Drosophila* Snazarus Regulates a Lipid Droplet Population at Plasma Membrane-Droplet Contacts in Adipocytes. *Dev. Cell* **2019**, *50*, 557–572.e555. [CrossRef] [PubMed]

172. Blumrich, A.; Vogler, G.; Dresen, S.; Diop, S.B.; Jaeger, C.; Leberer, S.; Grune, J.; Wirth, E.K.; Hoeft, B.; Renko, K.; et al. Fat-body brummer lipase determines survival and cardiac function during starvation in *Drosophila* melanogaster. *Iscience* **2021**, *24*, 102288. [CrossRef] [PubMed]

173. Bi, J.; Xiang, Y.; Chen, H.; Liu, Z.; Grönke, S.; Kühnlein, R.P.; Huang, X. Opposite and redundant roles of the two *Drosophila* perilipins in lipid mobilization. *J. Cell Sci.* **2012**, *125*, 3568–3577. [CrossRef]

174. Binh, T.D.; Nguyen, Y.D.H.; Pham, T.L.A.; Komori, K.; Nguyen, T.Q.C.; Taninaka, M.; Kamei, K. Dysfunction of lipid storage droplet-2 suppresses endoreplication and induces JNK pathway-mediated apoptotic cell death in *Drosophila* salivary glands. *Sci. Rep.* **2022**, *12*, 4302. [CrossRef]

175. Fauny, J.D.; Silber, J.; Zider, A. *Drosophila* Lipid Storage Droplet 2 gene (Lsd-2) is expressed and controls lipid storage in wing imaginal discs. *Dev. Dyn.* **2005**, *232*, 725–732. [CrossRef]

176. Binh, T.D.; Pham, T.L.A.; Men, T.T.; Dang, T.T.P.; Kamei, K. LSD-2 dysfunction induces dFoxO-dependent cell death in the wing of *Drosophila* melanogaster. *Biochem. Biophys. Res. Commun.* **2019**, *509*, 491–497. [CrossRef]

177. Goyal, L.; McCall, K.; Agapite, J.; Hartwieg, E.; Steller, H. Induction of apoptosis by *Drosophila* reaper, hid and grim through inhibition of IAP function. *EMBO J.* **2000**, *19*, 589–597. [CrossRef] [PubMed]

178. Eckelman, B.P.; Salvesen, G.S.; Scott, F.L. Human inhibitor of apoptosis proteins: Why XIAP is the black sheep of the family. *EMBO Rep.* **2006**, *7*, 988–994. [CrossRef] [PubMed]

179. Hanifeh, M.; Ataei, F. XIAP as a multifaceted molecule in Cellular Signaling. *Apoptosis* **2022**, *27*, 441–453. [CrossRef]

180. Lu, M.; Lin, S.C.; Huang, Y.; Kang, Y.J.; Rich, R.; Lo, Y.C.; Myszka, D.; Han, J.; Wu, H. XIAP induces NF-kappaB activation via the BIR1/TAB1 interaction and BIR1 dimerization. *Mol. Cell.* **2007**, *26*, 689–702. [CrossRef] [PubMed]
181. Yan, Y.; Wang, H.; Hu, M.; Jiang, L.; Wang, Y.; Liu, P.; Liang, X.; Liu, J.; Li, C.; Lindström-Battle, A.; et al. HDAC6 Suppresses Age-Dependent Ectopic Fat Accumulation by Maintaining the Proteostasis of PLIN2 in *Drosophila*. *Dev. Cell* **2017**, *43*, 99–111.e115. [CrossRef] [PubMed]
182. Katewa, S.D.; Akagi, K.; Bose, N.; Rakshit, K.; Camarella, T.; Zheng, X.; Hall, D.; Davis, S.; Nelson, C.S.; Brem, R.B.; et al. Peripheral Circadian Clocks Mediate Dietary Restriction-Dependent Changes in Lifespan and Fat Metabolism in *Drosophila*. *Cell Metab.* **2016**, *23*, 143–154. [CrossRef] [PubMed]
183. Katewa, S.D.; Kapahi, P. Dietary restriction and aging, 2009. *Aging Cell* **2010**, *9*, 105–112. [CrossRef]
184. Skorupa, D.A.; Dervisefendic, A.; Zwiener, J.; Pletcher, S.D. Dietary composition specifies consumption, obesity, and lifespan in *Drosophila* melanogaster. *Aging Cell* **2008**, *7*, 478–490. [CrossRef]
185. Solon-Biet, S.M.; McMahon, A.C.; Ballard, J.W.; Ruohonen, K.; Wu, L.E.; Cogger, V.C.; Warren, A.; Huang, X.; Pichaud, N.; Melvin, R.G.; et al. The ratio of macronutrients, not caloric intake, dictates cardiometabolic health, aging, and longevity in ad libitum-fed mice. *Cell Metab.* **2014**, *19*, 418–430. [CrossRef]
186. Puig, O.; Marr, M.T.; Ruhf, M.L.; Tjian, R. Control of cell number by *Drosophila* FOXO: Downstream and feedback regulation of the insulin receptor pathway. *Genes Dev.* **2003**, *17*, 2006–2020. [CrossRef] [PubMed]
187. Slack, C.; Giannakou, M.E.; Foley, A.; Goss, M.; Partridge, L. dFOXO-independent effects of reduced insulin-like signaling in *Drosophila*. *Aging Cell* **2011**, *10*, 735–748. [CrossRef]
188. Martins, R.; Lithgow, G.J.; Link, W. Long live FOXO: Unraveling the role of FOXO proteins in aging and longevity. *Aging Cell* **2016**, *15*, 196–207. [CrossRef] [PubMed]
189. Wang, M.C.; Bohmann, D.; Jasper, H. JNK signaling confers tolerance to oxidative stress and extends lifespan in *Drosophila*. *Dev. Cell* **2003**, *5*, 811–816. [CrossRef]
190. Yamamoto, R.; Tatar, M. Insulin receptor substrate chico acts with the transcription factor FOXO to extend *Drosophila* lifespan. *Aging Cell* **2011**, *10*, 729–732. [CrossRef]
191. Alic, N.; Giannakou, M.E.; Papatheodorou, I.; Hoddinott, M.P.; Andrews, T.D.; Bolukbasi, E.; Partridge, L. Interplay of dFOXO and two ETS-family transcription factors determines lifespan in *Drosophila* melanogaster. *PLoS Genet.* **2014**, *10*, e1004619. [CrossRef] [PubMed]
192. Min, K.J.; Yamamoto, R.; Buch, S.; Pankratz, M.; Tatar, M. *Drosophila* lifespan control by dietary restriction independent of insulin-like signaling. *Aging Cell* **2008**, *7*, 199–206. [CrossRef]
193. Gronke, S.; Clarke, D.F.; Broughton, S.; Andrews, T.D.; Partridge, L. Molecular evolution and functional characterization of *Drosophila* insulin-like peptides. *PLoS Genet.* **2010**, *6*, e1000857. [CrossRef]
194. Vereshchagina, N.; Wilson, C. Cytoplasmic activated protein kinase Akt regulates lipid-droplet accumulation in *Drosophila* nurse cells. *Development* **2006**, *133*, 4731–4735. [CrossRef]
195. DiAngelo, J.R.; Birnbaum, M.J. Regulation of fat cell mass by insulin in *Drosophila* melanogaster. *Mol. Cell. Biol.* **2009**, *29*, 6341–6352. [CrossRef]
196. Wang, B.; Moya, N.; Niessen, S.; Hoover, H.; Mihaylova, M.M.; Shaw, R.J.; Yates, J.R.; Fischer, W.H.; Thomas, J.B.; Montminy, M. A Hormone-Dependent Module Regulating Energy Balance. *Cell* **2011**, *145*, 596–606. [CrossRef]
197. Biteau, B.; Karpac, J.; Supoyo, S.; Degennaro, M.; Lehmann, R.; Jasper, H. Lifespan extension by preserving proliferative homeostasis in *Drosophila*. *PLoS Genet.* **2010**, *6*, e1001159. [CrossRef]
198. Wang, L.; Zeng, X.; Ryoo, H.D.; Jasper, H. Integration of UPRER and oxidative stress signaling in the control of intestinal stem cell proliferation. *PLoS Genet.* **2014**, *10*, e1004568. [CrossRef] [PubMed]
199. Wang, L.; Ryoo, H.D.; Qi, Y.; Jasper, H. PERK Limits *Drosophila* Lifespan by Promoting Intestinal Stem Cell Proliferation in Response to ER Stress. *PLoS Genet.* **2015**, *11*, e1005220. [CrossRef] [PubMed]
200. Luis, N.M.; Wang, L.F.; Ortega, M.; Deng, H.S.; Katewa, S.D.; Li, P.W.L.; Karpac, J.; Jasper, H.; Kapahi, P. Intestinal IRE1 Is Required for Increased Triglyceride Metabolism and Longer Lifespan under Dietary Restriction. *Cell Rep.* **2016**, *17*, 1207–1216. [CrossRef] [PubMed]
201. Bailey, A.P.; Koster, G.; Guillermier, C.; Hirst, E.M.; MacRae, J.I.; Lechene, C.P.; Postle, A.D.; Gould, A.P. Antioxidant Role for Lipid Droplets in a Stem Cell Niche of *Drosophila*. *Cell* **2015**, *163*, 340–353. [CrossRef]
202. Cermelli, S.; Guo, Y.; Gross, S.P.; Welte, M.A. The lipid-droplet proteome reveals that droplets are a protein-storage depot. *Curr. Biol.* **2006**, *16*, 1783–1795. [CrossRef] [PubMed]
203. Beller, M.; Riedel, D.; Jansch, L.; Dieterich, G.; Wehland, J.; Jackle, H.; Kuhnlein, R.P. Characterization of the *Drosophila* lipid droplet subproteome. *Mol. Cell. Proteom.* **2006**, *5*, 1082–1094. [CrossRef]
204. Kapahi, P.; Zid, B.M.; Harper, T.; Koslover, D.; Sapin, V.; Benzer, S. Regulation of lifespan in *Drosophila* by modulation of genes in the TOR signaling pathway. *Curr. Biol.* **2004**, *14*, 885–890. [CrossRef]
205. Taylor, R.C.; Dillin, A. Aging as an event of proteostasis collapse. *Cold Spring Harb. Perspect. Biol.* **2011**, *3*, a004440. [CrossRef]
206. Marshall, L.; Rideout, E.J.; Grewal, S.S. Nutrient/TOR-dependent regulation of RNA polymerase III controls tissue and organismal growth in *Drosophila*. *EMBO J.* **2012**, *31*, 1916–1930. [CrossRef] [PubMed]
207. Bjedov, I.; Toivonen, J.M.; Kerr, F.; Slack, C.; Jacobson, J.; Foley, A.; Partridge, L. Mechanisms of life span extension by rapamycin in the fruit fly *Drosophila* melanogaster. *Cell Metab.* **2010**, *11*, 35–46. [CrossRef]

208. Reeg, S.; Grune, T. Protein Oxidation in Toxicology. In *Studies on Experimental Toxicology and Pharmacology*; Oxidative Stress in Applied Basic Research and Clinical Practice; Humana Press: Cham, Switzerland, 2015; pp. 81–102. [CrossRef]

209. Simonsen, A.; Cumming, R.C.; Brech, A.; Isakson, P.; Schubert, D.R.; Finley, K.D. Promoting basal levels of autophagy in the nervous system enhances longevity and oxidant resistance in adult *Drosophila*. *Autophagy* **2008**, *4*, 176–184. [CrossRef]

210. Kovacs, T.; Billes, V.; Komlos, M.; Hotzi, B.; Manzeger, A.; Tarnoci, A.; Papp, D.; Szikszai, F.; Szinyakovics, J.; Racz, A.; et al. The small molecule AUTEN-99 (autophagy enhancer-99) prevents the progression of neurodegenerative symptoms. *Sci. Rep.* **2017**, *7*, 42014. [CrossRef] [PubMed]

211. Papp, D.; Kovács, T.; Billes, V.; Varga, M.; Tarnóci, A.; Hackler, L., Jr.; Puskás, L.G.; Liliom, H.; Tárnok, K.; Schlett, K.; et al. AUTEN-67, an autophagy-enhancing drug candidate with potent antiaging and neuroprotective effects. *Autophagy* **2016**, *12*, 273–286. [CrossRef]

212. Eisenberg, T.; Knauer, H.; Schauer, A.; Büttner, S.; Ruckenstuhl, C.; Carmona-Gutierrez, D.; Ring, J.; Schroeder, S.; Magnes, C.; Antonacci, L.; et al. Induction of autophagy by spermidine promotes longevity. *Nat. Cell Biol.* **2009**, *11*, 1305–1314. [CrossRef]

213. Pang, X.; Zhang, X.; Jiang, Y.; Su, Q.; Li, Q.; Li, Z. Autophagy: Mechanisms and Therapeutic Potential of Flavonoids in Cancer. *Biomolecules* **2021**, *11*, 135. [CrossRef]

214. Fantin, M.; Garelli, F.; Napoli, B.; Forgiarini, A.; Gumeni, S.; De Martin, S.; Montopoli, M.; Vantaggiato, C.; Orso, G. Flavonoids Regulate Lipid Droplets Biogenesis in *Drosophila* melanogaster. *Nat. Prod. Commun.* **2019**, *14*, 1934578X19852430. [CrossRef]

215. Wongchum, N.; Dechakhamphu, A. Xanthohumol prolongs lifespan and decreases stress-induced mortality in *Drosophila* melanogaster. *Comp. Biochem. Physiol. Part C Toxicol. Pharmacol.* **2021**, *244*, 108994. [CrossRef]

216. Gajender; Mazumder, A.; Sharma, A.; Azad, M.A.K. A Comprehensive Review of the Pharmacological Importance of Dietary Flavonoids as Hepatoprotective Agents. *Evid.-Based Complement. Altern. Med.* **2023**, *2023*, 4139117. [CrossRef]

217. Heijnen, C.G.; Haenen, G.R.; Oostveen, R.M.; Stalpers, E.M.; Bast, A. Protection of flavonoids against lipid peroxidation: The structure activity relationship revisited. *Free Radic. Res.* **2002**, *36*, 575–581. [CrossRef]

218. Partridge, L.; Alic, N.; Bjedov, I.; Piper, M.D. Ageing in *Drosophila*: The role of the insulin/Igf and TOR signalling network. *Exp. Gerontol.* **2011**, *46*, 376–381. [CrossRef]

219. Zid, B.M.; Rogers, A.N.; Katewa, S.D.; Vargas, M.A.; Kolipinski, M.C.; Lu, T.A.; Benzer, S.; Kapahi, P. 4E-BP extends lifespan upon dietary restriction by enhancing mitochondrial activity in *Drosophila*. *Cell* **2009**, *139*, 149–160. [CrossRef] [PubMed]

220. Vattem, K.M.; Wek, R.C. Reinitiation involving upstream ORFs regulates ATF4 mRNA translation in mammalian cells. *Proc. Natl. Acad. Sci. USA* **2004**, *101*, 11269–11274. [CrossRef] [PubMed]

221. Kang, M.J.; Vasudevan, D.; Kang, K.; Kim, K.; Park, J.E.; Zhang, N.; Zeng, X.; Neubert, T.A.; Marr, M.T., 2nd; Ryoo, H.D. 4E-BP is a target of the GCN2-ATF4 pathway during *Drosophila* development and aging. *J. Cell Biol.* **2017**, *216*, 115–129. [CrossRef]

222. Srivastava, A.; Lu, J.; Gadalla, D.S.; Hendrich, O.; Grönke, S.; Partridge, L. The Role of GCN2 Kinase in Mediating the Effects of Amino Acids on Longevity and Feeding Behaviour in *Drosophila*. *Front. Aging* **2022**, *3*, 944466. [CrossRef] [PubMed]

223. Mair, W.; Morantte, I.; Rodrigues, A.P.; Manning, G.; Montminy, M.; Shaw, R.J.; Dillin, A. Lifespan extension induced by AMPK and calcineurin is mediated by CRTC-1 and CREB. *Nature* **2011**, *470*, 404–408. [CrossRef]

224. Stenesen, D.; Suh, J.M.; Seo, J.; Yu, K.; Lee, K.S.; Kim, J.S.; Min, K.J.; Graff, J.M. Adenosine nucleotide biosynthesis and AMPK regulate adult life span and mediate the longevity benefit of caloric restriction in flies. *Cell Metab.* **2013**, *17*, 101–112. [CrossRef] [PubMed]

225. Johnson, E.C.; Kazgan, N.; Bretz, C.A.; Forsberg, L.J.; Hector, C.E.; Worthen, R.J.; Onyenwoke, R.; Brenman, J.E. Altered metabolism and persistent starvation behaviors caused by reduced AMPK function in *Drosophila*. *PLoS ONE* **2010**, *5*, e12799. [CrossRef]

226. Gutierrez, E.; Wiggins, D.; Fielding, B.; Gould, A.P. Specialized hepatocyte-like cells regulate *Drosophila* lipid metabolism. *Nature* **2007**, *445*, 275–280. [CrossRef]

227. Triggle, C.R.; Mohammed, I.; Bshesh, K.; Marei, I.; Ye, K.; Ding, H.; MacDonald, R.; Hollenberg, M.D.; Hill, M.A. Metformin: Is it a drug for all reasons and diseases? *Metabolism* **2022**, *133*, 155223. [CrossRef] [PubMed]

228. Anisimov, V.N.; Berstein, L.M.; Egormin, P.A.; Piskunova, T.S.; Popovich, I.G.; Zabezhinski, M.A.; Tyndyk, M.L.; Yurova, M.V.; Kovalenko, I.G.; Poroshina, T.E.; et al. Metformin slows down aging and extends life span of female SHR mice. *Cell Cycle* **2008**, *7*, 2769–2773. [CrossRef]

229. Cabreiro, F.; Au, C.; Leung, K.Y.; Vergara-Irigaray, N.; Cocheme, H.M.; Noori, T.; Weinkove, D.; Schuster, E.; Greene, N.D.; Gems, D. Metformin retards aging in *C. elegans* by altering microbial folate and methionine metabolism. *Cell* **2013**, *153*, 228–239. [CrossRef]

230. Slack, C.; Foley, A.; Partridge, L. Activation of AMPK by the putative dietary restriction mimetic metformin is insufficient to extend lifespan in *Drosophila*. *PLoS ONE* **2012**, *7*, e47699. [CrossRef] [PubMed]

231. Kalender, A.; Selvaraj, A.; Kim, S.Y.; Gulati, P.; Brûlé, S.; Viollet, B.; Kemp, B.E.; Bardeesy, N.; Dennis, P.; Schlager, J.J.; et al. Metformin, independent of AMPK, inhibits mTORC1 in a rag GTPase-dependent manner. *Cell Metab.* **2010**, *11*, 390–401. [CrossRef] [PubMed]

232. Funakoshi, M.; Tsuda, M.; Muramatsu, K.; Hatsuda, H.; Morishita, S.; Aigaki, T. A gain-of-function screen identifies wdb and lkb1 as lifespan-extending genes in *Drosophila*. *Biochem. Biophys. Res. Commun.* **2011**, *405*, 667–672. [CrossRef]

233. Choi, S.; Lim, D.S.; Chung, J. Feeding and Fasting Signals Converge on the LKB1-SIK3 Pathway to Regulate Lipid Metabolism in *Drosophila*. *PLoS Genet.* **2015**, *11*, e1005263. [CrossRef]

234. Grandison, R.C.; Piper, M.D.W.; Partridge, L. Amino-acid imbalance explains extension of lifespan by dietary restriction in *Drosophila*. *Nature* **2009**, *462*, 1061–1064. [CrossRef]
235. Kabil, H.; Kabil, O.; Banerjee, R.; Harshman, L.G.; Pletcher, S.D. Increased transsulfuration mediates longevity and dietary restriction in *Drosophila*. *Proc. Natl. Acad. Sci. USA* **2011**, *108*, 16831–16836. [CrossRef]
236. Jonsson, W.O.; Margolies, N.S.; Anthony, T.G. Dietary Sulfur Amino Acid Restriction and the Integrated Stress Response: Mechanistic Insights. *Nutrients* **2019**, *11*, 1349. [CrossRef]
237. Richie, J.P., Jr.; Leutzinger, Y.; Parthasarathy, S.; Malloy, V.; Orentreich, N.; Zimmerman, J.A. Methionine restriction increases blood glutathione and longevity in F344 rats. *Faseb J.* **1994**, *8*, 1302–1307. [CrossRef]
238. Miller, R.A.; Buehner, G.; Chang, Y.; Harper, J.M.; Sigler, R.; Smith-Wheelock, M. Methionine-deficient diet extends mouse lifespan, slows immune and lens aging, alters glucose, T4, IGF-I and insulin levels, and increases hepatocyte MIF levels and stress resistance. *Aging Cell* **2005**, *4*, 119–125. [CrossRef]
239. Ruckenstuhl, C.; Netzberger, C.; Entfellner, I.; Carmona-Gutierrez, D.; Kickenweiz, T.; Stekovic, S.; Gleixner, C.; Schmid, C.; Klug, L.; Sorgo, A.G.; et al. Lifespan Extension by Methionine Restriction Requires Autophagy-Dependent Vacuolar Acidification. *PLoS Genet.* **2014**, *10*, e1004347. [CrossRef]
240. Malloy, V.L.; Perrone, C.E.; Mattocks, D.A.; Ables, G.P.; Caliendo, N.S.; Orentreich, D.S.; Orentreich, N. Methionine restriction prevents the progression of hepatic steatosis in leptin-deficient obese mice. *Metabolism* **2013**, *62*, 1651–1661. [CrossRef]
241. Perrone, C.E.; Mattocks, D.A.; Jarvis-Morar, M.; Plummer, J.D.; Orentreich, N. Methionine restriction effects on mitochondrial biogenesis and aerobic capacity in white adipose tissue, liver, and skeletal muscle of F344 rats. *Metabolism* **2010**, *59*, 1000–1011. [CrossRef]
242. Nakata, K.; Kawase, M.; Ogino, S.; Kinoshita, C.; Murata, H.; Sakaue, T.; Ogata, K.; Ohmori, S. Effects of age on levels of cysteine, glutathione and related enzyme activities in livers of mice and rats and an attempt to replenish hepatic glutathione level of mouse with cysteine derivatives. *Mech. Ageing Dev.* **1996**, *90*, 195–207. [CrossRef] [PubMed]
243. Chakraborty, P.K.; Xiong, X.; Mustafi, S.B.; Saha, S.; Dhanasekaran, D.; Mandal, N.A.; McMeekin, S.; Bhattacharya, R.; Mukherjee, P. Role of cystathionine beta synthase in lipid metabolism in ovarian cancer. *Oncotarget* **2015**, *6*, 37367–37384. [CrossRef] [PubMed]
244. Negre-Salvayre, A.; Auge, N.; Ayala, V.; Basaga, H.; Boada, J.; Brenke, R.; Chapple, S.; Cohen, G.; Feher, J.; Grune, T.; et al. Pathological aspects of lipid peroxidation. *Free Radic. Res.* **2010**, *44*, 1125–1171. [CrossRef] [PubMed]
245. Negre-Salvayre, A.; Coatrieux, C.; Ingueneau, C.; Salvayre, R. Advanced lipid peroxidation end products in oxidative damage to proteins. Potential role in diseases and therapeutic prospects for the inhibitors. *Br. J. Pharmacol.* **2008**, *153*, 6–20. [CrossRef]
246. Johnson, D.C.; Dean, D.R.; Smith, A.D.; Johnson, M.K. Structure, function, and formation of biological iron-sulfur clusters. *Annu. Rev. Biochem.* **2005**, *74*, 247–281. [CrossRef]
247. Boal, A.K.; Yavin, E.; Barton, J.K. DNA repair glycosylases with a [4Fe-4S] cluster: A redox cofactor for DNA-mediated charge transport? *J. Inorg. Biochem.* **2007**, *101*, 1913–1921. [CrossRef]
248. Rouault, T.A. Mammalian iron-sulphur proteins: Novel insights into biogenesis and function. *Nat. Rev. Mol. Cell Biol.* **2015**, *16*, 45–55. [CrossRef] [PubMed]
249. Crooks, D.R.; Maio, N.; Lane, A.N.; Jarnik, M.; Higashi, R.M.; Haller, R.G.; Yang, Y.; Fan, T.W.; Linehan, W.M.; Rouault, T.A. Acute loss of iron-sulfur clusters results in metabolic reprogramming and generation of lipid droplets in mammalian cells. *J. Biol. Chem.* **2018**, *293*, 8297–8311. [CrossRef] [PubMed]
250. Koeppen, A.H. Friedreich's ataxia: Pathology, pathogenesis, and molecular genetics. *J. Neurol. Sci.* **2011**, *303*, 1–12. [CrossRef] [PubMed]
251. Martelli, A.; Napierala, M.; Puccio, H. Understanding the genetic and molecular pathogenesis of Friedreich's ataxia through animal and cellular models. *Dis. Model. Mech.* **2012**, *5*, 165–176. [CrossRef]
252. Puccio, H.; Simon, D.; Cossée, M.; Criqui-Filipe, P.; Tiziano, F.; Melki, J.; Hindelang, C.; Matyas, R.; Rustin, P.; Koenig, M. Mouse models for Friedreich ataxia exhibit cardiomyopathy, sensory nerve defect and Fe-S enzyme deficiency followed by intramitochondrial iron deposits. *Nat. Genet.* **2001**, *27*, 181–186. [CrossRef]
253. Navarro, J.A.; Ohmann, E.; Sanchez, D.; Botella, J.A.; Liebisch, G.; Moltó, M.D.; Ganfornina, M.D.; Schmitz, G.; Schneuwly, S. Altered lipid metabolism in a *Drosophila* model of Friedreich's ataxia. *Hum. Mol. Genet.* **2010**, *19*, 2828–2840. [CrossRef]
254. Bresgen, N.; Eckl, P.M. Oxidative stress and the homeodynamics of iron metabolism. *Biomolecules* **2015**, *5*, 808–847. [CrossRef] [PubMed]
255. Halliwell, B.; Gutteridge, J.M. Biologically relevant metal ion-dependent hydroxyl radical generation. An update. *FEBS Lett.* **1992**, *307*, 108–112. [CrossRef]
256. Schaur, R.J.; Siems, W.; Bresgen, N.; Eckl, P.M. 4-Hydroxy-nonenal-A Bioactive Lipid Peroxidation Product. *Biomolecules* **2015**, *5*, 2247–2337. [CrossRef]
257. Girard, V.; Goubard, V.; Querenet, M.; Seugnet, L.; Pays, L.; Nataf, S.; Dufourd, E.; Cluet, D.; Mollereau, B.; Davoust, N. Spen modulates lipid droplet content in adult *Drosophila* glial cells and protects against paraquat toxicity. *Sci. Rep.* **2020**, *10*, 20023. [CrossRef] [PubMed]
258. Long, M.; McWilliams, T.G. Lipid droplets promote efficient mitophagy. *Autophagy* **2022**, *19*, 724–725. [CrossRef] [PubMed]
259. Andersen, D.S.; Colombani, J.; Palmerini, V.; Chakrabandhu, K.; Boone, E.; Röthlisberger, M.; Toggweiler, J.; Basler, K.; Mapelli, M.; Hueber, A.O.; et al. The *Drosophila* TNF receptor Grindelwald couples loss of cell polarity and neoplastic growth. *Nature* **2015**, *522*, 482–486. [CrossRef]

260. Muliyil, S.; Levet, C.; Düsterhöft, S.; Dulloo, I.; Cowley, S.A.; Freeman, M. ADAM17-triggered TNF signalling protects the ageing *Drosophila* retina from lipid droplet-mediated degeneration. *Embo J.* **2020**, *39*, e104415. [CrossRef] [PubMed]

261. Lagrutta, L.C.; Layerenza, J.P.; Bronsoms, S.; Trejo, S.A.; Ves-Losada, A. Nuclear-lipid-droplet proteome: Carboxylesterase as a nuclear lipase involved in lipid-droplet homeostasis. *Heliyon* **2021**, *7*, e06539. [CrossRef]

262. Lian, J.; Nelson, R.; Lehner, R. Carboxylesterases in lipid metabolism: From mouse to human. *Protein Cell* **2018**, *9*, 178–195. [CrossRef] [PubMed]

263. Wang, H.; Wei, E.; Quiroga, A.D.; Sun, X.; Touret, N.; Lehner, R. Altered lipid droplet dynamics in hepatocytes lacking triacylglycerol hydrolase expression. *Mol. Biol. Cell* **2010**, *21*, 1991–2000. [CrossRef]

264. Lian, J.H.; Wei, E.H.; Wang, S.P.; Quiroga, A.D.; Li, L.N.; Di Pardo, A.; van der Veen, J.; Sipione, S.; Mitchell, G.A.; Lehner, R. Liver Specific Inactivation of Carboxylesterase 3/Triacylglycerol Hydrolase Decreases Blood Lipids Without Causing Severe Steatosis in Mice. *Hepatology* **2012**, *56*, 2154–2162. [CrossRef]

265. Fujimoto, T. Nuclear lipid droplets—How are they different from their cytoplasmic siblings? *J. Cell Sci.* **2022**, *135*, jcs259253. [CrossRef]

266. McPhee, M.J.; Salsman, J.; Foster, J.; Thompson, J.; Mathavarajah, S.; Dellaire, G.; Ridgway, N.D. Running 'LAPS' around nLD: Nuclear Lipid Droplet Form and Function. *Front. Cell Dev. Biol.* **2022**, *10*, 837406. [CrossRef]

267. Chen, H.; Zheng, X.; Xiao, D.; Zheng, Y. Age-associated de-repression of retrotransposons in the *Drosophila* fat body, its potential cause and consequence. *Aging Cell* **2016**, *15*, 542–552. [CrossRef]

268. Wood, J.G.; Jones, B.C.; Jiang, N.; Chang, C.; Hosier, S.; Wickremesinghe, P.; Garcia, M.; Hartnett, D.A.; Burhenn, L.; Neretti, N.; et al. Chromatin-modifying genetic interventions suppress age-associated transposable element activation and extend life span in *Drosophila*. *Proc. Natl. Acad. Sci. USA* **2016**, *113*, 11277–11282. [CrossRef] [PubMed]

269. Czech, B.; Hannon, G.J. Small RNA sorting: Matchmaking for Argonautes. *Nat. Rev. Genet.* **2011**, *12*, 19–31. [CrossRef]

270. Li, W.; Prazak, L.; Chatterjee, N.; Grüninger, S.; Krug, L.; Theodorou, D.; Dubnau, J. Activation of transposable elements during aging and neuronal decline in *Drosophila*. *Nat. Neurosci.* **2013**, *16*, 529–531. [CrossRef] [PubMed]

271. Li, Z.; Johnson, M.R.; Ke, Z.; Chen, L.; Welte, M.A. *Drosophila* lipid droplets buffer the H2Av supply to protect early embryonic development. *Curr. Biol.* **2014**, *24*, 1485–1491. [CrossRef]

272. Loyola, A.C.; Zhang, L.; Shang, R.; Dutta, P.; Li, J.; Li, W.X. Identification of methotrexate as a heterochromatin-promoting drug. *Sci. Rep.* **2019**, *9*, 11673. [CrossRef] [PubMed]

273. Wood, J.G.; Hillenmeyer, S.; Lawrence, C.; Chang, C.; Hosier, S.; Lightfoot, W.; Mukherjee, E.; Jiang, N.; Schorl, C.; Brodsky, A.S.; et al. Chromatin remodeling in the aging genome of *Drosophila*. *Aging Cell* **2010**, *9*, 971–978. [CrossRef]

274. Yang, L.; Ma, Z.; Wang, H.; Niu, K.; Cao, Y.; Sun, L.; Geng, Y.; Yang, B.; Gao, F.; Chen, Z.; et al. Ubiquitylome study identifies increased histone 2A ubiquitylation as an evolutionarily conserved aging biomarker. *Nat. Commun.* **2019**, *10*, 2191. [CrossRef]

275. Maleszewska, M.; Mawer, J.S.P.; Tessarz, P. Histone Modifications in Ageing and Lifespan Regulation. *Curr. Mol. Biol. Rep.* **2016**, *2*, 26–35. [CrossRef]

276. Feser, J.; Truong, D.; Das, C.; Carson, J.J.; Kieft, J.; Harkness, T.; Tyler, J.K. Elevated histone expression promotes life span extension. *Mol. Cell* **2010**, *39*, 724–735. [CrossRef]

277. Benayoun, B.A.; Pollina, E.A.; Brunet, A. Epigenetic regulation of ageing: Linking environmental inputs to genomic stability. *Nat. Rev. Mol. Cell Biol.* **2015**, *16*, 593–610. [CrossRef] [PubMed]

278. Muñoz-Najar, U.; Sedivy, J.M. Epigenetic control of aging. *Antioxid. Redox Signal.* **2011**, *14*, 241–259. [CrossRef] [PubMed]

279. Shah, P.P.; Donahue, G.; Otte, G.L.; Capell, B.C.; Nelson, D.M.; Cao, K.; Aggarwala, V.; Cruickshanks, H.A.; Rai, T.S.; McBryan, T.; et al. Lamin B1 depletion in senescent cells triggers large-scale changes in gene expression and the chromatin landscape. *Genes Dev.* **2013**, *27*, 1787–1799. [CrossRef]

280. Toiber, D.; Erdel, F.; Bouazoune, K.; Silberman, D.M.; Zhong, L.; Mulligan, P.; Sebastian, C.; Cosentino, C.; Martinez-Pastor, B.; Giacosa, S.; et al. SIRT6 Recruits SNF2H to DNA Break Sites, Preventing Genomic Instability through Chromatin Remodeling. *Mol. Cell* **2013**, *51*, 454–468. [CrossRef] [PubMed]

281. Kim, H.S.; Xiao, C.Y.; Wang, R.H.; Lahusen, T.; Xu, X.L.; Vassilopoulos, A.; Vazquez-Ortiz, G.; Jeong, W.I.; Park, O.; Ki, S.H.; et al. Hepatic-Specific Disruption of SIRT6 in Mice Results in Fatty Liver Formation Due to Enhanced Glycolysis and Triglyceride Synthesis. *Cell Metab.* **2010**, *12*, 224–236. [CrossRef]

282. Penrose, H.; Heller, S.; Cable, C.; Makboul, R.; Chadalawada, G.; Chen, Y.; Crawford, S.E.; Savkovic, S.D. Epidermal growth factor receptor mediated proliferation depends on increased lipid droplet density regulated via a negative regulatory loop with FOXO3/Sirtuin6. *Biochem. Biophys. Res. Commun.* **2016**, *469*, 370–376. [CrossRef]

283. Eckl, P.M.; Bresgen, N. Genotoxicity of lipid oxidation compounds. *Free Radic. Biol. Med.* **2017**, *111*, 244–252. [CrossRef]

284. Shubeita, G.T.; Tran, S.L.; Xu, J.; Vershinin, M.; Cermelli, S.; Cotton, S.L.; Welte, M.A.; Gross, S.P. Consequences of motor copy number on the intracellular transport of kinesin-1-driven lipid droplets. *Cell* **2008**, *135*, 1098–1107. [CrossRef]

285. Welte, M.A. Proteins under new management: Lipid droplets deliver. *Trends Cell Biol.* **2007**, *17*, 363–369. [CrossRef]

286. Welte, M.A.; Gross, S.P.; Postner, M.; Block, S.M.; Wieschaus, E.F. Developmental regulation of vesicle transport in *Drosophila* embryos: Forces and kinetics. *Cell* **1998**, *92*, 547–557. [CrossRef]

287. Bartsch, T.F.; Longoria, R.A.; Florin, E.L.; Shubeita, G.T. Lipid droplets purified from *Drosophila* embryos as an endogenous handle for precise motor transport measurements. *Biophys. J.* **2013**, *105*, 1182–1191. [CrossRef]

288. Larsen, K.S.; Xu, J.; Cermelli, S.; Shu, Z.; Gross, S.P. BicaudalD actively regulates microtubule motor activity in lipid droplet transport. *PLoS ONE* **2008**, *3*, e3763. [CrossRef] [PubMed]
289. Welte, M.A.; Cermelli, S.; Griner, J.; Viera, A.; Guo, Y.; Kim, D.H.; Gindhart, J.G.; Gross, S.P. Regulation of lipid-droplet transport by the perilipin homolog LSD2. *Curr. Biol.* **2005**, *15*, 1266–1275. [CrossRef] [PubMed]
290. Jäckle, H.; Jahn, R. Vesicle transport: Klarsicht clears up the matter. *Curr. Biol.* **1998**, *8*, R542–R544. [CrossRef]
291. Myat, M.M.; Andrew, D.J. Epithelial tube morphology is determined by the polarized growth and delivery of apical membrane. *Cell* **2002**, *111*, 879–891. [CrossRef]
292. Yu, Y.V.; Li, Z.; Rizzo, N.P.; Einstein, J.; Welte, M.A. Targeting the motor regulator Klar to lipid droplets. *BMC Cell Biol.* **2011**, *12*, 9. [CrossRef] [PubMed]
293. Guo, Y.; Jangi, S.; Welte, M.A. Organelle-specific control of intracellular transport: Distinctly targeted isoforms of the regulator Klar. *Mol. Biol. Cell* **2005**, *16*, 1406–1416. [CrossRef]
294. Ingle, L.; Wood, T.R.; Banta, A.M. A study of longevity, growth, reproduction and heart rate in Daphnia longispina as influenced by limitations in quantity of food. *J. Exp. Zool.* **1937**, *76*, 325–352. [CrossRef]
295. Fanestil, D.D.; Barrows, C.H., Jr. Aging in the rotifer. *J. Gerontol.* **1965**, *20*, 462–469.
296. Comfort, A. Effect of Delayed and resumed growth on the longevity of a fish (*Lebistes reticulatus*, peters) in captivity. *Gerontologia* **1963**, *49*, 150–155. [CrossRef]
297. Dilan, C.-B.; Begun, E.; Ahmet Tugrul, O.; Hulusi, K.; Michelle, A. Zebrafish Aging Models and Possible Interventions. In *Recent Advances in Zebrafish Researches*; Yusuf, B., Ed.; IntechOpen: Rijeka, Croatia, 2018; pp. 3–26.
298. Rudzinska, M.A. The influence of amount of food on the reproduction rate and longevity of a sectarian. (*Tokophyra infusionum*). *Science* **1951**, *113*, 10–11. [CrossRef]
299. Sutphin, G.L.; Kaeberlein, M. Dietary restriction by bacterial deprivation increases life span in wild-derived nematodes. *Exp. Gerontol.* **2008**, *43*, 130–135. [CrossRef]
300. Schulz, T.J.; Zarse, K.; Voigt, A.; Urban, N.; Birringer, M.; Ristow, M. Glucose Restriction Extends *Caenorhabditis elegans* Life Span by Inducing Mitochondrial Respiration and Increasing Oxidative Stress. *Cell Metab.* **2007**, *6*, 280–293. [CrossRef]
301. Suckow, B.K.; Suckow, M.A. Lifespan extension by the antioxidant curcumin in *Drosophila* melanogaster. *Int. J. Biomed. Sci.* **2006**, *2*, 402–405.
302. Mohammed, I.; Hollenberg, M.D.; Ding, H.; Triggle, C.R. A Critical Review of the Evidence That Metformin Is a Putative Anti-Aging Drug That Enhances Healthspan and Extends Lifespan. *Front. Endocrinol.* **2021**, *12*, 718942. [CrossRef] [PubMed]
303. Belenky, P.; Racette, F.G.; Bogan, K.L.; McClure, J.M.; Smith, J.S.; Brenner, C. Nicotinamide riboside promotes Sir2 silencing and extends lifespan via Nrk and Urh1/Pnp1/Meu1 pathways to NAD+. *Cell* **2007**, *129*, 473–484. [CrossRef]
304. Hofer, S.J.; Davinelli, S.; Bergmann, M.; Scapagnini, G.; Madeo, F. Caloric Restriction Mimetics in Nutrition and Clinical Trials. *Front. Nutr.* **2021**, *8*, 717343. [CrossRef]
305. Ross, M.H. Length of life and nutrition in the rat. *J. Nutr.* **1961**, *75*, 197–210. [CrossRef]
306. Weindruch, R.; Walford, R.L. Dietary restriction in mice beginning at 1 year of age: Effect on life-span and spontaneous cancer incidence. *Science* **1982**, *215*, 1415–1418. [CrossRef]
307. Nedergaard, J.; Bengtsson, T.; Cannon, B. Unexpected evidence for active brown adipose tissue in adult humans. *Am. J. Physiol.-Endocrinol. Metab.* **2007**, *293*, E444–E452. [CrossRef] [PubMed]
308. Darcy, J.; Tseng, Y.-H. ComBATing aging—Does increased brown adipose tissue activity confer longevity? *GeroScience* **2019**, *41*, 285–296. [CrossRef]
309. Zoico, E.; Rubele, S.; De Caro, A.; Nori, N.; Mazzali, G.; Fantin, F.; Rossi, A.; Zamboni, M. Brown and Beige Adipose Tissue and Aging. *Front. Endocrinol.* **2019**, *10*, 368. [CrossRef]
310. Fabbiano, S.; Suárez-Zamorano, N.; Rigo, D.; Veyrat-Durebex, C.; Stevanovic Dokic, A.; Colin, D.J.; Trajkovski, M. Caloric Restriction Leads to Browning of White Adipose Tissue through Type 2 Immune Signaling. *Cell Metab.* **2016**, *24*, 434–446. [CrossRef] [PubMed]
311. Baumeier, C.; Kaiser, D.; Heeren, J.; Scheja, L.; John, C.; Weise, C.; Eravci, M.; Lagerpusch, M.; Schulze, G.; Joost, H.-G.; et al. Caloric restriction and intermittent fasting alter hepatic lipid droplet proteome and diacylglycerol species and prevent diabetes in NZO mice. *Biochim. Biophys. Acta (BBA)—Mol. Cell Biol. Lipids* **2015**, *1851*, 566–576. [CrossRef] [PubMed]
312. Franceschi, C.; Garagnani, P.; Morsiani, C.; Conte, M.; Santoro, A.; Grignolio, A.; Monti, D.; Capri, M.; Salvioli, S. The Continuum of Aging and Age-Related Diseases: Common Mechanisms but Different Rates. *Front. Med.* **2018**, *5*, 61. [CrossRef]
313. van Deursen, J.M. The role of senescent cells in ageing. *Nature* **2014**, *509*, 439–446. [CrossRef] [PubMed]
314. Lizardo, D.Y.; Lin, Y.L.; Gokcumen, O.; Atilla-Gokcumen, G.E. Regulation of lipids is central to replicative senescence. *Mol. Biosyst.* **2017**, *13*, 498–509. [CrossRef]
315. Bartke, A.; Darcy, J. GH and ageing: Pitfalls and new insights. *Best Pract. Res. Clin. Endocrinol. Metab.* **2017**, *31*, 113–125. [CrossRef]
316. Vitale, G.; Salvioli, S.; Franceschi, C. Oxidative stress and the ageing endocrine system. *Nat. Rev. Endocrinol.* **2013**, *9*, 228–240. [CrossRef]
317. Vitale, G.; Cesari, M.; Mari, D. Aging of the endocrine system and its potential impact on sarcopenia. *Eur. J. Intern. Med.* **2016**, *35*, 10–15. [CrossRef]
318. Sadria, M.; Layton, A.T. Interactions among mTORC, AMPK and SIRT: A computational model for cell energy balance and metabolism. *Cell Commun. Signal.* **2021**, *19*, 57. [CrossRef] [PubMed]

319. Ralhan, I.; Chang, C.L.; Lippincott-Schwartz, J.; Ioannou, M.S. Lipid droplets in the nervous system. *J. Cell Biol.* **2021**, *220*, e202102136. [CrossRef]

320. Yang, D.S.; Stavrides, P.; Saito, M.; Kumar, A.; Rodriguez-Navarro, J.A.; Pawlik, M.; Huo, C.; Walkley, S.U.; Saito, M.; Cuervo, A.M.; et al. Defective macroautophagic turnover of brain lipids in the TgCRND8 Alzheimer mouse model: Prevention by correcting lysosomal proteolytic deficits. *Brain* **2014**, *137*, 3300–3318. [CrossRef]

321. Derk, J.; Bermudez Hernandez, K.; Rodriguez, M.; He, M.; Koh, H.; Abedini, A.; Li, H.; Fenyo, D.; Schmidt, A.M. Diaphanous 1 (DIAPH1) is Highly Expressed in the Aged Human Medial Temporal Cortex and Upregulated in Myeloid Cells during Alzheimer's Disease. *J. Alzheimers Dis.* **2018**, *64*, 995–1007. [CrossRef] [PubMed]

322. Hamilton, L.K.; Dufresne, M.; Joppe, S.E.; Petryszyn, S.; Aumont, A.; Calon, F.; Barnabe-Heider, F.; Furtos, A.; Parent, M.; Chaurand, P.; et al. Aberrant Lipid Metabolism in the Forebrain Niche Suppresses Adult Neural Stem Cell Proliferation in an Animal Model of Alzheimer's Disease. *Cell Stem Cell* **2015**, *17*, 397–411. [CrossRef] [PubMed]

323. Qi, G.; Mi, Y.; Shi, X.; Gu, H.; Brinton, R.D.; Yin, F. ApoE4 Impairs Neuron-Astrocyte Coupling of Fatty Acid Metabolism. *Cell Rep.* **2021**, *34*, 108572. [CrossRef]

324. Brekk, O.R.; Honey, J.R.; Lee, S.; Hallett, P.J.; Isacson, O. Cell type-specific lipid storage changes in Parkinson's disease patient brains are recapitulated by experimental glycolipid disturbance. *Proc. Natl. Acad. Sci. USA* **2020**, *117*, 27646–27654. [CrossRef]

325. Cole, N.B.; Murphy, D.D.; Grider, T.; Rueter, S.; Brasaemle, D.; Nussbaum, R.L. Lipid droplet binding and oligomerization properties of the Parkinson's disease protein alpha-synuclein. *J. Biol. Chem.* **2002**, *277*, 6344–6352. [CrossRef] [PubMed]

326. Yako, T.; Otsu, W.; Nakamura, S.; Shimazawa, M.; Hara, H. Lipid Droplet Accumulation Promotes RPE Dysfunction. *Int. J. Mol. Sci.* **2022**, *23*, 1790. [CrossRef]

327. Arbaizar-Rovirosa, M.; Gallizioli, M.; Pedragosa, J.; Lozano, J.J.; Casal, C.; Pol, A.; Planas, A.M. Age-dependent lipid droplet-rich microglia worsen stroke outcome in old mice. *bioRxiv* **2022**, 2022.2003.2014.484305. [CrossRef]

328. Plakkal Ayyappan, J.; Paul, A.; Goo, Y.H. Lipid droplet-associated proteins in atherosclerosis (Review). *Mol. Med. Rep.* **2016**, *13*, 4527–4534. [CrossRef]

329. Sukhorukov, V.N.; Khotina, V.A.; Chegodaev, Y.S.; Ivanova, E.; Sobenin, I.A.; Orekhov, A.N. Lipid Metabolism in Macrophages: Focus on Atherosclerosis. *Biomedicines* **2020**, *8*, 262. [CrossRef]

330. Goldberg, I.J.; Reue, K.; Abumrad, N.A.; Bickel, P.E.; Cohen, S.; Fisher, E.A.; Galis, Z.S.; Granneman, J.G.; Lewandowski, E.D.; Murphy, R.; et al. Deciphering the Role of Lipid Droplets in Cardiovascular Disease: A Report from the 2017 National Heart, Lung, and Blood Institute Workshop. *Circulation* **2018**, *138*, 305–315. [CrossRef] [PubMed]

331. Huang, W.; Gao, F.; Zhang, Y.; Chen, T.; Xu, C. Lipid Droplet-Associated Proteins in Cardiomyopathy. *Ann. Nutr. Metab.* **2022**, *78*, 1–13. [CrossRef]

332. Al Saedi, A.; Debruin, D.A.; Hayes, A.; Hamrick, M. Lipid metabolism in sarcopenia. *Bone* **2022**, *164*, 116539. [CrossRef]

333. Conte, M.; Vasuri, F.; Trisolino, G.; Bellavista, E.; Santoro, A.; Degiovanni, A.; Martucci, E.; D'Errico-Grigioni, A.; Caporossi, D.; Capri, M.; et al. Increased Plin2 expression in human skeletal muscle is associated with sarcopenia and muscle weakness. *PLoS ONE* **2013**, *8*, e73709. [CrossRef]

334. Weyand, C.M.; Wu, B.; Goronzy, J.J. The metabolic signature of T cells in rheumatoid arthritis. *Curr. Opin. Rheumatol.* **2020**, *32*, 159–167. [CrossRef] [PubMed]

335. Naiff, P.F.; Kuckelhaus, S.A.S.; Corazza, D.; Leite, L.M.; Couto, S.; deOliveira, M.S.; Santiago, L.M.; Silva, L.F.; Oliveira, L.A.; Grisi, D.C.; et al. Quantification of lipid bodies in monocytes from patients with periodontitis. *Clin. Exp. Dent. Res.* **2021**, *7*, 93–100. [CrossRef] [PubMed]

336. Rho, J.H.; Kim, H.J.; Joo, J.Y.; Lee, J.Y.; Lee, J.H.; Park, H.R. Periodontal Pathogens Promote Foam Cell Formation by Blocking Lipid Efflux. *J. Dent. Res.* **2021**, *100*, 1367–1377. [CrossRef]

337. McGee-Lawrence, M.E.; Carpio, L.R.; Schulze, R.J.; Pierce, J.L.; McNiven, M.A.; Farr, J.N.; Khosla, S.; Oursler, M.J.; Westendorf, J.J. Hdac3 Deficiency Increases Marrow Adiposity and Induces Lipid Storage and Glucocorticoid Metabolism in Osteochondroprogenitor Cells. *J. Bone Miner. Res.* **2016**, *31*, 116–128. [CrossRef] [PubMed]

338. Rendina-Ruedy, E.; Rosen, C.J. Lipids in the Bone Marrow: An Evolving Perspective. *Cell Metab.* **2020**, *31*, 219–231. [CrossRef] [PubMed]

339. Lippiello, L.; Walsh, T.; Fienhold, M. The association of lipid abnormalities with tissue pathology in human osteoarthritic articular cartilage. *Metabolism* **1991**, *40*, 571–576. [CrossRef]

340. Lee, S.W.; Rho, J.H.; Lee, S.Y.; Chung, W.T.; Oh, Y.J.; Kim, J.H.; Yoo, S.H.; Kwon, W.Y.; Bae, J.Y.; Seo, S.Y.; et al. Dietary fat-associated osteoarthritic chondrocytes gain resistance to lipotoxicity through PKCK2/STAMP2/FSP27. *Bone Res.* **2018**, *6*, 20. [CrossRef] [PubMed]

341. Mustonen, A.M.; Nieminen, P. Fatty Acids and Oxylipins in Osteoarthritis and Rheumatoid Arthritis-a Complex Field with Significant Potential for Future Treatments. *Curr. Rheumatol. Rep.* **2021**, *23*, 41. [CrossRef]

342. Tong, X.; Liu, S.; Stein, R.; Imai, Y. Lipid Droplets' Role in the Regulation of beta-Cell Function and beta-Cell Demise in Type 2 Diabetes. *Endocrinology* **2022**, *163*, bqac007. [CrossRef]

343. Tong, X.; Stein, R. Lipid Droplets Protect Human beta-Cells From Lipotoxicity-Induced Stress and Cell Identity Changes. *Diabetes* **2021**, *70*, 2595–2607. [CrossRef]

344. He, Y.; Su, Y.; Duan, C.; Wang, S.; He, W.; Zhang, Y.; An, X.; He, M. Emerging role of aging in the progression of NAFLD to HCC. *Ageing Res. Rev.* **2023**, *84*, 101833. [CrossRef]

345. Stephens, N.A.; Skipworth, R.J.; Macdonald, A.J.; Greig, C.A.; Ross, J.A.; Fearon, K.C. Intramyocellular lipid droplets increase with progression of cachexia in cancer patients. *J. Cachexia Sarcopenia Muscle* **2011**, *2*, 111–117. [CrossRef] [PubMed]
346. Nardi, F.; Fitchev, P.; Brooks, K.M.; Franco, O.E.; Cheng, K.; Hayward, S.W.; Welte, M.A.; Crawford, S.E. Lipid droplet velocity is a microenvironmental sensor of aggressive tumors regulated by V-ATPase and PEDF. *Lab. Investig.* **2019**, *99*, 1822–1834. [CrossRef] [PubMed]
347. Hamsanathan, S.; Gurkar, A.U. Lipids as Regulators of Cellular Senescence. *Front. Physiol.* **2022**, *13*, 796850. [CrossRef]
348. Ogrodnik, M.; Zhu, Y.; Langhi, L.G.P.; Tchkonia, T.; Kruger, P.; Fielder, E.; Victorelli, S.; Ruswhandi, R.A.; Giorgadze, N.; Pirtskhalava, T.; et al. Obesity-Induced Cellular Senescence Drives Anxiety and Impairs Neurogenesis. *Cell Metab.* **2019**, *29*, 1061–1077.e1068. [CrossRef]
349. Chee, W.Y.; Kurahashi, Y.; Kim, J.; Miura, K.; Okuzaki, D.; Ishitani, T.; Kajiwara, K.; Nada, S.; Okano, H.; Okada, M. beta-catenin-promoted cholesterol metabolism protects against cellular senescence in naked mole-rat cells. *Commun. Biol.* **2021**, *4*, 357. [CrossRef]
350. Justesen, J.; Stenderup, K.; Ebbesen, E.N.; Mosekilde, L.; Steiniche, T.; Kassem, M. Adipocyte tissue volume in bone marrow is increased with aging and in patients with osteoporosis. *Biogerontology* **2001**, *2*, 165–171. [CrossRef]
351. Verma, S.; Rajaratnam, J.H.; Denton, J.; Hoyland, J.A.; Byers, R.J. Adipocytic proportion of bone marrow is inversely related to bone formation in osteoporosis. *J. Clin. Pathol.* **2002**, *55*, 693–698. [CrossRef] [PubMed]
352. Rosen, C.J.; Bouxsein, M.L. Mechanisms of disease: Is osteoporosis the obesity of bone? *Nat. Clin. Pract. Rheumatol.* **2006**, *2*, 35–43. [CrossRef]
353. Rendina-Ruedy, E.; Guntur, A.R.; Rosen, C.J. Intracellular lipid droplets support osteoblast function. *Adipocyte* **2017**, *6*, 250–258. [CrossRef]
354. Diascro, D.D.; Vogel, R.L.; Johnson, T.E.; Witherup, K.M.; Pitzenberger, S.M.; Rutledge, S.J.; Prescott, D.J.; Rodan, G.A.; Schmidt, A. High fatty acid content in rabbit serum is responsible for the differentiation of osteoblasts into adipocyte-like cells. *J. Bone Miner. Res.* **1998**, *13*, 96–106. [CrossRef] [PubMed]
355. Sebo, Z.L.; Rendina-Ruedy, E.; Ables, G.P.; Lindskog, D.M.; Rodeheffer, M.S.; Fazeli, P.K.; Horowitz, M.C. Bone Marrow Adiposity: Basic and Clinical Implications. *Endocr. Rev.* **2019**, *40*, 1187–1206. [CrossRef]
356. Maurin, A.C.; Chavassieux, P.M.; Frappart, L.; Delmas, P.D.; Serre, C.M.; Meunier, P.J. Influence of mature adipocytes on osteoblast proliferation in human primary cocultures. *Bone* **2000**, *26*, 485–489. [CrossRef]
357. Elbaz, A.; Wu, X.; Rivas, D.; Gimble, J.M.; Duque, G. Inhibition of fatty acid biosynthesis prevents adipocyte lipotoxicity on human osteoblasts in vitro. *J. Cell. Mol. Med.* **2010**, *14*, 982–991. [CrossRef]
358. Backesjo, C.M.; Li, Y.; Lindgren, U.; Haldosen, L.A. Activation of Sirt1 decreases adipocyte formation during osteoblast differentiation of mesenchymal stem cells. *Cells Tissues Organs* **2009**, *189*, 93–97. [CrossRef]
359. Najt, C.P.; Khan, S.A.; Heden, T.D.; Witthuhn, B.A.; Perez, M.; Heier, J.L.; Mead, L.E.; Franklin, M.P.; Karanja, K.K.; Graham, M.J.; et al. Lipid Droplet-Derived Monounsaturated Fatty Acids Traffic via PLIN5 to Allosterically Activate SIRT1. *Mol. Cell* **2020**, *77*, 810–824 e818. [CrossRef]
360. Han, L.; Zhou, R.; Niu, J.; McNutt, M.A.; Wang, P.; Tong, T. SIRT1 is regulated by a PPARgamma-SIRT1 negative feedback loop associated with senescence. *Nucleic Acids Res.* **2010**, *38*, 7458–7471. [CrossRef]
361. Moerman, E.J.; Teng, K.; Lipschitz, D.A.; Lecka-Czernik, B. Aging activates adipogenic and suppresses osteogenic programs in mesenchymal marrow stroma/stem cells: The role of PPAR-gamma2 transcription factor and TGF-beta/BMP signaling pathways. *Aging Cell* **2004**, *3*, 379–389. [CrossRef]
362. Gong, J.; Sun, Z.; Wu, L.; Xu, W.; Schieber, N.; Xu, D.; Shui, G.; Yang, H.; Parton, R.G.; Li, P. Fsp27 promotes lipid droplet growth by lipid exchange and transfer at lipid droplet contact sites. *J. Cell Biol.* **2011**, *195*, 953–963. [CrossRef]
363. Wang, F.; Ren, S.Y.; Chen, J.F.; Liu, K.; Li, R.X.; Li, Z.F.; Hu, B.; Niu, J.Q.; Xiao, L.; Chan, J.R.; et al. Myelin degeneration and diminished myelin renewal contribute to age-related deficits in memory. *Nat. Neurosci.* **2020**, *23*, 481–486. [CrossRef]
364. Farokhian, F.; Yang, C.; Beheshti, I.; Matsuda, H.; Wu, S. Age-Related Gray and White Matter Changes in Normal Adult Brains. *Aging Dis.* **2017**, *8*, 899–909. [CrossRef]
365. Capilla-Gonzalez, V.; Cebrian-Silla, A.; Guerrero-Cazares, H.; Garcia-Verdugo, J.M.; Quiñones-Hinojosa, A. Age-related changes in astrocytic and ependymal cells of the subventricular zone. *Glia* **2014**, *62*, 790–803. [CrossRef] [PubMed]
366. Kabaso, D.; Coskren, P.J.; Henry, B.I.; Hof, P.R.; Wearne, S.L. The electrotonic structure of pyramidal neurons contributing to prefrontal cortical circuits in macaque monkeys is significantly altered in aging. *Cereb. Cortex* **2009**, *19*, 2248–2268. [CrossRef]
367. O'Brien, J.S.; Sampson, E.L. Lipid composition of the normal human brain: Gray matter, white matter, and myelin. *J. Lipid Res.* **1965**, *6*, 537–544. [CrossRef]
368. Yang, C.; Wang, X.; Wang, J.; Wang, X.; Chen, W.; Lu, N.; Siniossoglou, S.; Yao, Z.; Liu, K. Rewiring Neuronal Glycerolipid Metabolism Determines the Extent of Axon Regeneration. *Neuron* **2020**, *105*, 276–292.e275. [CrossRef]
369. Kaushik, S.; Rodriguez-Navarro, J.A.; Arias, E.; Kiffin, R.; Sahu, S.; Schwartz, G.J.; Cuervo, A.M.; Singh, R. Autophagy in hypothalamic AgRP neurons regulates food intake and energy balance. *Cell Metab.* **2011**, *14*, 173–183. [CrossRef] [PubMed]
370. Ioannou, M.S.; Jackson, J.; Sheu, S.H.; Chang, C.L.; Weigel, A.V.; Liu, H.; Pasolli, H.A.; Xu, C.S.; Pang, S.; Matthies, D.; et al. Neuron-Astrocyte Metabolic Coupling Protects against Activity-Induced Fatty Acid Toxicity. *Cell* **2019**, *177*, 1522–1535.e1514. [CrossRef]

371. Martinez-Vicente, M.; Talloczy, Z.; Wong, E.; Tang, G.; Koga, H.; Kaushik, S.; de Vries, R.; Arias, E.; Harris, S.; Sulzer, D.; et al. Cargo recognition failure is responsible for inefficient autophagy in Huntington's disease. *Nat. Neurosci.* **2010**, *13*, 567–576. [CrossRef]
372. Marschallinger, J.; Iram, T.; Zardeneta, M.; Lee, S.E.; Lehallier, B.; Haney, M.S.; Pluvinage, J.V.; Mathur, V.; Hahn, O.; Morgens, D.W.; et al. Lipid-droplet-accumulating microglia represent a dysfunctional and proinflammatory state in the aging brain. *Nat. Neurosci.* **2020**, *23*, 194–208. [CrossRef]
373. Shimabukuro, M.K.; Langhi, L.G.; Cordeiro, I.; Brito, J.M.; Batista, C.M.; Mattson, M.P.; Mello Coelho, V. Lipid-laden cells differentially distributed in the aging brain are functionally active and correspond to distinct phenotypes. *Sci. Rep.* **2016**, *6*, 23795. [CrossRef]
374. Bresgen, N.; Jaksch, H.; Bauer, H.C.; Eckl, P.; Krizbai, I.; Tempfer, H. Astrocytes are more resistant than cerebral endothelial cells toward geno- and cytotoxicity mediated by short-term oxidative stress. *J. Neurosci. Res.* **2006**, *84*, 1821–1828. [CrossRef] [PubMed]
375. Schroeter, M.L.; Mertsch, K.; Giese, H.; Muller, S.; Sporbert, A.; Hickel, B.; Blasig, I.E. Astrocytes enhance radical defence in capillary endothelial cells constituting the blood-brain barrier. *FEBS Lett.* **1999**, *449*, 241–244. [CrossRef]
376. Husain, M.A.; Laurent, B.; Plourde, M. APOE and Alzheimer's Disease: From Lipid Transport to Physiopathology and Therapeutics. *Front. Neurosci.* **2021**, *15*, 630502. [CrossRef]
377. Corder, E.H.; Saunders, A.M.; Strittmatter, W.J.; Schmechel, D.E.; Gaskell, P.C.; Small, G.W.; Roses, A.D.; Haines, J.L.; Pericak-Vance, M.A. Gene dose of apolipoprotein E type 4 allele and the risk of Alzheimer's disease in late onset families. *Science* **1993**, *261*, 921–923. [CrossRef]
378. Liu, L.; MacKenzie, K.R.; Putluri, N.; Maletic-Savatic, M.; Bellen, H.J. The Glia-Neuron Lactate Shuttle and Elevated ROS Promote Lipid Synthesis in Neurons and Lipid Droplet Accumulation in Glia via APOE/D. *Cell Metab.* **2017**, *26*, 719–737.e716. [CrossRef]
379. Xu, L.; Pu, J. Alpha-Synuclein in Parkinson's Disease: From Pathogenetic Dysfunction to Potential Clinical Application. *Park. Dis.* **2016**, *2016*, 1720621. [CrossRef]
380. Polymeropoulos, M.H.; Lavedan, C.; Leroy, E.; Ide, S.E.; Dehejia, A.; Dutra, A.; Pike, B.; Root, H.; Rubenstein, J.; Boyer, R.; et al. Mutation in the alpha-synuclein gene identified in families with Parkinson's disease. *Science* **1997**, *276*, 2045–2047. [CrossRef]
381. Loov, C.; Scherzer, C.R.; Hyman, B.T.; Breakefield, X.O.; Ingelsson, M. alpha-Synuclein in Extracellular Vesicles: Functional Implications and Diagnostic Opportunities. *Cell. Mol. Neurobiol.* **2016**, *36*, 437–448. [CrossRef]
382. Gasser, T.; Hardy, J.; Mizuno, Y. Milestones in PD genetics. *Mov. Disord.* **2011**, *26*, 1042–1048. [CrossRef]
383. Islimye, E.; Girard, V.; Gould, A.P. Functions of Stress-Induced Lipid Droplets in the Nervous System. *Front. Cell Dev. Biol.* **2022**, *10*, 863907. [CrossRef]
384. Taran, A.S.; Shuvalova, L.D.; Lagarkova, M.A.; Alieva, I.B. Huntington's Disease-An Outlook on the Interplay of the HTT Protein, Microtubules and Actin Cytoskeletal Components. *Cells* **2020**, *9*, 1514. [CrossRef]
385. Churkina Taran, A.S.; Shakhov, A.S.; Kotlobay, A.A.; Alieva, I.B. Huntingtin and Other Neurodegeneration-Associated Proteins in the Development of Intracellular Pathologies: Potential Target Search for Therapeutic Intervention. *Int. J. Mol. Sci.* **2022**, *23*, 5533. [CrossRef]
386. Gruber, A.; Hornburg, D.; Antonin, M.; Krahmer, N.; Collado, J.; Schaffer, M.; Zubaite, G.; Lüchtenborg, C.; Sachsenheimer, T.; Brügger, B.; et al. Molecular and structural architecture of polyQ aggregates in yeast. *Proc. Natl. Acad. Sci. USA* **2018**, *115*, E3446–E3453. [CrossRef]
387. Cyske, Z.; Gaffke, L.; Pierzynowska, K.; Węgrzyn, G. Tubulin Cytoskeleton in Neurodegenerative Diseases-not Only Primary Tubulinopathies. *Cell. Mol. Neurobiol.* **2022**. *Online ahead of print.* [CrossRef]
388. Ketut, S.; Pande, D.; Made Siswadi, S.; Kuswardhani, R.A.T. Age is an Important Risk Factor for Type 2 Diabetes Mellitus and Cardiovascular Diseases. In *Glucose Tolerance*; Sureka, C., Ed.; IntechOpen: Rijeka, Croatia, 2012; p. Ch. 5.
389. Han, S.K.; Baik, S.K.; Kim, M.Y. Non-alcoholic fatty liver disease: Definition and subtypes. *Clin. Mol. Hepatol.* **2022**, *29*, S5–S16. [CrossRef]
390. Guo, X.; Yin, X.; Liu, Z.; Wang, J. Non-Alcoholic Fatty Liver Disease (NAFLD) Pathogenesis and Natural Products for Prevention and Treatment. *Int. J. Mol. Sci.* **2022**, *23*, 5489. [CrossRef]
391. Buzzetti, E.; Pinzani, M.; Tsochatzis, E.A. The multiple-hit pathogenesis of non-alcoholic fatty liver disease (NAFLD). *Metabolism* **2016**, *65*, 1038–1048. [CrossRef]
392. Ogrodnik, M.; Miwa, S.; Tchkonia, T.; Tiniakos, D.; Wilson, C.L.; Lahat, A.; Day, C.P.; Burt, A.; Palmer, A.; Anstee, Q.M.; et al. Cellular senescence drives age-dependent hepatic steatosis. *Nat. Commun.* **2017**, *8*, 15691. [CrossRef]
393. Lau, L.F. CCN1/CYR61: The very model of a modern matricellular protein. *Cell. Mol. Life Sci.* **2011**, *68*, 3149–3163. [CrossRef]
394. Kim, K.H.; Won, J.H.; Cheng, N.; Lau, L.F. The matricellular protein CCN1 in tissue injury repair. *J. Cell Commun. Signal.* **2018**, *12*, 273–279. [CrossRef] [PubMed]
395. Ju, L.; Sun, Y.; Xue, H.; Chen, L.; Gu, C.; Shao, J.; Lu, R.; Luo, X.; Wei, J.; Ma, X.; et al. CCN1 promotes hepatic steatosis and inflammation in non-alcoholic steatohepatitis. *Sci. Rep.* **2020**, *10*, 3201. [CrossRef]
396. Kim, K.H.; Chen, C.C.; Monzon, R.I.; Lau, L.F. Matricellular protein CCN1 promotes regression of liver fibrosis through induction of cellular senescence in hepatic myofibroblasts. *Mol. Cell. Biol.* **2013**, *33*, 2078–2090. [CrossRef]
397. Quan, T.; Qin, Z.; Voorhees, J.J.; Fisher, G.J. Cysteine-rich protein 61 (CCN1) mediates replicative senescence-associated aberrant collagen homeostasis in human skin fibroblasts. *J. Cell Biochem.* **2012**, *113*, 3011–3018. [CrossRef]

398. Qin, Z.; Robichaud, P.; He, T.; Fisher, G.J.; Voorhees, J.J.; Quan, T. Oxidant exposure induces cysteine-rich protein 61 (CCN1) via c-Jun/AP-1 to reduce collagen expression in human dermal fibroblasts. *PLoS ONE* **2014**, *9*, e115402. [CrossRef]

399. Borkham-Kamphorst, E.; Steffen, B.T.; Van de Leur, E.; Haas, U.; Tihaa, L.; Friedman, S.L.; Weiskirchen, R. CCN1/CYR61 overexpression in hepatic stellate cells induces ER stress-related apoptosis. *Cell. Signal.* **2016**, *28*, 34–42. [CrossRef]

400. Cheng, N.Y.; Kim, K.H.; Lau, L.F. Senescent hepatic stellate cells promote liver regeneration through IL-6 and ligands of CXCR2. *JCI Insight* **2022**, *7*, e158207. [CrossRef]

401. Han, D.; Ybanez, M.D.; Ahmadi, S.; Yeh, K.; Kaplowitz, N. Redox regulation of tumor necrosis factor signaling. *Antioxid. Redox Signal.* **2009**, *11*, 2245–2263. [CrossRef]

402. Duvigneau, J.C.; Luis, A.; Gorman, A.M.; Samali, A.; Kaltenecker, D.; Moriggl, R.; Kozlov, A.V. Crosstalk between inflammatory mediators and endoplasmic reticulum stress in liver diseases. *Cytokine* **2019**, *124*, 154577. [CrossRef]

403. Long, Z.; Cao, M.; Su, S.H.; Wu, G.Y.; Meng, F.S.; Wu, H.; Liu, J.Z.; Yu, W.H.; Atabai, K.; Wang, X. Inhibition of hepatocyte nuclear factor 1b induces hepatic steatosis through DPP4/NOX1-mediated regulation of superoxide. *Free Radic. Biol. Med.* **2017**, *113*, 71–83. [CrossRef]

404. Bellanné-Chantelot, C.; Clauin, S.; Chauveau, D.; Collin, P.; Daumont, M.; Douillard, C.; Dubois-Laforgue, D.; Dusselier, L.; Gautier, J.F.; Jadoul, M.; et al. Large genomic rearrangements in the hepatocyte nuclear factor-1beta (TCF2) gene are the most frequent cause of maturity-onset diabetes of the young type 5. *Diabetes* **2005**, *54*, 3126–3132. [CrossRef]

405. Bonnycastle, L.L.; Willer, C.J.; Conneely, K.N.; Jackson, A.U.; Burrill, C.P.; Watanabe, R.M.; Chines, P.S.; Narisu, N.; Scott, L.J.; Enloe, S.T.; et al. Common variants in maturity-onset diabetes of the young genes contribute to risk of type 2 diabetes in Finns. *Diabetes* **2006**, *55*, 2534–2540. [CrossRef]

406. Barak, P.; Rai, A.; Rai, P.; Mallik, R. Quantitative optical trapping on single organelles in cell extract. *Nat. Methods* **2013**, *10*, 68–70. [CrossRef]

407. Rai, P.; Kumar, M.; Sharma, G.; Barak, P.; Das, S.; Kamat, S.S.; Mallik, R. Kinesin-dependent mechanism for controlling triglyceride secretion from the liver. *Proc. Natl. Acad. Sci. USA* **2017**, *114*, 12958–12963. [CrossRef]

408. Thiam, A.R.; Antonny, B.; Wang, J.; Delacotte, J.; Wilfling, F.; Walther, T.C.; Beck, R.; Rothman, J.E.; Pincet, F. COPI buds 60-nm lipid droplets from reconstituted water-phospholipid-triacylglyceride interfaces, suggesting a tension clamp function. *Proc. Natl. Acad. Sci. USA* **2013**, *110*, 13244–13249. [CrossRef]

409. Kumar, M.; Ojha, S.; Rai, P.; Joshi, A.; Kamat, S.S.; Mallik, R. Insulin activates intracellular transport of lipid droplets to release triglycerides from the liver. *J. Cell Biol.* **2019**, *218*, 3697–3713. [CrossRef]

410. Singh, R.; Kaushik, S.; Wang, Y.; Xiang, Y.; Novak, I.; Komatsu, M.; Tanaka, K.; Cuervo, A.M.; Czaja, M.J. Autophagy regulates lipid metabolism. *Nature* **2009**, *458*, 1131–1135. [CrossRef]

411. Singh, R.; Cuervo, A.M. Lipophagy: Connecting autophagy and lipid metabolism. *Int. J. Cell Biol.* **2012**, *2012*, 282041. [CrossRef]

412. Korovila, I.; Höhn, A.; Jung, T.; Grune, T.; Ott, C. Reduced Liver Autophagy in High-Fat Diet Induced Liver Steatosis in New Zealand Obese Mice. *Antioxidants* **2021**, *10*, 501. [CrossRef]

413. Greenberg, A.S.; Coleman, R.A.; Kraemer, F.B.; McManaman, J.L.; Obin, M.S.; Puri, V.; Yan, Q.W.; Miyoshi, H.; Mashek, D.G. The role of lipid droplets in metabolic disease in rodents and humans. *J. Clin. Investig.* **2011**, *121*, 2102–2110. [CrossRef]

414. Unger, R.H.; Orci, L. Lipoapoptosis: Its mechanism and its diseases. *Biochim. Biophys. Acta* **2002**, *1585*, 202–212. [CrossRef]

415. Akazawa, Y.; Nakao, K. Lipotoxicity pathways intersect in hepatocytes: Endoplasmic reticulum stress, c-Jun N-terminal kinase-1, and death receptors. *Hepatol. Res.* **2016**, *46*, 977–984. [CrossRef]

416. Gaggini, M.; Ndreu, R.; Michelucci, E.; Rocchiccioli, S.; Vassalle, C. Ceramides as Mediators of Oxidative Stress and Inflammation in Cardiometabolic Disease. *Int. J. Mol. Sci.* **2022**, *23*, 2719. [CrossRef]

417. Poli, G.; Schaur, R.J.; Siems, W.G.; Leonarduzzi, G. 4-hydroxynonenal: A membrane lipid oxidation product of medicinal interest. *Med. Res. Rev.* **2008**, *28*, 569–631. [CrossRef]

418. Gueraud, F.; Atalay, M.; Bresgen, N.; Cipak, A.; Eckl, P.M.; Huc, L.; Jouanin, I.; Siems, W.; Uchida, K. Chemistry and biochemistry of lipid peroxidation products. *Free Radic. Res.* **2010**, *44*, 1098–1124. [CrossRef]

419. Dixon, S.J.; Lemberg, K.M.; Lamprecht, M.R.; Skouta, R.; Zaitsev, E.M.; Gleason, C.E.; Patel, D.N.; Bauer, A.J.; Cantley, A.M.; Yang, W.S.; et al. Ferroptosis: An iron-dependent form of nonapoptotic cell death. *Cell* **2012**, *149*, 1060–1072. [CrossRef] [PubMed]

420. Xiong, F.; Zhou, Q.; Huang, X.; Cao, P.; Wang, Y. Ferroptosis plays a novel role in nonalcoholic steatohepatitis pathogenesis. *Front. Pharmacol.* **2022**, *13*, 1055793. [CrossRef]

421. Kazankov, K.; Jørgensen, S.M.D.; Thomsen, K.L.; Møller, H.J.; Vilstrup, H.; George, J.; Schuppan, D.; Grønbæk, H. The role of macrophages in nonalcoholic fatty liver disease and nonalcoholic steatohepatitis. *Nat. Rev. Gastroenterol. Hepatol.* **2019**, *16*, 145–159. [CrossRef]

422. Tiniakos, D.G.; Vos, M.B.; Brunt, E.M. Nonalcoholic fatty liver disease: Pathology and pathogenesis. *Annu. Rev. Pathol.* **2010**, *5*, 145–171. [CrossRef]

423. Takahashi, Y.; Fukusato, T. Histopathology of nonalcoholic fatty liver disease/nonalcoholic steatohepatitis. *World J. Gastroenterol.* **2014**, *20*, 15539–15548. [CrossRef]

424. Matsumoto, A.; Naito, M.; Itakura, H.; Ikemoto, S.; Asaoka, H.; Hayakawa, I.; Kanamori, H.; Aburatani, H.; Takaku, F.; Suzuki, H.; et al. Human macrophage scavenger receptors: Primary structure, expression, and localization in atherosclerotic lesions. *Proc. Natl. Acad. Sci. USA* **1990**, *87*, 9133–9137. [CrossRef]

425. Moore, K.J.; Kunjathoor, V.V.; Koehn, S.L.; Manning, J.J.; Tseng, A.A.; Silver, J.M.; McKee, M.; Freeman, M.W. Loss of receptor-mediated lipid uptake via scavenger receptor A or CD36 pathways does not ameliorate atherosclerosis in hyperlipidemic mice. *J. Clin. Investig.* **2005**, *115*, 2192–2201. [CrossRef]

426. Gerrity, R.G. The role of the monocyte in atherogenesis: I. Transition of blood-borne monocytes into foam cells in fatty lesions. *Am. J. Pathol.* **1981**, *103*, 181–190. [PubMed]

427. Tabas, I. Nonoxidative modifications of lipoproteins in atherogenesis. *Annu. Rev. Nutr.* **1999**, *19*, 123–139. [CrossRef]

428. Pohl, A.; Devaux, P.F.; Herrmann, A. Function of prokaryotic and eukaryotic ABC proteins in lipid transport. *Biochim. Biophys. Acta* **2005**, *1733*, 29–52. [CrossRef] [PubMed]

429. Cuchel, M.; Rader, D.J. Macrophage reverse cholesterol transport: Key to the regression of atherosclerosis? *Circulation* **2006**, *113*, 2548–2555. [CrossRef]

430. Chang, T.Y.; Chang, C.C.; Lin, S.; Yu, C.; Li, B.L.; Miyazaki, A. Roles of acyl-coenzyme A:cholesterol acyltransferase-1 and -2. *Curr. Opin. Lipidol.* **2001**, *12*, 289–296. [CrossRef]

431. Yano, H.; Fujiwara, Y.; Horlad, H.; Pan, C.; Kai, K.; Niino, D.; Ohsawa, K.; Higashi, M.; Nosaka, K.; Okuno, Y.; et al. Blocking cholesterol efflux mechanism is a potential target for antilymphoma therapy. *Cancer Sci.* **2022**, *113*, 2129–2143. [CrossRef]

432. Hynynen, R.; Suchanek, M.; Spandl, J.; Bäck, N.; Thiele, C.; Olkkonen, V.M. OSBP-related protein 2 is a sterol receptor on lipid droplets that regulates the metabolism of neutral lipids. *J. Lipid Res.* **2009**, *50*, 1305–1315. [CrossRef]

433. Guyard, V.; Monteiro-Cardoso, V.F.; Omrane, M.; Sauvanet, C.; Houcine, A.; Boulogne, C.; Ben Mbarek, K.; Vitale, N.; Faklaris, O.; El Khallouki, N.; et al. ORP5 and ORP8 orchestrate lipid droplet biogenesis and maintenance at ER-mitochondria contact sites. *J. Cell Biol.* **2022**, *221*, 2107. [CrossRef]

434. Anderson, A.; Campo, A.; Fulton, E.; Corwin, A.; Jerome, W.G., 3rd; O'Connor, M.S. 7-Ketocholesterol in disease and aging. *Redox Biol.* **2020**, *29*, 101380. [CrossRef]

435. Ghzaiel, I.; Nury, T.; Zarrouk, A.; Vejux, A.; Lizard, G. Oxiapoptophagy in Age-Related Diseases. Comment on Ouyang et al. 7-Ketocholesterol Induces Oxiapoptophagy and Inhibits Osteogenic Differentiation in MC3T3-E1 Cells. *Cells* **2022**, *11*, 2882. [CrossRef]

436. Monier, S.; Samadi, M.; Prunet, C.; Denance, M.; Laubriet, A.; Athias, A.; Berthier, A.; Steinmetz, E.; Jürgens, G.; Nègre-Salvayre, A.; et al. Impairment of the cytotoxic and oxidative activities of 7 beta-hydroxycholesterol and 7-ketocholesterol by esterification with oleate. *Biochem. Biophys. Res. Commun.* **2003**, *303*, 814–824. [CrossRef]

437. Nury, T.; Zarrouk, A.; Yammine, A.; Mackrill, J.J.; Vejux, A.; Lizard, G. Oxiapoptophagy: A type of cell death induced by some oxysterols. *Br. J. Pharmacol.* **2021**, *178*, 3115–3123. [CrossRef]

438. Ouyang, J.; Xiao, Y.; Ren, Q.; Huang, J.; Zhou, Q.; Zhang, S.; Li, L.; Shi, W.; Chen, Z.; Wu, L. 7-Ketocholesterol Induces Oxiapoptophagy and Inhibits Osteogenic Differentiation in MC3T3-E1 Cells. *Cells* **2022**, *11*, 2882. [CrossRef]

439. Samadi, A.; Sabuncuoglu, S.; Samadi, M.; Isikhan, S.Y.; Chirumbolo, S.; Peana, M.; Lay, I.; Yalcinkaya, A.; Bjørklund, G. A Comprehensive Review on Oxysterols and Related Diseases. *Curr. Med. Chem.* **2021**, *28*, 110–136. [CrossRef]

440. Lee-Rueckert, M.; Lappalainen, J.; Kovanen, P.T.; Escola-Gil, J.C. Lipid-Laden Macrophages and Inflammation in Atherosclerosis and Cancer: An Integrative View. *Front. Cardiovasc. Med.* **2022**, *9*, 777822. [CrossRef]

441. Odegaard, J.I.; Chawla, A. Alternative macrophage activation and metabolism. *Annu. Rev. Pathol.* **2011**, *6*, 275–297. [CrossRef]

442. Lathe, R.; Sapronova, A.; Kotelevtsev, Y. Atherosclerosis and Alzheimer–diseases with a common cause? Inflammation, oxysterols, vasculature. *BMC Geriatr.* **2014**, *14*, 36. [CrossRef]

443. Razani, B.; Feng, C.; Coleman, T.; Emanuel, R.; Wen, H.; Hwang, S.; Ting, J.P.; Virgin, H.W.; Kastan, M.B.; Semenkovich, C.F. Autophagy links inflammasomes to atherosclerotic progression. *Cell Metab.* **2012**, *15*, 534–544. [CrossRef]

444. Robichaud, S.; Rasheed, A.; Pietrangelo, A.; Kim, A.D.; Boucher, D.M.; Emerton, C.; Vijithakumar, V.; Gharibeh, L.; Fairman, G.; Mak, E.; et al. Autophagy Is Differentially Regulated in Leukocyte and Nonleukocyte Foam Cells during Atherosclerosis. *Circ. Res.* **2022**, *130*, 831–847. [CrossRef]

445. Feil, S.; Fehrenbacher, B.; Lukowski, R.; Essmann, F.; Schulze-Osthoff, K.; Schaller, M.; Feil, R. Transdifferentiation of vascular smooth muscle cells to macrophage-like cells during atherogenesis. *Circ. Res.* **2014**, *115*, 662–667. [CrossRef]

446. Banke, N.H.; Wende, A.R.; Leone, T.C.; O'Donnell, J.M.; Abel, E.D.; Kelly, D.P.; Lewandowski, E.D. Preferential oxidation of triacylglyceride-derived fatty acids in heart is augmented by the nuclear receptor PPARalpha. *Circ. Res.* **2010**, *107*, 233–241. [CrossRef]

447. Barger, P.M.; Brandt, J.M.; Leone, T.C.; Weinheimer, C.J.; Kelly, D.P. Deactivation of peroxisome proliferator-activated receptor-alpha during cardiac hypertrophic growth. *J. Clin. Investig.* **2000**, *105*, 1723–1730. [CrossRef]

448. Ding, G.; Fu, M.; Qin, Q.; Lewis, W.; Kim, H.W.; Fukai, T.; Bacanamwo, M.; Chen, Y.E.; Schneider, M.D.; Mangelsdorf, D.J.; et al. Cardiac peroxisome proliferator-activated receptor gamma is essential in protecting cardiomyocytes from oxidative damage. *Cardiovasc. Res.* **2007**, *76*, 269–279. [CrossRef]

449. Amen, T.; Kaganovich, D. Small Molecule Screen Reveals Joint Regulation of Stress Granule Formation and Lipid Droplet Biogenesis. *Front. Cell Dev. Biol.* **2020**, *8*, 606111. [CrossRef]

450. Sharma, S.; Adrogue, J.V.; Golfman, L.; Uray, I.; Lemm, J.; Youker, K.; Noon, G.P.; Frazier, O.H.; Taegtmeyer, H. Intramyocardial lipid accumulation in the failing human heart resembles the lipotoxic rat heart. *FASEB J.* **2004**, *18*, 1692–1700. [CrossRef]

451. Cui, X.; Wang, J.; Zhang, Y.; Wei, J.; Wang, Y. Plin5, a New Target in Diabetic Cardiomyopathy. *Oxidative Med. Cell. Longev.* **2022**, *2022*, 2122856. [CrossRef]

452. Zhang, X.; Xu, W.; Xu, R.; Wang, Z.; Zhang, X.; Wang, P.; Peng, K.; Li, M.; Li, J.; Tan, Y.; et al. Plin5 Bidirectionally Regulates Lipid Metabolism in Oxidative Tissues. *Oxidative Med. Cell. Longev.* **2022**, *2022*, 4594956. [CrossRef]
453. Wang, H.; Sreenivasan, U.; Hu, H.; Saladino, A.; Polster, B.M.; Lund, L.M.; Gong, D.W.; Stanley, W.C.; Sztalryd, C. Perilipin 5, a lipid droplet-associated protein, provides physical and metabolic linkage to mitochondria. *J. Lipid Res.* **2011**, *52*, 2159–2168. [CrossRef]
454. Holzem, K.M.; Vinnakota, K.C.; Ravikumar, V.K.; Madden, E.J.; Ewald, G.A.; Dikranian, K.; Beard, D.A.; Efimov, I.R. Mitochondrial structure and function are not different between nonfailing donor and end-stage failing human hearts. *Faseb J.* **2016**, *30*, 2698–2707. [CrossRef]
455. Pollak, N.M.; Schweiger, M.; Jaeger, D.; Kolb, D.; Kumari, M.; Schreiber, R.; Kolleritsch, S.; Markolin, P.; Grabner, G.F.; Heier, C.; et al. Cardiac-specific overexpression of perilipin 5 provokes severe cardiac steatosis via the formation of a lipolytic barrier. *J. Lipid Res.* **2013**, *54*, 1092–1102. [CrossRef]
456. Wang, H.; Sreenivasan, U.; Gong, D.W.; O'Connell, K.A.; Dabkowski, E.R.; Hecker, P.A.; Ionica, N.; Konig, M.; Mahurkar, A.; Sun, Y.; et al. Cardiomyocyte-specific perilipin 5 overexpression leads to myocardial steatosis and modest cardiac dysfunction. *J. Lipid Res.* **2013**, *54*, 953–965. [CrossRef]
457. Kuramoto, K.; Okamura, T.; Yamaguchi, T.; Nakamura, T.Y.; Wakabayashi, S.; Morinaga, H.; Nomura, M.; Yanase, T.; Otsu, K.; Usuda, N.; et al. Perilipin 5, a lipid droplet-binding protein, protects heart from oxidative burden by sequestering fatty acid from excessive oxidation. *J. Biol. Chem.* **2012**, *287*, 23852–23863. [CrossRef]
458. Zhu, Y.; Zhang, X.; Zhang, L.; Zhang, M.; Li, L.; Luo, D.; Zhong, Y. Perilipin5 protects against lipotoxicity and alleviates endoplasmic reticulum stress in pancreatic beta-cells. *Nutr. Metab.* **2019**, *16*, 50. [CrossRef]
459. Cinato, M.; Mardani, I.; Miljanovic, A.; Drevinge, C.; Laudette, M.; Bollano, E.; Henricsson, M.; Tolo, J.; Bauza Thorbrugge, M.; Levin, M.; et al. Cardiac Plin5 interacts with SERCA2 and promotes calcium handling and cardiomyocyte contractility. *Life Sci. Alliance* **2023**, *6*, e202201690. [CrossRef]
460. Cruz, A.L.S.; Barreto, E.A.; Fazolini, N.P.B.; Viola, J.P.B.; Bozza, P.T. Lipid droplets: Platforms with multiple functions in cancer hallmarks. *Cell Death Dis.* **2020**, *11*, 105. [CrossRef]
461. Ukraintseva, S.V.; Yashin, A.I. Opposite Phenotypes of Cancer and Aging Arise from Alternative Regulation of Common Signaling Pathways. *Ann. N. Y. Acad. Sci.* **2003**, *1010*, 489–492. [CrossRef]
462. Antunes, P.; Cruz, A.; Barbosa, J.; Bonifácio, V.D.B.; Pinto, S.N. Lipid Droplets in Cancer: From Composition and Role to Imaging and Therapeutics. *Molecules* **2022**, *27*, 991. [CrossRef]
463. Li, Z.; Liu, H.; Luo, X. Lipid droplet and its implication in cancer progression. *Am. J. Cancer Res.* **2020**, *10*, 4112–4122.
464. Lung, J.; Hung, M.-S.; Wang, T.-Y.; Chen, K.-L.; Luo, C.-W.; Jiang, Y.-Y.; Wu, S.-Y.; Lee, L.-W.; Lin, P.-Y.; Chen, F.-F.; et al. Lipid Droplets in Lung Cancers Are Crucial for the Cell Growth and Starvation Survival. *Int. J. Mol. Sci.* **2022**, *23*, 12533. [CrossRef]
465. Shyu, P.; Wong, X.F.A.; Crasta, K.; Thibault, G. Dropping in on lipid droplets: Insights into cellular stress and cancer. *Biosci. Rep.* **2018**, *38*, BSR20180764. [CrossRef]
466. Castelli, S.; De Falco, P.; Ciccarone, F.; Desideri, E.; Ciriolo, M.R. Lipid Catabolism and ROS in Cancer: A Bidirectional Liaison. *Cancers* **2021**, *13*, 5484. [CrossRef]
467. Zhang, J.; Liu, Z.; Lian, Z.; Liao, R.; Chen, Y.; Qin, Y.; Wang, J.; Jiang, Q.; Wang, X.; Gong, J. Monoacylglycerol Lipase: A Novel Potential Therapeutic Target and Prognostic Indicator for Hepatocellular Carcinoma. *Sci. Rep.* **2016**, *6*, 35784. [CrossRef]
468. Patel, D.; Salloum, D.; Saqcena, M.; Chatterjee, A.; Mroz, V.; Ohh, M.; Foster, D.A. A Late G1 Lipid Checkpoint That Is Dysregulated in Clear Cell Renal Carcinoma Cells. *J. Biol. Chem.* **2017**, *292*, 936–944. [CrossRef]
469. Qi, W.; Weber, C.R.; Wasland, K.; Roy, H.; Wali, R.; Joshi, S.; Savkovic, S.D. Tumor suppressor FOXO3 mediates signals from the EGF receptor to regulate proliferation of colonic cells. *Am. J. Physiol. -Gastrointest. Liver Physiol.* **2011**, *300*, G264–G272. [CrossRef] [PubMed]
470. Yu, W.; Bozza, P.T.; Tzizik, D.M.; Gray, J.P.; Cassara, J.; Dvorak, A.M.; Weller, P.F. Co-compartmentalization of MAP kinases and cytosolic phospholipase A2 at cytoplasmic arachidonate-rich lipid bodies. *Am. J. Pathol.* **1998**, *152*, 759–769. [PubMed]
471. Yu, W.; Cassara, J.; Weller, P.F. Phosphatidylinositide 3-kinase localizes to cytoplasmic lipid bodies in human polymorphonuclear leukocytes and other myeloid-derived cells. *Blood* **2000**, *95*, 1078–1085. [CrossRef]
472. Mittal, D.; Gubin, M.M.; Schreiber, R.D.; Smyth, M.J. New insights into cancer immunoediting and its three component phases—Elimination, equilibrium and escape. *Curr. Opin. Immunol.* **2014**, *27*, 16–25. [CrossRef]
473. Wong, R.S.Y. Apoptosis in cancer: From pathogenesis to treatment. *J. Exp. Clin. Cancer Res.* **2011**, *30*, 87. [CrossRef]
474. Bozza, P.T.; Bakker-Abreu, I.; Navarro-Xavier, R.A.; Bandeira-Melo, C. Lipid body function in eicosanoid synthesis: An update. *Prostaglandins Leukot. Essent. Fat. Acids (PLEFA)* **2011**, *85*, 205–213. [CrossRef]
475. Accioly, M.T.; Pacheco, P.; Maya-Monteiro, C.M.; Carrossini, N.; Robbs, B.K.; Oliveira, S.S.; Kaufmann, C.; Morgado-Diaz, J.A.; Bozza, P.T.; Viola, J.P. Lipid bodies are reservoirs of cyclooxygenase-2 and sites of prostaglandin-E2 synthesis in colon cancer cells. *Cancer Res.* **2008**, *68*, 1732–1740. [CrossRef]
476. Johnson, A.M.; Kleczko, E.K.; Nemenoff, R.A. Eicosanoids in Cancer: New Roles in Immunoregulation. *Front. Pharmacol.* **2020**, *11*, 595498. [CrossRef]
477. Finetti, F.; Travelli, C.; Ercoli, J.; Colombo, G.; Buoso, E.; Trabalzini, L. Prostaglandin E2 and Cancer: Insight into Tumor Progression and Immunity. *Biology* **2020**, *9*, 434. [CrossRef]

478. Veglia, F.; Tyurin, V.A.; Mohammadyani, D.; Blasi, M.; Duperret, E.K.; Donthireddy, L.; Hashimoto, A.; Kapralov, A.; Amoscato, A.; Angelini, R.; et al. Lipid bodies containing oxidatively truncated lipids block antigen cross-presentation by dendritic cells in cancer. *Nat. Commun.* **2017**, *8*, 2122. [CrossRef]

479. Hakumaki, J.M.; Kauppinen, R.A. 1H NMR visible lipids in the life and death of cells. *Trends Biochem. Sci.* **2000**, *25*, 357–362. [CrossRef] [PubMed]

480. Blankenberg, F.G. In vivo detection of apoptosis. *J. Nucl. Med.* **2008**, *49* (Suppl. S2), 81S–95S. [CrossRef] [PubMed]

481. Henique, C.; Mansouri, A.; Fumey, G.; Lenoir, V.; Girard, J.; Bouillaud, F.; Prip-Buus, C.; Cohen, I. Increased mitochondrial fatty acid oxidation is sufficient to protect skeletal muscle cells from palmitate-induced apoptosis. *J. Biol. Chem.* **2010**, *285*, 36818–36827. [CrossRef]

482. Choi, S.E.; Jung, I.R.; Lee, Y.J.; Lee, S.J.; Lee, J.H.; Kim, Y.; Jun, H.S.; Lee, K.W.; Park, C.B.; Kang, Y. Stimulation of lipogenesis as well as fatty acid oxidation protects against palmitate-induced INS-1 beta-cell death. *Endocrinology* **2011**, *152*, 816–827. [CrossRef] [PubMed]

483. Boren, J.; Brindle, K.M. Apoptosis-induced mitochondrial dysfunction causes cytoplasmic lipid droplet formation. *Cell Death Differ.* **2012**, *19*, 1561–1570. [CrossRef]

484. Wang, J.-B.; Qi, L.-L.; Zheng, S.-D.; Wu, T.-X. Curcumin induces apoptosis through the mitochondria-mediated apoptotic pathway in HT-29 cells. *J. Zhejiang Univ. Sci. B* **2009**, *10*, 93–102. [CrossRef]

485. Roy, M.; Chakraborty, S.; Siddiqi, M.; Bhattacharya, R.K. Induction of Apoptosis in Tumor Cells by Natural Phenolic Compounds. *Asian Pac. J. Cancer Prev. APJCP* **2002**, *3*, 61–67.

486. Jiang, M.C.; Yang-Yen, H.F.; Yen, J.J.; Lin, J.K. Curcumin induces apoptosis in immortalized NIH 3T3 and malignant cancer cell lines. *Nutr. Cancer* **1996**, *26*, 111–120. [CrossRef]

487. Cao, J.; Liu, Y.; Jia, L.; Zhou, H.M.; Kong, Y.; Yang, G.; Jiang, L.P.; Li, Q.J.; Zhong, L.F. Curcumin induces apoptosis through mitochondrial hyperpolarization and mtDNA damage in human hepatoma G2 cells. *Free Radic. Biol. Med.* **2007**, *43*, 968–975. [CrossRef]

488. Zhang, I.; Cui, Y.; Amiri, A.; Ding, Y.; Campbell, R.E.; Maysinger, D. Pharmacological inhibition of lipid droplet formation enhances the effectiveness of curcumin in glioblastoma. *Eur. J. Pharm. Biopharm.* **2016**, *100*, 66–76. [CrossRef]

489. Jin, C.; Yuan, P. Implications of lipid droplets in lung cancer: Associations with drug resistance. *Oncol. Lett.* **2020**, *20*, 2091–2104. [CrossRef] [PubMed]

490. Lossi, L. The concept of intrinsic versus extrinsic apoptosis. *Biochem. J.* **2022**, *479*, 357–384. [CrossRef]

491. Song, J.H.; Tse, M.C.L.; Bellail, A.; Phuphanich, S.; Khuri, F.; Kneteman, N.M.; Hao, C. Lipid Rafts and Nonrafts Mediate Tumor Necrosis Factor–Related Apoptosis-Inducing Ligand–Induced Apoptotic and Nonapoptotic Signals in Non–Small Cell Lung Carcinoma Cells. *Cancer Res.* **2007**, *67*, 6946–6955. [CrossRef]

492. Zembroski, A.S.; Andolino, C.; Buhman, K.K.; Teegarden, D. Proteomic Characterization of Cytoplasmic Lipid Droplets in Human Metastatic Breast Cancer Cells. *Front. Oncol.* **2021**, *11*, 576326. [CrossRef] [PubMed]

493. Sofi, F.; Abbate, R.; Gensini, G.F.; Casini, A. Accruing evidence on benefits of adherence to the Mediterranean diet on health: An updated systematic review and meta-analysis. *Am. J. Clin. Nutr.* **2010**, *92*, 1189–1196. [CrossRef] [PubMed]

494. Russo, M.A.; Sansone, L.; Polletta, L.; Runci, A.; Rashid, M.M.; De Santis, E.; Vernucci, E.; Carnevale, I.; Tafani, M. Sirtuins and resveratrol-derived compounds: A model for understanding the beneficial effects of the Mediterranean diet. *Endocr. Metab. Immune Disord. Drug Targets* **2014**, *14*, 300–308. [CrossRef]

495. Park, S.J.; Ahmad, F.; Philp, A.; Baar, K.; Williams, T.; Luo, H.; Ke, H.; Rehmann, H.; Taussig, R.; Brown, A.L.; et al. Resveratrol ameliorates aging-related metabolic phenotypes by inhibiting cAMP phosphodiesterases. *Cell* **2012**, *148*, 421–433. [CrossRef]

 *biomolecules* 

Article

Effect of mitoTEMPO on Redox Reactions in Different Body Compartments upon Endotoxemia in Rats

Adelheid Weidinger [1,†] , Andras T. Meszaros [1,2,†] , Sergiu Dumitrescu [1] and Andrey V. Kozlov [1,*]

1   Ludwig Boltzmann Institute for Traumatology, The Research Center in Cooperation with AUVA, 1200 Vienna, Austria
2   Department of Visceral, Transplant and Thoracic Surgery, Medical University of Innsbruck, 6020 Innsbruck, Austria
*   Correspondence: andrey.kozlov@trauma.lbg.ac.at; Tel.: +43-059393-41980
†   These authors contributed equally to this work.

Abstract: Mitochondrial ROS (mitoROS) control many reactions in cells. Biological effects of mitoROS *in vivo* can be investigated by modulation via mitochondria-targeted antioxidants (mtAOX, mitoTEMPO). The aim of this study was to determine how mitoROS influence redox reactions in different body compartments in a rat model of endotoxemia. We induced inflammatory response by lipopolysaccharide (LPS) injection and analyzed effects of mitoTEMPO in blood, abdominal cavity, bronchoalveolar space, and liver tissue. MitoTEMPO decreased the liver damage marker aspartate aminotransferase; however, it neither influenced the release of cytokines (e.g., tumor necrosis factor, IL-4) nor decreased ROS generation by immune cells in the compartments examined. In contrast, *ex vivo* mitoTEMPO treatment substantially reduced ROS generation. Examination of liver tissue revealed several redox paramagnetic centers sensitive to *in vivo* LPS and mitoTEMPO treatment and high levels of nitric oxide (NO) in response to LPS. NO levels in blood were lower than in liver, and were decreased by *in vivo* mitoTEMPO treatment. Our data suggest that (i) inflammatory mediators are not likely to directly contribute to ROS-mediated liver damage and (ii) mitoTEMPO is more likely to affect the redox status of liver cells reflected in a redox change of paramagnetic molecules. Further studies are necessary to understand these mechanisms.

Keywords: reactive oxygen species; systemic inflammatory response syndrome; cytokines; mitochondria-targeted antioxidants; bronchoalveolar system; peritoneal lavage; mitoTEMPO

Citation: Weidinger, A.; Meszaros, A.T.; Dumitrescu, S.; Kozlov, A.V. Effect of mitoTEMPO on Redox Reactions in Different Body Compartments upon Endotoxemia in Rats. *Biomolecules* **2023**, *13*, 794. https://doi.org/10.3390/biom13050794

Academic Editors: José Pedraza Chaverri, Mark Rinnerthaler and Markus Ralser

Received: 22 February 2023
Revised: 27 April 2023
Accepted: 29 April 2023
Published: 5 May 2023

1. Introduction

Systemic inflammation and consequent systemic inflammatory response syndrome (SIRS) is a complex cascade of pro- and anti-inflammatory processes. It is associated with release of inflammatory mediators such as tumor necrosis factor (TNF)-α and interleukins (IL-1, IL-6, IL-4, MCP1 and others) [1] as well as by excessive generation of nitric oxide (NO) by inducible nitric oxide synthase (iNOS), thought to be a mediator of distant organ damage [2]. The exact mechanism of cellular damage is not fully clarified, although the theory of a deleterious interplay of inflammatory mediators complemented by immune cell cascade is widely accepted [1]. The question remains of the extent to which immune cells and their products play a causative role in the distant organ damage.

The inflammatory response is often accompanied by multiple organ dysfunction syndrome (MODS), which is often associated with excessive generation of reactive oxygen and nitrogen species (RONS). It has been shown that this process is orchestrated by mitochondrial ROS (mitoROS), mainly superoxide ($O_2^{\bullet-}$), a primary ROS generated by the electron transfer system in the mitochondrial inner membrane [3]. However, it is very difficult to dissect direct effects of ROS *in vivo* due to their very short lifetime. A common and precise way to access these biological effects *in vivo* is the examination of effects of so-called mitochondria-targeted antioxidants (mtAOX). These molecules comprise two structural parts. The first is positively charged and hydrophobic. This part is considered as chemically inert, and brings the entire molecule to the negatively charged mitochondrial matrix. The second part has specific ROS-scavenging moieties [4]. TEMPO is the most often used antioxidant part of the molecule.

Generally, the major reason for increased generation of mitochondrial ROS is a shift in the redox equilibrium within the cells. These disturbances are often accompanied by changes in intracellular redox active centers and deviation of electrons from their physiological pathways (redox shuttles) to one-electron reduction of oxygen-yielding superoxide radicals [5]. Such redox-motivated electron transfer appears in the mitochondrial electron transport chain (ETC) or electron transport system linked to cytochrome P450 in the endoplasmic reticulum. One-electron reduction of oxygen appears under pathological circumstances such as hypoxia or inflammation. The major sites of leakage of electrons in mitochondria are Complex I and III of the electron transfer system [6]. Another important component of the respiratory chain is ubiquinone (Q10), which exerts both pro- and antioxidant capacities [7]. A number of these centers containing enzymes of mitochondria, such as P450, Q10, and iron, can be detected in a specific redox state by electron paramagnetic resonance spectroscopy (EPR).

Spatially, there are two general sources of ROS that are considered potentially damaging for tissues. Extracellular ROS are mainly generated by immune cells [8], while intracellular ROS production leads to cell damage from intracellular sources [2]. Thus, it can be expected that the antioxidant mitoTEMPO acts both in intra- and extra-cellular environments owing to the almost ubiquitous presence of mitochondria.

To this end, the effects of mtAOX on ROS generation and on various redox processes were evaluated in a lipopolysaccharide (LPS)-induced rodent SIRS model. LPS induces acute systemic inflammatory response in rodents [9] and humans [10]. This model mirrors certain aspects of septic shock in humans, though the correlation between rodent LPS models and clinical septic shock is poor [11]. Nonetheless, this model has provided the majority of mechanistic insights into systemic inflammation [10], such as the release of cytokines [12], induction of apoptosis [13], signal transduction [14], and others.

In the present study, we set out to characterize ROS generation in three major pathologically relevant sites of oxidative stress: (1) in the intravascular compartment (blood), (2) in the intraabdominal compartment, and (3) in the bronchoalveolar system. To the best of our knowledge, there are no studies on the effects of mitoTEMPO in both body compartments and in a reference tissue (liver) susceptible to inflammatory damage.

2. Materials and Methods

2.1. Chemicals

All chemicals were obtained from Sigma-Aldrich (St. Louis, MO, USA) unless otherwise noted.

2.2. Animals

The *in vivo* experiments were performed on male Sprague Dawley rats (250–300 g/ 390–540 g; Animal Research Laboratories, Himberg, Austria/Charles River, Germany) which were kept under controlled standard animal housing conditions at least for 7 days prior to usage in experiments with free access to standard laboratory rodent food and water. The rats with high weight were used in experiments first; small animals were kept approximately 2 weeks longer in the animal research facility of the institute. All interventions were conducted in compliance with the Guide for the Care and Use of Laboratory Animals published by the National Institute of Health with approval from the Animal Protocol Review Board of the city government of Vienna, Austria (no. M58003956-2011-9, no. 815758/2014/16). To prevent unnecessary pain, Buprenorphin (Richter Pharma AG, Wels, Austria, 0.05 mg/kg body weight) was injected subcutaneously at the time of the LPS treatment and 10 h thereafter. At the end of experiments, rats were anesthetized by inhalation of a mixture of 3% isoflurane and oxygen (Vapor 2000, Dräger, Austria).

2.3. Lipopolysaccharide and mtAOX Treatment

Rats were injected with lipopolysaccharide (LPS) from Escherichia coli serotype 026:B6 (activity $\geq$ 500,000 EU/mg) at a dose of 2.5 mg/kg body weight dissolved in saline (Fresenius Kabi, Graz, Austria). Samples were collected at 16 h after LPS injection. In a separate set of experiments (Figure 1), animals were divided into four groups. All rats were injected with the same dose of LPS (Escherichia coli serotype 026:B6, 8 mg/kg body weight, activity $\geq$ 10,000 EU/mg) dissolved in saline. Control animals were injected with saline only. Treatment with mitoTEMPO (50 nmol/kg) was administered intraperitoneally at 1 h before LPS treatment and 11 h after LPS treatment. The LPS solution was vortexed for 1 min and sonicated for 30 s before application, then injected in the penis vein under isoflurane anesthesia (Vapor 2000, Dräger, Austria) in volumes ranging from 0.5 to 0.75 mL.

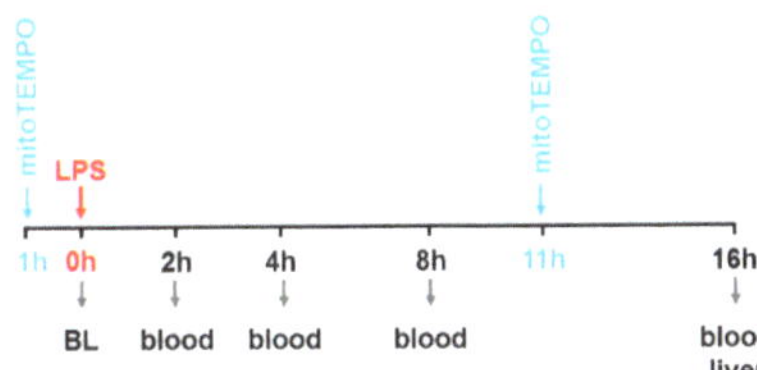

Figure 1. Experimental protocol of LPS and mitoTEMPO treatment followed by blood and tissue sampling. Treatment with mitoTEMPO was administered intraperitoneally at 1 h before LPS treatment and 11 h after LPS treatment. Blood samples were collected at 0/2/4/8/16 h and tissue samples at 16 h.

2.4. Sampling of Blood and Liver Tissue

After a small skin cut was made, the left femoral artery was dissected and catheterized using a 24 GA i.v. cannula (BD Neoflon, Becton Dickinson Infusion Therapy AB, Helsingborg, Sweden). Next, 10–12 mL of whole blood was collected in a 50 mL Falcon tube prefilled with 200 µL of sodium heparin (1000 IU/mL, Gilvasan Pharma GmbH, Austria) for subsequent processing and for the in vitro part of the study.

For determination of mononitrosyl–hemoglobin complex (NO-Hb) levels, blood samples were taken into Minicollect tubes coated with lithium heparin (Greiner Bio-One GmbH, Kremsmünster, Austria). After centrifugation at 4 °C and 1600× g for 10 min, plasma was removed and the erythrocyte pellet was collected in 1 mL plastic syringes. Subsequently, erythrocyte pellets were shock-frozen in liquid nitrogen and stored at −80 °C until EPR measurement.

Following blood sampling, animals were euthanized by decapitation. Subsequently, the liver was excised and transferred immediately to a beaker filled with ice cold Ringer solution (Fresenius Kabi, Austria). After cooling, the liver was cut into small pieces on a Petri dish placed on ice. Tissue samples to a volume of 0.4 mL were filled in 1 mL plastic syringes, shock-frozen in liquid nitrogen, and stored at −80 °C for further measurements. Liver tissue for measurement of mitochondrial respiration was used freshly.

2.5. Measurement of Mitochondrial Respiration

Respiratory parameters of mitochondria were monitored using high-resolution respirometry (Oxygraph-2k, Oroboros Instruments, Innsbruck, Austria) as previously described [15]. Rat liver homogenate was incubated in buffer containing 105 mM KCl, 5 mM KH_2PO_4, 20 mM Tris-HCl, 0.5 mM EDTA, and 5 mg/mL fatty acid-free bovine serum albumin (pH 7.2, 37 °C). Respiration was stimulated by the addition of 5 mM glutamate and 5 mM malate. Transition to State-3 respiration was induced by the addition of 1 mM adenosine diphosphate. Oxygen consumption rates were obtained by calculating the negative time derivative of the measured oxygen concentration. The respiratory control ratio was calculated by dividing State 3 respiration by State 2 respiration.

2.6. Electron Paramagnetic Resonance Spectroscopy

EPR is the most appropriate method to detect redox active centers directly in untreated tissues. It can be used to determine redox-active iron-sulfur species [16], copper containing compounds [17], free radicals and ferrous ions [18], and nitric oxide [19]. A particular advantage of this method is that the analytic procedure can be performed directly in frozen tissue biopsies at liquid nitrogen temperature [19]. This ensures the absence of artefacts due to processing of tissues, such as homogenization, extraction etc.

EPR spectra were recorded at liquid nitrogen temperature (-196 °C) with a Magnettech MiniScope MS 200 EPR spectrometer (Magnettech Ltd., Berlin, Germany) at a modulation frequency of 100 kHz and microwave frequency of 9.429 GHz. The settings for NO-Hb spectra in blood samples were microwave power 8.3 mW and modulation amplitude 5 G. NO-Hb complexes were recorded at 3300 ± 200 G and quantified by double-integrating the EPR spectra. The settings for NO-Hb and Fe-NO complex in liver samples were microwave power 30 mW and modulation amplitude 6 G. Liver spectra were recorded at 3200 ± 500 G. The settings for p450 (g = 2.25), free radicals (g = 2.002), and succinate dehydrogenase (g = 1.94) were microwave power 1 mW and modulation amplitude 5 G). The intensity of the signals was recorded at 3200 ± 1000 G.

2.7. Analysis of Total Nitric Oxide

Total nitric oxide levels (NOx) were analyzed with Sievers 280i-NO Analyzer (General Electric Company, Boston, MA, USA) as previously described [20]. Plasma samples were injected through a septum into the glass vessel, where NO species were converted by a redox active reagent (VCl3) to NO(g). Subsequent reaction with ozone causing photon emission was detected as chemiluminescence intensity.

2.8. Sampling of Peritoneal and Bronchoalveolar Lavage

Bronchoalveolar lavage was collected as described elsewhere [21]. A 19-gauge needle hub was inserted into the trachea and 3 mL of ice-cold phosphate-buffered saline (PBS) and 10% fetal bovine serum (FBS) were administered and aspirated slowly through the needle hub. After transfer into a 15 mL tube (Greiner Bio-One, Kremsmünster, Austria), the fluid was kept on ice. The procedure was repeated four times.

For collection of peritoneal cells, we used a protocol adapted from [22]. Briefly, the peritoneum was exposed and 5 mL of ice-cold PBS with 3% FBS was injected into the peritoneal cavity through a needle. A gentle massage to the abdomen was applied to dislodge any attached cells into the PBS solution. The suspension was collected through a 22-gauge needle into a syringe and transferred into a 15 mL tube, then kept on ice. The above-described procedure was repeated five times.

2.9. Isolation of Cells

The suspension of bronchoalveolar and peritoneal cells was centrifuged at $400 \times g$ for 10 min at 4 °C, the supernatant was discarded, and the cell number was counted with a Cell-Dyn 3700 Hematology Analyzer (Abbott Laboratories, Lake Bluff, IL, USA). Cell counts in blood and in peritoneal and bronchoalveolar fluids showed mainly polymorphonuclear leukocytes (PMN, mainly neutrophil granulocytes; see Supplementary Figure S1). After the blood count, blood samples were centrifuged at $300 \times g$ for 10 min at 4 °C, then the plasma and buffy coat were removed and the remaining blood pellet was incubated with a lysis buffer containing 168 mM NH4Cl, 10 mM KHCO3, and 973 µM EDTA for 10 min at 4 °C. Cells were then washed twice with 10 °C cold PBS and centrifuged at $400 \times g$ for 8 min before finally performing the counts. For the in vitro mtAOX treatment, cell suspensions were incubated at 37 °C for 45 min with mitoTEMPO (500 nM) or vehicle (NaCl).

2.10. Ex Vivo ROS Measurement

The H_2O_2 production in the cell suspension was assessed by N-acetyl-3,7-dihydroxy phenoxazine (Amplex Red, Life Technologies, Eugene, OR, USA). Amplex Red is a sensitive and chemically stable fluorogenic probe, and produces fluorescent resolufin with H_2O_2 in a horseradish peroxidase (HRP)-catalyzed oxidation with excitation/emission maxima at 563/587. The reaction stoichiometry of Amplex Red and H_2O_2 is 1:1 [21]. Using a fluorescence plate reader (POLARstar Omega 3MG, Labtech, Germany), white blood cells, bronchoalveolar lavage, and peritoneal lavage cells (15×10^3 cells per well in Krebs buffer) were incubated in black 96-well microplates (Greiner Cellstar) at 37 °C with Amplex Red (10 µM) and HRP ($0.2\ U\ mL^{-1}$). The fluorescence intensity was recorded for 30 min with an excitation of 544 nm/emission of 590 nm and gain at 1200. The slope was calculated from a 10 min interval between the 5th and 15th minute of the measurement.

2.11. Statistics

Statistical evaluation of the experimental results was performed using GraphPad Prism(Version 9.4.1., GraphPad Software, Boston, MA, USA). The results are shown as the mean ± standard error of the mean (SEM). Statistical significance was evaluated by ANOVA followed by Post Hoc, Holm-Šídák's multiple comparisons test unless indicated otherwise in the figure legends; n numbers are indicated in the figure legends. Statistical significance is indicated as follows: $* p < 0.05$; $** p < 0.01$; $*** p < 0.001$; $**** p < 0.0001$.

3. Results

3.1. LPS Treatment Leads to Tissue Damage in an mtAOX-Dependent Way

As a first step, we validated our rat endotoxemia model (Figure 1).

By 16 h post-LPS challenge, an increase in aspartate aminotransferase AST (Figure 2a) could be observed. MitoTEMPO pretreatment, however, could ameliorate this parenchymal damage. According to the literature, an early peak of tumor necrosis factor (TNF)-α can be seen at 2 h post-LPS in the plasma [2]. In this model, *in vivo* treatment with mitoTEMPO neither influenced the release of early acute phase cytokines (TNF, IL-1) into the circulation (Figure 2b,c) nor substantially decreased the levels of cytokines such as IL-4 and MCP-1 linked to the activation of monocytes (Figure 1d,e). In addition, we found that mitochondrial function in the liver is affected by LPS challenge (Figure 1f). As with the AST release, addition of mitoTEMPO normalized the respiratory control ratio (Figure 1f).

3.2. In Vivo mitoTEMPO Treatment Does Not Influence Ex Vivo ROS Generation

Next, we examined the *ex vivo* ROS generation by immune cells isolated from peripheral blood, peritoneal fluid, and lavage from the bronchoalveolar system. However, we did not observe any change in ROS generation in response to *in vivo* mitoTEMPO treatment in any of the studied compartments (Figure 3a–c). Further, to test whether the capacity of cells to generate ROS is affected by mitoTEMPO at all, we activated cells *ex vivo* by phorbol 12-myristate 13-acetate (PMA); again, we did not find any difference in this case (Figure 3d–f). Treatment with mitoTEMPO did not influence the relative counts of immune cells in any of the three compartments (Supplementary Figure S1). These data suggest that either ROS generation by immune cells is not regulated by mitoROS or mitoTEMPO did not affect immune cells *in vivo*.

3.3. Ex Vivo mitoTEMPO Application Reduces ROS Production

To investigate the direct effects of mitoTEMPO on ROS production of immune cells, cells were similarly extracted from all body compartments as described above. In this case, mitoTEMPO was applied *ex vivo*. In contrast to the *in vivo* treatment, mitoTEMPO substantially reduced the rate of ROS generation (Figure 4a–c). Interestingly, this inhibition disappeared when cells were additionally activated *ex vivo* by PMA, with the exception of bronchoalveolar lavage cells (Figure 4d–f).

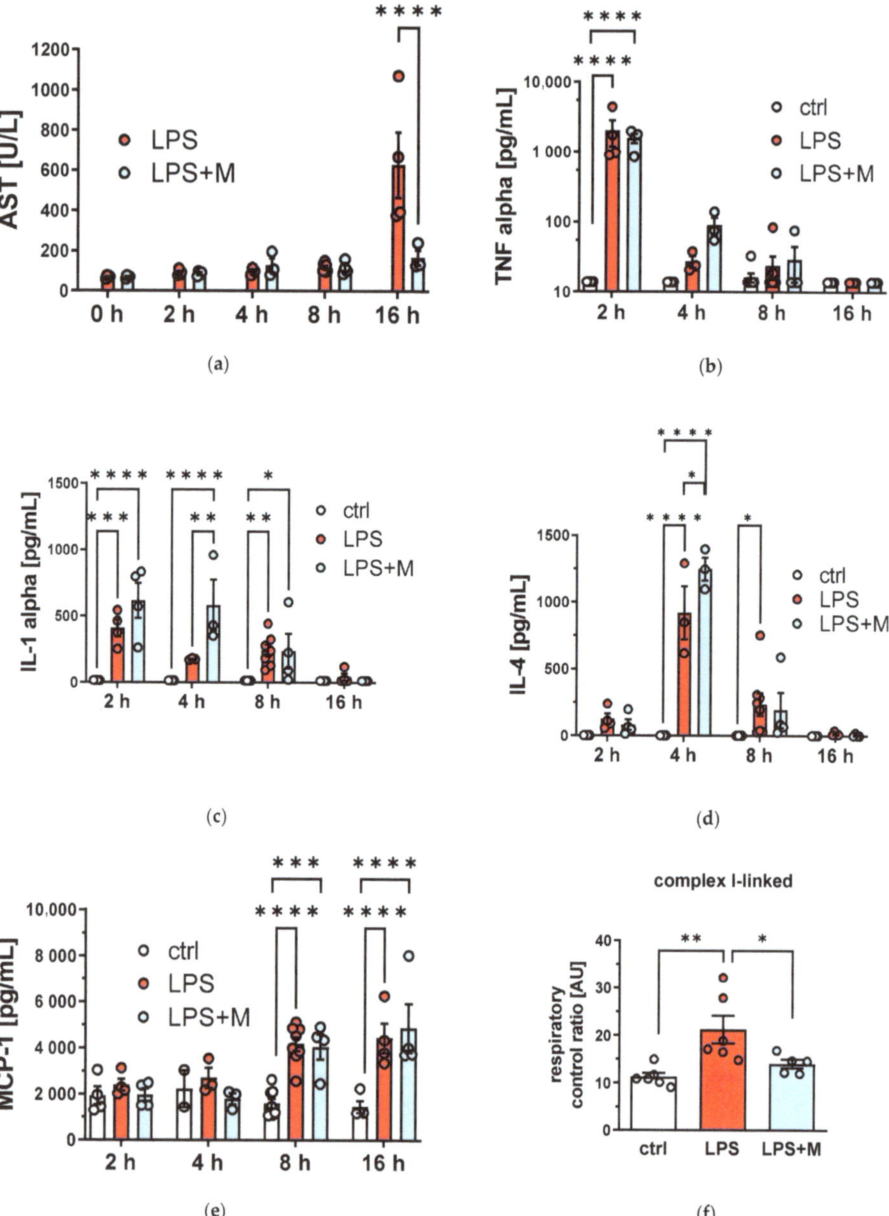

Figure 2. Effect of mitochondria-targeted antioxidant (mitoTEMPO, M) on liver function and humoral immune response after LPS treatment: (**a**) time course of AST (**b**), TNF-alpha (**c**), IL1-alpha (**d**) IL-4 (**e**), and MCP1. The animals were treated with 106 U/kg (approx. 2 mg/kg) LPS or saline (control group) and observed for up to 16 h. Samples were taken at 2/4/8/16 h. (**f**) Mitochondrial respiratory control ratio. Liver samples were taken at 16 h. The data are presented as mean ± SEM. Statistical evaluation was performed by ANOVA followed by Holm–Šídák's multiple comparisons test. n = 3–8. Statistical significance is indicated as follows: * $p < 0.05$; ** $p < 0.01$; *** $p < 0.001$; **** $p < 0.0001$.

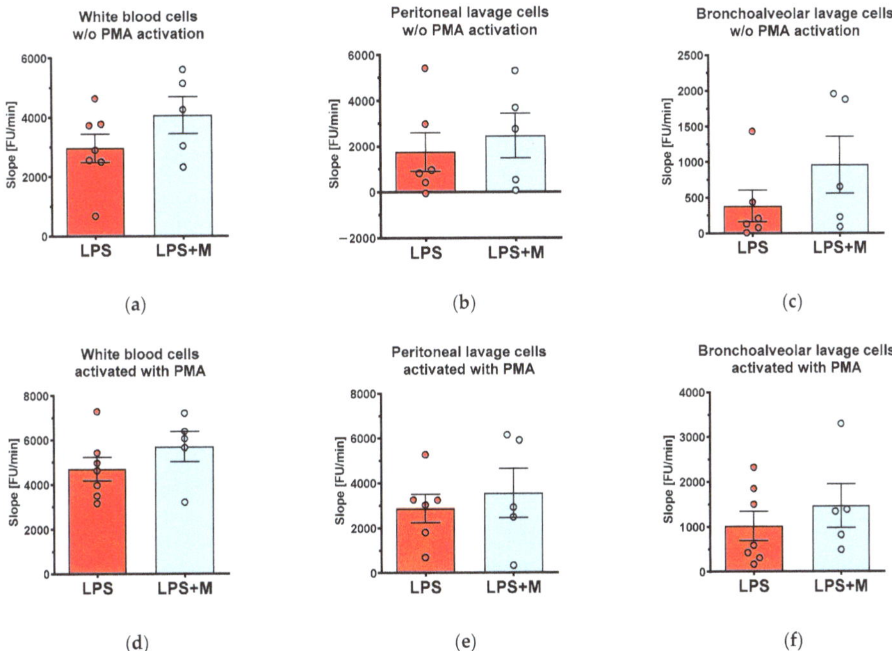

Figure 3. *In vivo* effects of the mitochondria-targeted antioxidant mitoTEMPO (M) on ROS generation in blood and bronchoalveolar lavage after LPS treatment: blood (**a,d**) and peritoneal (**b,e**), and bronchoalveolar (**c,f**) lavage were collected under standardized conditions as described in the methods section. ROS generation was measured in cells without activation with PMA (**a–c**) and after *ex vivo* activation with PMA (**d–f**). The data are presented as mean ± SEM. n = 5–7. Statistical evaluation was performed by ANOVA followed by Holm-Šídák's multiple comparisons test.

3.4. EPR Reveals mitoTEMPO-Dependent Decrease in Free Radical and NO Signals

Investigation of redox-sensitive intracellular paramagnetic centers using EPR technique (Figure 5a) revealed that their redox state is responsive to both LPS and mitoTEMPO. Both substances increased the intensity of the signal coming from p450 (Figure 5b), while the free radical signal, which originates predominantly from mitochondrial Q10, was increased by LPS (Figure 5c), though this increase was diminished by mitoTEMPO (Figure 5c). The signal g = 1.94, which is usually attributed to succinate dehydrogenase of mitochondria, was increased by LPS but did not change in response to mitoTEMPO (Figure 5d). In LPS-treated rats, the concentration of mononitrosyl-hemoglobin-complexes (NO-Hb) determined in liver tissue (Figure 5f) was higher than in the blood (Figure 5e), supporting the assumption that liver is a (the) prominent NO source, eventually higher than the vasculature and the blood itself. Increased concentrations of NO-Hb were attenuated by mitoTEMPO (Figure 5e,f). Similar changes were observed with liver dinitrosyl iron complexes (Fe-NO) reporting intracellular NO levels (Figure 5g). In addition, NOx levels in plasma after LPS treatment, determined by ozone-chemiluminescence technology, were increased (Figure 5h). This increase could be reduced by mitoTEMPO treatment (Figure 5h).

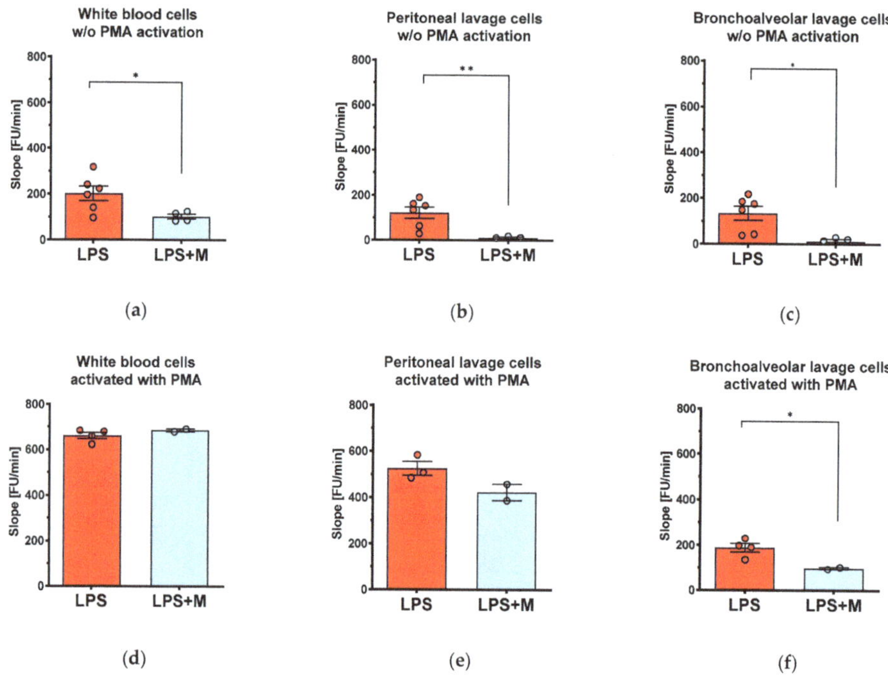

Figure 4. *Ex vivo* effects of mitochondria-targeted antioxidant mitoTEMPO (M) on ROS generation in blood and peritoneal and bronchoalveolar lavage exposed to LPS. Blood (**a,d**) and peritoneal (**b,e**), and bronchoalveolar (**c,f**) lavage were collected under standardized conditions as described in the methods section. ROS generation was measured in cells without activation with PMA (**a–c**) and after *ex vivo* activation with PMA (**d–f**). The data are presented as mean ± SEM. n = 2–6. Statistical evaluation was performed by ANOVA followed by Holm-Šídák's multiple comparisons test. Statistical significance is indicated as follows: * $p < 0.05$; ** $p < 0.01$.

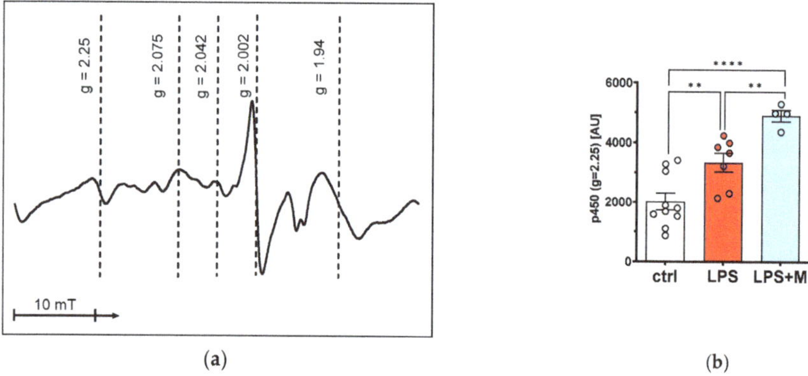

Figure 5. *Cont.*

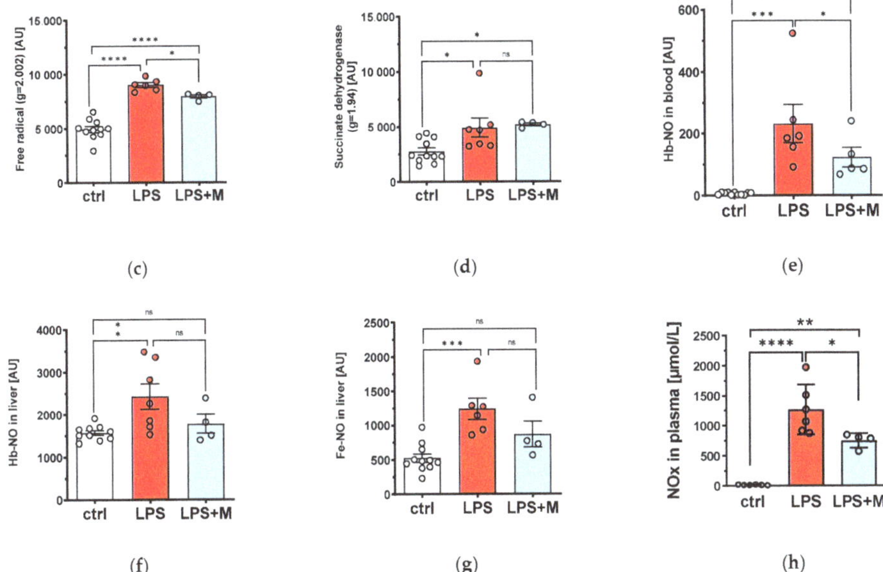

Figure 5. Redox state of paramagnetic centers: EPR spectrum of liver tissue (**a**); changes in the intensity of the cyt P450 signal (g = 2.25) (**b**); changes in the intensity of tissue free radical signal (g = 2.002) (**c**); changes in the intensity of succinate dehydrogenase signal (g = 1.94) (**d**); NO-Hb levels in circulating blood (**e**) and blood in liver vessels (**f**); Fe-NO levels in liver cells (**g**); NOx level in plasma (**h**). The data are presented as mean ± SEM. Statistical evaluation was performed by ANOVA followed by Holm-Šídák's multiple comparisons test. n = 4–10. Statistical significance is indicated as follows: * $p < 0.05$; ** $p < 0.01$; *** $p < 0.001$; **** $p < 0.0001$.

4. Discussion

In the present study, we employed a rodent model, which is extensively described in the literature, to investigate systemic inflammation and consequent organ damage. Bacterial endotoxin (LPS) treatment is known to lead to a dose-dependent increase in mortality along with elevation of organ damage markers and nitric oxide levels, as well as substantial changes in plasma cytokine pattern [23,24]. It is generally accepted that an inflammatory activation of cytokines orchestrated by and acting through ROS play a central role in end-organ damage [25].

Mitochondria-targeted antioxidants such as mitoTEMPO are widely employed to elucidate functional aspects of ROS-dependent mechanisms. Because we have already investigated the effect of mitoTEMPO on control tissues in previous studies [2,26], we did not include this group here. Consequently, the conclusions are limited to comparisons between control and LPS on one hand, and on the other LPS and LPS with mitoTEMPO.

In the present study, mitoTEMPO treatment successfully mitigated the tissue damage to the liver parenchyma, as evidenced by AST levels. This is in line with previous publications [2]. We observed that the respiratory control ratio is increased in response to LPS and normalized upon addition of mitoTEMPO. It has already been reported that upon inflammatory conditions the mitochondrial respiratory function can respond by a decrease or an increase in the capacity of the electron transport chain [27]. In our previous studies on rodents, we observed an increase in the State 3 respiration [28], as we show here, while in experiments with peritonitis in pigs the respiratory activity was decreased [29]. The reason that mitochondrial function is upregulated in certain cases and downregulated in others is not completely clear. However, the normalization of mitochondrial function to its control values by mitoTEMPO suggests that the changes in mitochondrial function observed here reflect pathological changes in the liver and that these changes are mediated by mitoROS.

According to the literature, an early peak of TNF-α can be seen at 2 h post-LPS [2]. TNF-α and IL-1 are central cytokines in the acute systemic inflammatory response, on the one hand exerting direct effects on hepatocytes, inducing apoptosis and necrosis [30], and on the other hand sending later cytokines into action. Our data suggest that fast (2 h and earlier) release of TNF/IL-1 at the beginning of acute inflammatory response orchestrates the immune response rather than directly contributing to the liver damage. This may indeed be possible, as the first wave of TNF/IL-1 is released from membrane-bound pools of these inflammatory mediators and not from a pool which needs to be synthetized de novo. This does not preclude the fact that in the later second wave of their release, when they are upregulated on the genetic level, their levels will be sensitive to mitoTEMPO (Figure 6).

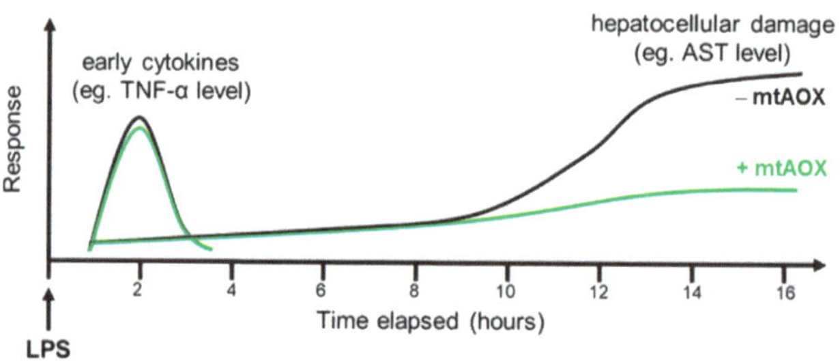

Figure 6. Schematic presentation of the time course of early cytokine response and hepatocellular damage after intravenous LPS challenge with and without mtAOX treatment in rats. LPS-induced hepatocellular damage, reflected by serum AST levels, is shown on the same timescale. AST can be seen rising slightly with the time following LPS injection, reaching significance compared to baseline levels at 8 h post-LPS. MtAOX treatment shows a strong reduction in circulating AST levels 16 h post-LPS, indicating a protective effect of mtAOX, and conversely a key role of mtROS in organ damage induced by endotoxemia.

Assuming effective intramitochondrial radical scavenging properties of mtAOX, we can conclude the following based on the experimental data. Beneficial effects of mtAOX on hepatocellular injury by 16 h are not directly linked to TNF-α, as mtAOX prevented AST release despite no alterations in TNF levels. Either later cytokines in the cascade are ROS-sensitive, or another process is contributing to cellular damage.

To determine whether ROS production in different body compartments can be modulated by mitoTEMPO, we examined ROS generation by immune cells in the blood, peritoneal fluid, and lavage from the bronchoalveolar system. We predominantly obtained PMNs from the three body compartments. Considering the central role of circulating PMNs in this systemic inflammation model, which is not TNF-α-dependent, an explanation for the protective effects could be a direct effect of mtAOX on PMNs. Although these cells have only a few functional mitochondria, a growing body of evidence suggests important functions of this organelle in initiation and migration [31], and Dikalov [32] and Daiber [33] have previously reported a feed-forward loop between mtROS and nicotinamide adenine dinucleotide phosphate (NADPH) oxidase activity. Theoretically, by inhibiting mtROS less ROS is formed by NADPH oxidase, resulting in less parenchymal cellular damage. Thus, we designed an experiment to investigate a direct effect of mtAOX on PMNs. There was no difference between PMA-induced ROS production of treated and non-treated groups, implying no causative role of reduced oxidative bursts of circulating immune cells capable of eliciting oxidative damage in the target organ in this setting.

We failed to reproducibly determine ROS generation of controls, and analyzed only the cells obtained from animals treated with LPS. In contrast to our expectations, *in vivo* injection of mitoTEMPO did not influence the rate of ROS generation in any of the examined compartments. Interestingly, the *ex vivo* treatment of the same cells with mitoTEMPO substantially reduced ROS generation. We assume that mitoTEMPO is distributed inside the body within the tissues; however, at the time when PMNs generate ROS it is possible that mitoTEMPO may not be present in sufficiently high concentrations in the corresponding compartments. In addition, it is possible that these cells might not be completely activated, as external treatment with PMA induced an additional release of ROS. The ROS release induced by PMA was not sensitive to mitoTEMPO. These data suggest that effect of mitoTEMPO in terms of ROS release is not mediated by activated immune cells.

Our observation that mitoTEMPO attenuated ROS generation *ex vivo* when exposure occurred prior to PMA treatment suggests that the generation of extracellular ROS in all three compartments upon treatment with LPS is controlled by mitoROS. This may not be the case when cells are activated by PMA. Indeed, PMA may activate immune cells in a mitoROS-independent manner by directly activating protein kinase C (PKC), and subsequently NADPH oxidase. However, LPS is known to act via toll-like receptors (TLR), which trigger the activation of NADPH-oxidases over mitochondrial ROS [34].

When analyzing the EPR spectra of the liver we observed specific changes of redox sensitive paramagnetic centers, particularly those involved in mitochondrial ROS generation and scavenging. We observed an increase in the Q10 signal in response to LPS, which was attenuated by mitoTEMPO. Q10 is an antioxidant in its reduced state, and it is able to scavenge ROS to yield stable free radicals. The latter can be detected by EPR. *In vivo* treatment with mitoTEMPO reduced this signal to its normal levels. Induction of iNOS is a sign of ongoing inflammation, and it has previously been shown that iNOS is upregulated in our model [2]. In our experiments, we observed increased NO levels in liver tissue and blood. The concentration of NO in liver tissue was higher than in the blood, suggesting that NO is formed predominantly in the liver (and probably other tissues) and then released into blood. *In vivo* treatment with mitoTEMPO reduced the NO levels elevated by LPS.

Based on the data presented here, and according to the literature [9], it can be concluded that classical early pro-inflammatory cytokines (TNF-α, IL-6) alone are not direct mediators of tissue damage *in vivo*. MtAOX treatment prevented hepatocellular damage despite obvious elevation of TNF-α in the circulation. Our data suggest that inflammatory mediators orchestrate systemic immune response, rather than directly contributing to ROS-mediated liver damage. Because ROS can originate in extracellular and intracellular compartments, it is expected that the mitochondrial antioxidant mitoTEMPO should prevent ROS-induced organ damage. This is supported by the fact that mitochondrial ROS regulate the release of intracellular and extracellular ROS from other sources [3]. Consequently, mitoTEMPO influences the redox status of liver cells, which can contribute to its local beneficial effect, though it may not be sufficient to reduce the systemic ROS generation by immune cells. The latter is likely due to the major portion of mitoTEMPO being absorbed by tissues and only a small portion being absorbed by immune cells. This may be explained by the volume/absorption surface of the tissues, which is much higher than the surface of immune cells.

Supplementary Materials: The following supporting information can be downloaded at: https:// www.mdpi.com/article/10.3390/biom13050794/s1, Supplementary Figure S1: Effect of mitoTEMPO on cell counts in blood and peritoneal and bronchoalveolar fluids.

Author Contributions: Conceptualization, A.V.K.; methodology, A.T.M. and S.D.; validation, A.T.M., A.W., S.D. and A.V.K.; formal analysis, A.T.M., A.W. and S.D.; investigation, A.T.M. and S.D.; resources, A.V.K.; data curation, A.W.; writing—original draft preparation, A.W. and A.V.K.; writing—review and editing, A.T.M., A.W., S.D. and A.V.K.; supervision, A.V.K.; project administration, A.W. All authors have read and agreed to the published version of the manuscript.

Funding: This research received no external funding.

Institutional Review Board Statement: All interventions were conducted in compliance with the Guide for the Care and Use of Laboratory Animals published by the National Institute of Health with approval from the Animal Protocol Review Board (M58003956-2011-9, no. 815758/2014/16).

Informed Consent Statement: Not applicable.

Data Availability Statement: The data of this study are available on request from the corresponding author, [A.V.K.].

Conflicts of Interest: The authors declare no conflict of interest.

References

1. Jarczak, D.; Nierhaus, A. Cytokine Storm—Definition, Causes, and Implications. *Int. J. Mol. Sci.* **2022**, *23*, 11740. [CrossRef]
2. Weidinger, A.; Müllebner, A.; Paier-Pourani, J.; Banerjee, A.; Miller, I.; Lauterböck, L.; Duvigneau, J.C.; Skulachev, V.P.; Redl, H.; Kozlov, A.V. Vicious Inducible Nitric Oxide Synthase-Mitochondrial Reactive Oxygen Species Cycle Accelerates Inflammatory Response and Causes Liver Injury in Rats. *Antioxid. Redox Signal.* **2015**, *22*, 572–586. [CrossRef] [PubMed]
3. Kozlov, A.V.; Lancaster, J.R.; Meszaros, A.T.; Weidinger, A. Mitochondria-Meditated Pathways of Organ Failure upon Inflammation. *Redox Biol.* **2017**, *13*, 170–181. [CrossRef]
4. Skulachev, V.P. Solution of the Problem of Energy Coupling in Terms of Chemiosmotic Theory. *J. Bioenerg.* **1972**, *3*, 25–38. [CrossRef] [PubMed]
5. Turrens, J.F. Mitochondrial Formation of Reactive Oxygen Species. *J. Physiol.* **2003**, *552*, 335–344. [CrossRef] [PubMed]
6. Murphy, M.P. How Mitochondria Produce Reactive Oxygen Species. *Biochem. J.* **2009**, *417*, 1–13. [CrossRef]
7. Gille, L.; Staniek, K.; Rosenau, T.; Duvigneau, J.C.; Kozlov, A. V Tocopheryl Quinones and Mitochondria. *Mol. Nutr. Food Res.* **2010**, *54*, 601–615. [CrossRef] [PubMed]
8. Jaeschke, H.; Smith, C.W. Mechanisms of Neutrophil-Induced Parenchymal Cell Injury. *J. Leukoc. Biol.* **1997**, *61*, 647–653. [CrossRef]
9. Juskewitch, J.E.; Knudsen, B.E.; Platt, J.L.; Nath, K.A.; Knutson, K.L.; Brunn, G.J.; Grande, J.P. LPS-Induced Murine Systemic Inflammation Is Driven by Parenchymal Cell Activation and Exclusively Predicted by Early MCP-1 Plasma Levels. *Am. J. Pathol.* **2012**, *180*, 32–40. [CrossRef]
10. Brooks, D.; Barr, L.C.; Wiscombe, S.; McAuley, D.F.; Simpson, A.J.; Rostron, A.J. Human Lipopolysaccharide Models Provide Mechanistic and Therapeutic Insights into Systemic and Pulmonary Inflammation. *Eur. Respir. J.* **2020**, *56*, 1901298. [CrossRef]
11. Libert, C.; Ayala, A.; Bauer, M.; Cavaillon, J.-M.M.; Deutschman, C.; Frostell, C.; Knapp, S.; Kozlov, A.V.; Wang, P.; Osuchowski, M.F.; et al. Part II: Minimum Quality Threshold in Preclinical Sepsis Studies (MQTiPSS) for Types of Infections and Organ Dysfunction Endpoints. *Shock* **2019**, *51*, 23–32. [CrossRef] [PubMed]
12. Flad, H.-D.; Loppnow, H.; Rietschel, E.T.; Ulmer, A.J. Agonists and Antagonists for Lipopolysaccharide-Induced Cytokines. *Immunobiology* **1993**, *187*, 303–316. [CrossRef] [PubMed]
13. Bannerman, D.D.; Goldblum, S.E. Mechanisms of Bacterial Lipopolysaccharide-Induced Endothelial Apoptosis. *Am. J. Physiol.—Lung Cell. Mol. Physiol.* **2003**, *284*, L899–L914. [CrossRef] [PubMed]
14. Beutler, B.; Kruys, V. Lipopolysaccharide Signal Transduction, Regulation of Tumor Necrosis Factor Biosynthesis, and Signaling by Tumor Necrosis Factor Itself. *J. Cardiovasc. Pharmacol.* **1995**, *25* (Suppl. 2), S1–S8. [CrossRef] [PubMed]
15. Weidinger, A.; Dungel, P.; Perlinger, M.; Singer, K.; Ghebes, C.; Duvigneau, J.C.; Müllebner, A.; Schäfer, U.; Redl, H.; Kozlov, A.V. Experimental Data Suggesting That Inflammation Mediated Rat Liver Mitochondrial Dysfunction Results from Secondary Hypoxia Rather than from Direct Effects of Inflammatory Mediators. *Front. Physiol.* **2013**, *4*, 138. [CrossRef]
16. Stich, T.A. Characterization of Paramagnetic Iron-Sulfur Clusters Using Electron Paramagnetic Resonance Spectroscopy. *Methods Mol. Biol.* **2021**, *2353*, 259–280. [CrossRef]
17. Jakubowska, M.A.; Pyka, J.; Michalczyk-Wetula, D.; Baczyński, K.; Cieśla, M.; Susz, A.; Ferdek, P.E.; Płonka, B.K.; Fiedor, L.; Płonka, P.M. Electron Paramagnetic Resonance Spectroscopy Reveals Alterations in the Redox State of Endogenous Copper and Iron Complexes in Photodynamic Stress-Induced Ischemic Mouse Liver. *Redox Biol.* **2020**, *34*, 101566. [CrossRef]
18. Kozlov, A.V.; Yegorov, D.Y.; Vladimirov, Y.A.; Azizova, O.A. Intracellular Free Iron in Liver Tissue and Liver Homogenate: Studies with Electron Paramagnetic Resonance on the Formation of Paramagnetic Complexes with Desferal and Nitric Oxide. *Free Radic. Biol. Med.* **1992**, *13*, 9–16. [CrossRef]
19. Dumitrescu, S.D.; Meszaros, A.T.; Puchner, S.; Weidinger, A.; Boros, M.; Redl, H.; Kozlov, A.V. EPR Analysis of Extra- and Intracellular Nitric Oxide in Liver Biopsies. *Magn. Reson. Med.* **2017**, *77*, 2372–2380. [CrossRef]
20. Pelletier, M.M.; Kleinbongard, P.; Ringwood, L.; Hito, R.; Hunter, C.J.; Schechter, A.N.; Gladwin, M.T.; Dejam, A. The Measurement of Blood and Plasma Nitrite by Chemiluminescence: Pitfalls and Solutions. *Free Radic. Biol. Med.* **2006**, *41*, 541–548. [CrossRef]
21. Song, J.-A.; Yang, H.-S.; Lee, J.; Kwon, S.; Jung, K.J.; Heo, J.-D.; Cho, K.-H.; Song, C.W.; Lee, K. Standardization of Bronchoalveolar Lavage Method Based on Suction Frequency Number and Lavage Fraction Number Using Rats. *Toxicol. Res.* **2010**, *26*, 203–208. [CrossRef] [PubMed]
22. Ray, A.; Dittel, B.N. Isolation of Mouse Peritoneal Cavity Cells. *J. Vis. Exp.* **2010**, *35*, 1488. [CrossRef]
23. Suffredini, A.F.; Hockstein, H.D.; McMahon, F.G. Dose-Related Inflammatory Effects of Intravenous Endotoxin in Humans: Evaluation of a New Clinical Lot of Escherichia Coli O:113 Endotoxin. *J. Infect. Dis.* **1999**, *179*, 1278–1282. [CrossRef] [PubMed]
24. An, N.; Song, Y.; Zhang, X.; Ci, X.; Li, H.; Cao, Y.; Zhang, M.; Cui, J.; Deng, X. Pretreatment of Mice with Rifampicin Prolongs Survival of Endotoxic Shock by Modulating the Levels of Inflammatory Cytokines. *Immunopharmacol. Immunotoxicol.* **2008**, *30*, 437–446. [CrossRef] [PubMed]

25. Mittal, M.; Siddiqui, M.R.; Tran, K.; Reddy, S.P.; Malik, A.B. Reactive Oxygen Species in Inflammation and Tissue Injury. *Antioxid. Redox Signal.* **2014**, *20*, 1126–1167. [CrossRef]

26. Weidinger, A.; Birgisdóttir, L.; Schäffer, J.; Meszaros, A.T.; Zavadskis, S.; Müllebner, A.; Hecker, M.; Duvigneau, J.C.; Sommer, N.; Kozlov, A.V. Systemic Effects of MitoTEMPO upon Lipopolysaccharide Challenge Are Due to Its Antioxidant Part, While Local Effects in the Lung Are Due to Triphenylphosphonium. *Antioxidants* **2022**, *11*, 323. [CrossRef]

27. Jeger, V.; Djafarzadeh, S.; Jakob, S.M.; Takala, J. Mitochondrial Function in Sepsis. *Eur. J. Clin. Investig.* **2013**, *43*, 532–542. [CrossRef]

28. Kozlov, A.V.; Staniek, K.; Haindl, S.; Piskernik, C.; Ohlinger, W.; Gille, L.; Nohl, H.; Bahrami, S.; Redl, H. Different Effects of Endotoxic Shock on the Respiratory Function of Liver and Heart Mitochondria in Rats. *Am. J. Physiol. Gastrointest. Liver Physiol.* **2006**, *290*, G543–G549. [CrossRef]

29. Kozlov, A.V.; van Griensven, M.; Haindl, S.; Kehrer, I.; Duvigneau, J.C.; Hartl, R.T.; Ebel, T.; Jafarmadar, M.; Calzia, E.; Gnaiger, E.; et al. Peritoneal Inflammation in Pigs Is Associated with Early Mitochondrial Dysfunction in Liver and Kidney. *Inflammation* **2010**, *33*, 295–305. [CrossRef]

30. Zang, G.Q.; Zhou, X.Q.; Yu, H.; Xie, Q.; Zhao, G.M.; Wang, B.; Guo, Q.; Xiang, Y.Q.; Liao, D. Effect of Hepatocyte Apoptosis Induced by TNF-α on Acute Severe Hepatitis in Mouse Models. *World J. Gastroenterol.* **2000**, *6*, 688–692. [CrossRef]

31. Zhang, Q.; Raoof, M.; Chen, Y.; Sumi, Y.; Sursal, T.; Junger, W.; Brohi, K.; Itagaki, K.; Hauser, C.J. Circulating Mitochondrial DAMPs Cause Inflammatory Responses to Injury. *Nature* **2010**, *464*, 104–107. [CrossRef] [PubMed]

32. Dikalov, S.I.; Li, W.; Doughan, A.K.; Blanco, R.R.; Zafari, A.M. Mitochondrial Reactive Oxygen Species and Calcium Uptake Regulate Activation of Phagocytic NADPH Oxidase. *Am. J. Physiol. Regul. Integr. Comp. Physiol.* **2012**, *302*, R1134–R1142. [CrossRef]

33. Kröller-Schön, S.; Steven, S.; Kossmann, S.; Scholz, A.; Daub, S.; Oelze, M.; Xia, N.; Hausding, M.; Mikhed, Y.; Zinssius, E.; et al. Molecular Mechanisms of the Crosstalk between Mitochondria and NADPH Oxidase through Reactive Oxygen Species-Studies in White Blood Cells and in Animal Models. *Antioxid. Redox Signal.* **2014**, *20*, 247–266. [CrossRef] [PubMed]

34. Nguyen, G.T.; Green, E.R.; Mecsas, J. Neutrophils to the ROScue: Mechanisms of NADPH Oxidase Activation and Bacterial Resistance. *Front. Cell. Infect. Microbiol.* **2017**, *7*, 373. [CrossRef] [PubMed]

Article

Timing of Chromosome DNA Integration throughout the Yeast Cell Cycle

Valentina Tosato [1,2], Beatrice Rossi [1], Jason Sims [3] and Carlo V. Bruschi [1,4,*]

[1] Yeast Molecular Genetics, ICGEB-International Center for Genetic Engineering and Biotechnology, AREA Science Park, Padriciano 99, 34149 Trieste, Italy
[2] Department of Chemical and Pharmaceutical Sciences, University of Trieste, Via Giorgieri 1, 34127 Trieste, Italy
[3] St. Anna Children's Cancer Research Institute, Zimmermannplatz 10, 1090 Vienna, Austria
[4] Department of Cell Biology, University of Salzburg, Hellbrunner Straße 34, 5020 Salzburg, Austria
* Correspondence: chabravo@gmail.com

Abstract: The dynamic mechanism of cell uptake and genomic integration of exogenous linear DNA still has to be completely clarified, especially within each phase of the cell cycle. We present a study of integration events of double-stranded linear DNA molecules harboring at their ends sequence homologies to the host's genome, all throughout the cell cycle of the model organism *Saccharomyces cerevisiae*, comparing the efficiency of chromosomal integration of two types of DNA cassettes tailored for site-specific integration and bridge-induced translocation. Transformability increases in S phase regardless of the sequence homologies, while the efficiency of chromosomal integration during a specific cycle phase depends upon the genomic targets. Moreover, the frequency of a specific translocation between chromosomes XV and VIII strongly increased during DNA synthesis under the control of Pol32 polymerase. Finally, in the null *POL32* double mutant, different pathways drove the integration in the various phases of the cell cycle and bridge-induced translocation was possible outside the S phase even without Pol32. The discovery of this cell-cycle dependent regulation of specific pathways of DNA integration, associated with an increase of ROS levels following translocation events, is a further demonstration of a sensing ability of the yeast cell in determining a cell-cycle-related choice of DNA repair pathways under stress.

Keywords: BIT; cell cycle; DNA integration; Pol32; yeast

check for **updates**

Citation: Tosato, V.; Rossi, B.; Sims, J.; Bruschi, C.V. Timing of Chromosome DNA Integration throughout the Yeast Cell Cycle. *Biomolecules* **2023**, *13*, 614. https://doi.org/10.3390/biom13040614

Academic Editor: Jürg Bähler

Received: 13 January 2023
Revised: 22 March 2023
Accepted: 23 March 2023
Published: 29 March 2023

1. Introduction

Bridge-Induced Translocation (BIT) is a unique molecular system to generate specific chromosomal translocations in *Saccharomyces cerevisiae* by exploiting its endogenous recombination machinery [1]. The whole process is based on a selectable DNA cassette bearing at its ends sequence homologies to two genomic loci on different chromosomes, providing the molecular substrate for integration into the yeast genome via homologous recombination. Chromosome translocants are generated among an ensemble of genomic rearrangements comprising ectopic integrations, intra-chromosomal deletions, recombination with the endogenous 2 μ plasmid and non-specific translocations promoted by DNA micro-homology [1]. Moreover, a few of the correct translocants may suffer heavy secondary gross chromosomal rearrangements (GCRs), which seem to affect their life span [2]. In this respect, at the end of their chronological life (CLS), these translocants undergo the loss of the translocated chromosome in a strongly locus-dependent manner [2]. In general, the efficiency of BIT does not reflect the efficiency of overall recombination events since only cells that survive the recombination event can be recovered and analyzed. BIT efficiency is strongly locus-dependent and, with targeting DNA sequence homologies of 65 nt, fluctuates between 5 and 15% of all transformants. A two-step integration model for BIT has been proposed [3]. in which the translocation efficiency is inversely related to the distance of a replication-dependent "D-loop" from the telomere and a massive deregulation of gene expression occurs around the translocation breakpoints [4]. In this view, the BIT system resembles a DNA site-specific integration (SSI) event, but with the target sequence homologies belonging to two different chromosomes. Although yeast DNA transformation protocols are extensively used in *S. cerevisiae* studies, very little is known about the molecular mechanisms underlying the transformation process, especially the timing of DNA uptake and transport into the nucleus [5]. Only very few reports exist concerning the transformation of synchronized yeast cells with circular [6] or linear [7]. DNA and it appears that synchronization of yeast cells in S phase might specifically improve gene targeting and, therefore, gene replacement, especially in poorly transformable yeast strains [7]. It is suggested that HR responds to replication stress through replicative and repair activities that operate at different stages of the cell cycle and in distinct subnuclear structures [8]. However, a correlation between the transformation efficiency throughout the cell cycle and the different integration mechanisms of linear DNA molecules is still missing. Furthermore, elucidating the dynamics of translocation induction from a cell cycle perspective could be an initial milestone in dissecting the effects of chromosomal translocations on aging and oxidative stress. These findings could shed light on how translocation events at each cell cycle stage have different effects on cell adaptation to reactive oxygen species (ROS) and impact the cell life span.

In order to clarify the above issues, in the present work we arrested *S. cerevisiae* cells in G1, S and G2/M phases, using the α-factor (α-f), hydroxyurea (HU) and nocodazole (NOC), respectively, and we successively transformed the synchronized cells with exogenous linear DNA cassettes. For each transformation experiment, two different types of linear DNA cassettes, the Site-Specific Integration (SSI) and the Bridge Induced Translocation (BIT) cassette, were tested. Moreover, we compared the efficiency of targeting four different genomic loci. The results indicate that BIT efficiency mainly depends on the phase of cell cycle and suggest that for each translocation experiment, an "optimum" timing of integration can be characterized. We previously hypothesized that, during BIT, after the insertion of the first homologous end, a double-strand break (DSB) is generated promoting its repair by Break-Induced Replication (BIR) [9]. The DNA polymerase delta subunit Pol32, dispensable for replication and gene conversion (GC), is the pivotal enzyme for BIR [10] although chromatin remodeling is also important for accurate DNA replication [11]. In this work we demonstrated that the deletion of *POL32* (*YJR043C*) prevents linear DNA integration and BIT translocation during DNA synthesis leading to the conclusion that, at least in S phase, Pol32 is a fundamental player in the non-reciprocal translocation outcome. Nevertheless, BIT is still possible in the null mutant, with translocation paradoxically accounting for the majority of the homologous integration events. This enforces the idea that there are two pathways for the completion of the chromosomal DNA bridge, one that is Pol32-dependent and is restricted to the S phase only, and another one that is Pol32-independent and occurs in the other phases of the cell cycle. Further analyses of the synchronized mutants and of the ectopic healings of the DSBs will help to disclose whether this time-specific BIR is restricted to G2, preceding mitosis, and if other possible molecular mechanisms such as Single Strand Annealing (SSA) could be partly responsible for the chromosomal repair in absence of Pol32.

2. Materials and Methods

2.1. Yeast Strains and Media

The diploid strain San1 was used to generate the translocations XV–VIII/IX–XVI and was used as the control strain throughout this work. San1 was obtained by mating YPH250 (a, *ade2-101o leu2-Δ1 lys2-801a his3-Δ200 trp1-Δ1 ura3-52*, ATCC 96519) with Fas20 (α, *ade1 ade2 ade8 can1r leu2 trp1 ura3-52*) [12]. A derivative of San1 hemizygous for the mating type locus, San1VTΔMAT, was generated in this work for the G1 arrest with the α-f. To generate this strain, primers for the deletion of the MATα locus were chosen following Lee et al. [13]. The two primers contained a stretch of DNA homology with kanamycin necessary for the POP-OUT of the recyclable marker. The POP-OUT technology, developed in our laboratory [14], allows the complete deletion of the MATα locus and its substitution with a Flip Recognition Target (FRT) scar. When α-f was used to synchronize the cells in G1, the experiments were always performed using San1VTΔMAT as the reference strain. This strain performed as San1 in terms of efficiency of transformability, but showed minimal differences in the distribution of integration events with a bias toward ectopic integrations. These differences were measured and reported in the analysis of translocation outcomes while they were totally negligible in SSI experiments (where the correct integration always represents the majority of the events). Thus, San1VTΔMAT has been used as the reference strain in the translocation experiments while it has been omitted in transformability and SSI experiments since it performed exactly like San1. The understanding of the moderate predisposition toward BIT ectopic integrations of San1VTΔMAT is presently under investigation (personal communication).

The POP-OUT technology [14] was used twice to delete both copies of *POL32* in San1 and to generate the double deletant strain (*pol32Δ/pol32Δ*), which is indicated in this work as Δ*POL32*. All the primers synthesized and utilized for this work are listed in Table S1. The strains were all grown at 30 °C in rich medium (YPD, Difco, BD, Milano, Italy) and geneticin (G418, final concentration 200 μg/mL) was added when required.

2.2. Molecular Biology Techniques and Microscopy

Standard recombinant DNA techniques were carried out according to [15]. Translocation breakpoints (TBPs) and integration junctions were sequenced by BMR Genomics (Padua, Italy). The cassettes to generate the translocations between the chromosomes XV–VIII and XVI–IX were obtained by PCR using the bacterial plasmid template pFa6AKanMX4 [16] using the primers reported in Table S1 and the High Fidelity TAQ polymerase (Kapa Biosystems), according to optimized protocols [1,17].

DAPI (4′,6-diamidino-2-phenylindole) staining was performed as previously described [17] to check either the Δ*POL32* mutant or the cell cycle phases after the arrest treatments and using a Leica DMBL photomicroscope equipped with a computer-driven CCD camera at 60× and 100× magnifications.

2.3. Transformation and Synchronization Procedure

Transformation of *S. cerevisiae* was performed with the lithium acetate (LiAc) methodology following the EUROFAN guidelines for PCR-based gene replacements [16]. In particular, transformation of the S-phase-arrested cells was performed as follows. San1 cells were grown overnight up to stationary phase; then, in a 500 mL flask, 100 mL of YPD was inoculated with up to 5×10^6 cells/mL and were left to grow for 90 min. At this point, hydroxyurea (HU) was added at a final concentration of 100 mM. After 120 min of incubation, the resulting S-phase-arrested cells were counted, washed twice with sterile water and used for transformation either with the linear cassettes or the plasmid YCp50. The arrest was confirmed by microscopy, ranging between 94% and 95% of the cell population both in San1 and in the Δ*POL32* mutant.

For the arrest in G2/M phase, the same protocol was followed, using nocodazole (NOC) instead of HU, at a final concentration of 15 μg/mL in YPD plus 1% DMSO. After the arrest in G2/M phase, the cells were counted, washed twice with water and used for transformation. The arrest was almost complete with a rapid block of nuclear division easily detectable with a microscope (Figure S1).

For the arrest in G1, we optimized the protocol described by Lieberman [18] using the strain San1VTΔMAT. The cells were inoculated in 100 mL YPD at a concentration up to 5×10^6 cells/mL and left to grow for 60 min; at that point, α-f was added to a final concentration of 5 μg/mL. The cells were then incubated again with vigorous shaking for another 60 min after which an additional treatment with α-f (5 μg/mL) was performed. Sixty minutes after this second treatment, the cells were definitively blocked in G1 showing the typical "schmoo" morphology (Figure S1). They were counted, washed twice with water, and used for transformation when more than 95% had the "schmoo" morphology.

The cells used for the transformation experiments were grown up to a density of $1.4–1.6 \times 10^7$ cells/mL. More precise cell culture details for each set of transformations are reported in Table S2. When the transformability efficiency was low, such as in *ssu1* of G2/M-arrested cells, the same transformation protocol was repeated until a comparable number of transformants was collected. The same number of transformants (40 for each locus) was then analyzed (Table S2B).

The average amount of DNA cassettes used for each transformation was 15 µg.

The strain transformability control was performed with 400 ng of the centromeric plasmid YCp50 [19]. The transformants were always selected as G418-resistant clones; the presence of translocations was monitored by colony PCR and sequencing as previously described [1,17] with the primers reported in Table S1.

2.4. Measurement of ROS

To measure the reactive oxygen species (ROS) levels, we used the dihydroethidium (DHE) method based on the oxidation of DHE to ethidium and 2-hydroxy ethidium, which both give a fluorescence with an excitation wavelength of 485 nm and emission of 595 nm, detectable by flow cytometry. The cells were grown overnight to a plateau phase, where they exhibit 100% survival, washed two times with PBS, resuspended in a PBS solution with 5 µg/mL dihydroethdium (DHE), and then incubated at 30 °C for 15 min in a shaker kept in darkness. The fluorescence was measured in a BD Biosciences FACScalibur flow cytometer, counting 100,000 cells for each sample. A control staining was done with Propidium Iodide (PI) by incubating cells from the same cultures in PBS with 10 µg/mL of PI and counting them in the same way [2].

2.5. p-Value Calculation

The exact hypergeometric probability of obtaining the experimental results reported in Table S2 was determined using the Freeman–Halton extension of the Fisher exact probability test for a three-row by two-column contingency table. Category 1 comprised the untreated cells (asynchronous) and the treated cells (cells blocked in a specific cell cycle phase). Category 2 included three groups of different integration events (translocation, ectopic, one-side integration). Other details are reported in the legend of Table S2.

Relative Efficiency (Er) Calculation

As reported previously [9], the frequencies (v) are represented by the total number of obtained transformants on selective medium per number of treated cells. The efficiency of transformation (E) is obtained by dividing v by the total DNA amount (µg) used in the transformation experiment. Each efficiency value (of transformation, translocation or SSI) of the linear cassettes was compared with the efficiency of transformability with the circular plasmid YCp50 of the same strain in the same phase of the cell cycle. The resulting efficiency numbers were then divided by the transformation efficiency of an asynchronous population and the relative efficiency (Er) was obtained.

3. Results

3.1. Integration of Bridge-Induced Translocation and Site-Specific DNA Cassettes

At first, we decided to test whether a strong link existed between the cell cycle phases and the transformation efficiency with BIT or SSI cassettes.

San1 diploid strain cells of *S. cerevisiae* were synchronized in G1, S and G2/M phases (Figure S1) and then transformed (Figure 1) with two different types of linear DNA cassettes: the BIT or the SSI cassette (Figures 2 and 3, respectively). We used two different BIT DNA cassettes: one carrying a 65 bp homology to the *ADH1* promoter (chromosome XV) at one end, and to the *DUR3* gene (chromosome VIII) at the other (Figure 2A). The second carries a 65 bp homology to the terminator region of *SSU1* (chromosome XVI) at one end, and to the promoter region of *SUC2* (chromosome IX) at the other [17] (Figure 2B). A diploid strain (San1) was used since the particular loci of these translocations may generate dead cells in haploids.

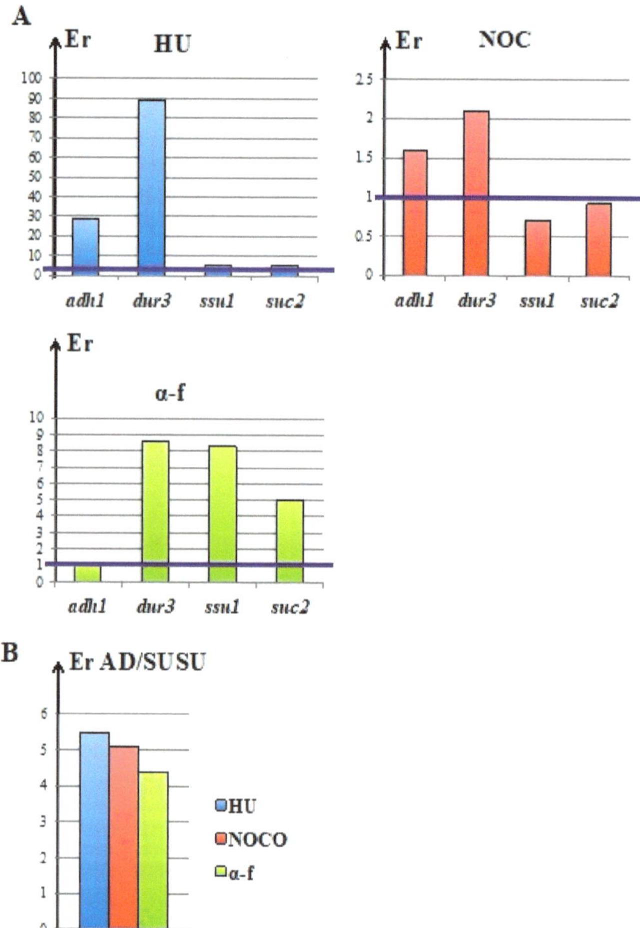

Figure 1. Transformability efficiency. (**A**) Relative efficiency of transformability (Er, see M & M for the mathematical calculation) with four different SSI DNA cassettes homologous to *adh1*, *dur3*, *ssu1* and *suc2* loci when yeast cells have been blocked with HU (blue), NOC (red) or α-f (green). The violet line represents the efficiency of transformability of an asynchronous population given a value of 1. In the case of α-f-blocked cells, the strain San1VTΔMAT has been used as a wild type strain and reference strain for transformations in asynchronous conditions. (**B**) Comparison between the efficiency of transformability between the AD and the SUSU translocation cassettes in cells blocked with HU (blue), NOC (red) or α-f (green). All the raw data used to create the Figure are listed in Table S2.

Several translocants were obtained using AD and SUSU cassettes (Table S2A). Almost all of them had an extremely high increases in ROS levels with a physiological adaptation allowing a medium or long CLS [2]. However, since DHE staining could show false positive results due to necrotic cells, a control stain was done with Propidium Iodide that enters damaged membranes of necrotic cells, to discriminate the false positives from the actual increase in ROS (Figure S2). The result obtained with one SUSU translocant named SUSU5 indeed showed that the increase in ROS was not due to necrotic cell staining.

To obtain the recombination frequency at the selected loci involved in these two translocation events, we tested the transformation efficiency with four different SSI cassettes (Figure 1A). For each SSI cassette, one homology is the BIT-cassette integration site while the other is located only few hundred nucleotides away from the first one (Figure 3, Table S1). As a DNA transformability control, the data were compared with the transformation efficiency of the centromeric plasmid YCp50 [18] (Table S2). Moreover, all the data were normalized in each experiment with the data obtained with San1 asynchronous cells grown in YPD rich medium. The transformability of strain San1VTΔMAT was also assessed and reported as the control for cells blocked in G1 when substantial differences with San1 were detected as described in the following paragraphs.

A

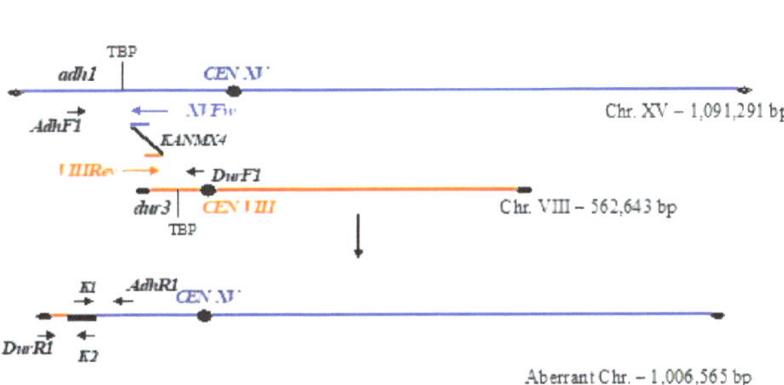

B

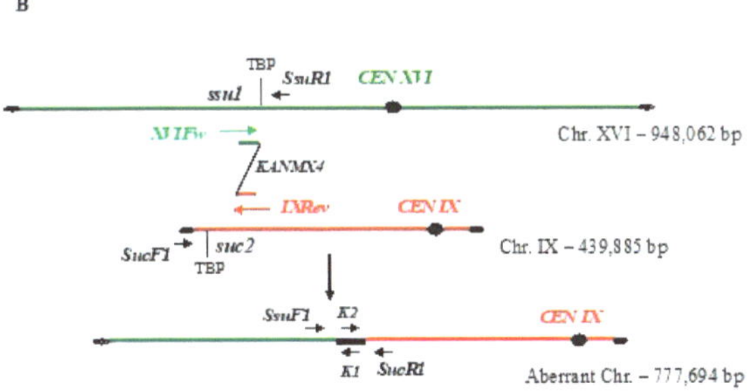

Figure 2. Schematic representation of the two Bridge-Induced Translocation events studied in this work. (**A**) BIT between chromosomes VIII and XV. The amplified selectable DNA cassette has two 65 bp ends homologous to the *dur* (VIII) and *adh* (XV) loci. Primers XVFw and VIIIRev, used to amplify the BIT cassette, are indicated with an arrow in blue and orange colors, respectively; primers used to check the integration of the translocated chromosomes (AdhF1, AdhR1, DurF1, DurR1) are reported with an arrow in black along their respective chromosomes. AdhR1 and DurR1 primers were used to amplify the TBP (translocation breakpoint) between the two chromosomes. (**B**) BIT between chromosomes IX and XVI. The amplified selectable DNA cassette has two 65 bp ends homologous to the ssu1 and suc2 loci. Primers XVIF and IXRev used to amplify the BIT cassette are indicated with green and red colors, respectively; primers used to check the integration of the translocated chromosomes (SsuF1, SsuR1, SucF1, SucR1) are reported in black along their respective chromosomes. SsuF1 and SucR1 primers were used to amplify the translocation breakpoint (TBP) between the two chromosomes. CEN: centromere. K1 and K2: primers used to check proper integration. Only the resulting translocant chromosomes without other fragments or secondary rearrangements coming from the two BIT events are reported in this figure.

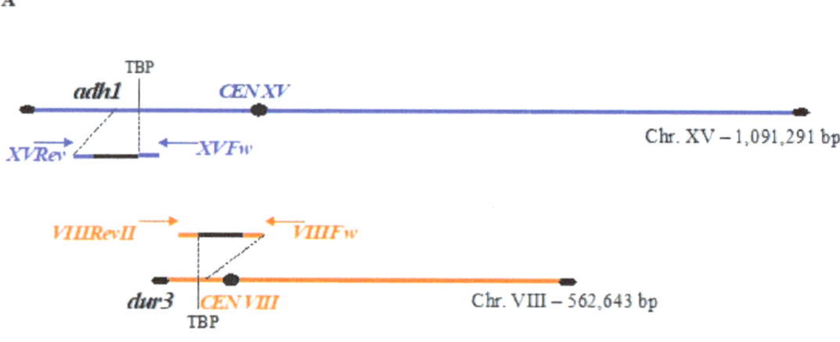

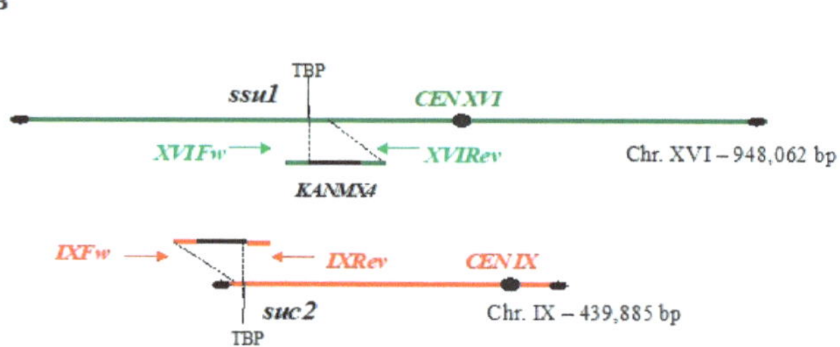

Figure 3. Schematic representation of the four events of Site-Specific Integration studied in this work. The precise position of each DNA cassette integrated on chromosome XV, VIII, XVI and IX (next to loci *adh1*, *dur3*, *ssu1* and *suc2*, respectively) is shown together with the primers used for the diagnostic amplifications (sequences in Table S1). (**A**) *adh1* (chromosome XV, blue color) and *dur3* loci (chromosome VIII, orange color) are shown. (**B**) *ssu1* (chromosome XVI, green color) and *suc2* (chromosome IX, red color) loci are shown. TBP: translocation breakpoint. CEN: centromere.

3.2. The Locus and the Cell Cycle Phase During Integration Strongly Affected Transformation Efficiency with Linear DNA Molecules

In Figure 1 we considered the total number of transformants that were obtained in each phase of the cell cycle regardless of the outcome of integration. In particular, Figure 1A shows the relative efficiency of transformation (Er) of yeast cells synchronized at different times (HU = S phase; NOC = G2/M phase; α-f = G0/G1, see Figure S1) with four different SSI DNA cassettes. These results suggest that the S-phase arrest generally favors cell transformation, although with considerable targeting variations among the four loci, while G2/M arrest seemed to decrease the possibility to obtain survivors after the transformation, despite the doubled amount of DNA in this phase of the cell cycle (Figure S1). The high number of transformation experiments that are required to get a comparable number of transformants after NOC treatment (Table S2B) supports this indication. The potential effect of dimetylsulfoxyde (DMSO) addition to the medium was checked and evaluated in each nocodazole-related experiment. The conclusion was that DMSO did not affect the efficiency of yeast transformation either with a circular or with a linear DNA molecule. The relative efficiencies of transformation (Er) with two different BIT cassettes (AD, which bridges chromosomes XV and VIII and SUSU, which bridges chromosomes XVI and IX) were calculated, compared and reported in Figure 1B. The AD is noticeably more efficient than SUSU in producing viable transformants meaning that targeting *adh1* and *dur3* loci allows a higher number of life-compatible chromosomal rearrangements. Further computation details and the raw data are reported in Table S2. Taken together, all these data reveal a leading position of the *dur3* genomic locus, which can deeply affect transformation efficiency all throughout the cell cycle.

3.3. The S Phase May Have a Leading Role in Translocation Success

Past and recent data obtained in our laboratory [3,9,17] suggested that DNA synthesis might have an important role in the BIT translocation resolution and in the induction of secondary complex chromosome rearrangements following the primary BIT event. Among these secondary gross rearrangements, the loss of the acentric chromosomal end and the generation of a partial trisomy of the centric chromosome were the most frequent ones [17]. To investigate the specific role of S phase in BIT translocation, the two collections of BIT transformants obtained with cells blocked in S phase and described in the previous paragraph, were analyzed by colony PCR with specific primers (Figure 2, Table S1). These data were compared with those obtained from an asynchronous San1 population. The total number of transformants analyzed for each group is reported in Table S2. The results, summarized in Figure 4, indicate that during S phase, the efficiency of translocation of the AD cassette was almost 3-fold higher than for the asynchronous cell population, whereas the efficiency of the SUSU cassette was slightly decreased. The DNA bridges of the sixteen translocants obtained in S phase with the AD cassette were sequenced and did not reveal any error within the *dur3* end. We also noticed that when S-phase-arrested cells were transformed with the AD cassette, the percentage of the one-end-only (either *adh1* or *dur3*) integration event decreased whereas when the cells were transformed with the SUSU cassette, the integration frequency at the *suc2* locus increased (Figure 4). Therefore, we can conclude that the target loci have a pivotal role in promoting translocation during DNA synthesis. Moreover, integration during S phase could also be partially explained by the induction of DNA damage as a consequence of HU treatment [20].

To better understand the role of the S phase in the integration within a specific locus, we also used PCR to analyze the collection of transformants obtained with the set of four SSI DNA cassettes (Figure 5). For each cassette, among all the transformants obtained, 40 of them (Table S2B) were checked by colony PCR either in asynchronous (YPD) or synchronous cells. Again, the locus *dur3* showed a different profile with respect to the other loci, being characterized by the totality of integration events during synthesis. Since the locus *adh-up*, which is the one responsible for the AD integration on chromosome XV, also showed a significant targeting in S phase (Figure 5), these data taken together justify the noticeable increased frequency of AD translocation during synthesis (Figures 1B and 4). In contrast, the integration of the two homologous sequences involved in the SUSU translocation (*ssu1*up, *suc2*down; Figure 5) rarely occurred during S phase (2.5%), confirming the decreased percentage of SUSU translocation events highlighted in Figures 1B and 4.

3.4. Nocodazole and α-Factor Strongly Decreased the Occurrence of Homologous Integration and Chromosomal Translocation

The integration efficiency of BIT (AD, SUSU) and SSI cassettes during G2/M phase was analyzed as previously described for S phase using the primers reported in Figures 2 and 3 (primers location) and Table S1 (primers sequence)

The results reported in Figure 5 suggested than when yeast cells are blocked with NOC, the BIT is generally repressed, as demonstrated by the complete absence of translocants found in the AD translocation and by a marginal 1.8% of translocants recovered with the SUSU cassette (Figure 4). Moreover, ectopic integrations were predominant in the G2/M phase regardless of the type of cassette. Not only translocation events, but also homology-driven SSI events were not favored as shown by the bars labelled as NOC in Figure 5.

Therefore, these data suggest that arrest in the G2/M phase seems to result in low homologous integration and frequent ectopic events while DNA synthesis promotes integration events. These results correlate well with the known poor resection in yeast of single-stranded DNA fragments and to the highly ordered chromatin structure in the G2/M phase [21] in contrast to the proficient formation of 3' ends during synthesis, which actively promotes homologous recombination [22].

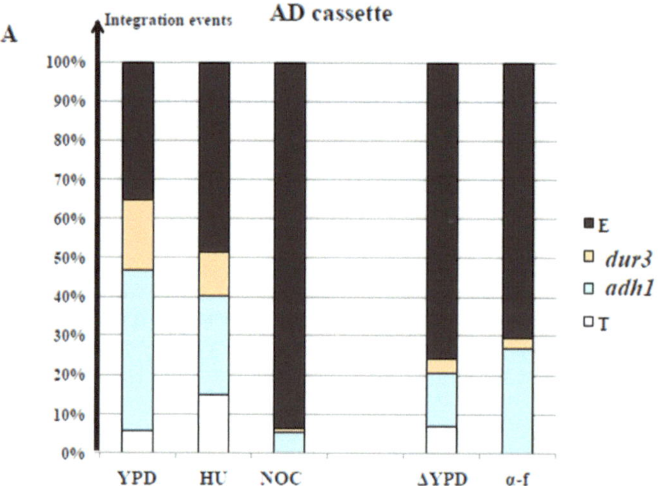

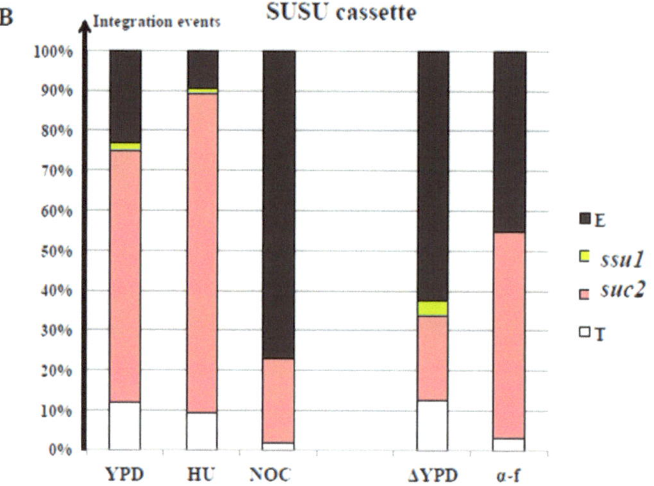

Figure 4. Percentage of integration events obtained using the XV–VIII and XVI–IX BIT cassettes. In this Figure, 100% represents the total amount of integration events. The raw data of the percentages used to plot the bars of the figure are reported in Table S2. (**A**) The three columns on the left show the percentage of translocation events (T, white), integration in adh1 only (*ADH1*, blue), in *dur3* only (*DUR3*, orange) and of ectopic events (E, dark grey) obtained using the XV-VIII BIT cassette in an asynchronous cell population of San1 (indicated as YPD) and in populations synchronized either with HU or with NOC. The distribution of the events in cells synchronized with α-f is separately reported because only these data have been compared with the asynchronous strain San1VTΔMAT (also reported in the figure and indicated as ΔYPD). (**B**) The three columns on the left show the percentage of the translocation event (T, white), integration in ssu1 (*SSU1*, green), suc2 (*SUC2*, red) and ectopic events (E, dark grey) obtained using the XVI–IX BIT DNA cassette in an asynchronous cell population (indicated as YPD) and in yeast cells synchronized either with HU or NOC. The data obtained with cells blocked using α-f are shown on the right and are compared with the strain San1VTΔMAT (indicated as ΔYPD).

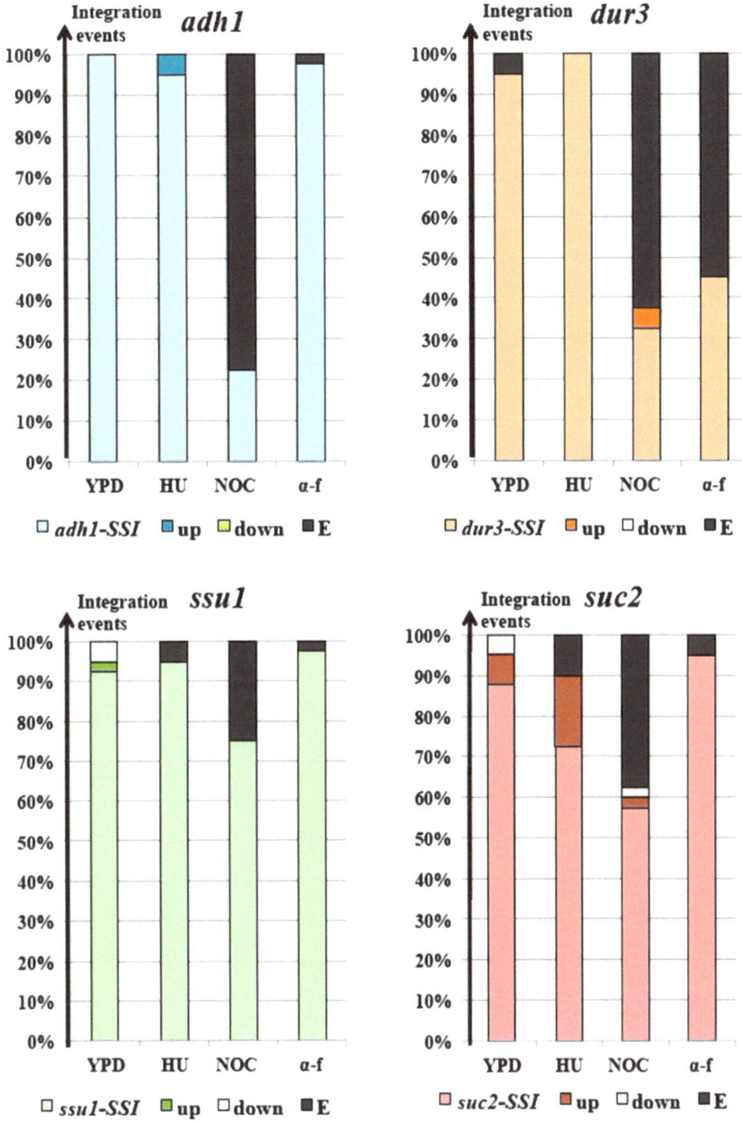

Figure 5. Percentage of integration events obtained using the SSI cassettes. For each locus the percentage of the integration events obtained in an asynchronous cell population grown in YPD (indicated as YPD) and in yeast cells synchronized with HU, NOC or α-f is reported. For each SSI DNA cassette (*adh1, dur3, ssu1, suc2*) each plot shows (i) the percentage of complete SSI events (indicated as locus-SSI) when both, upstream and downstream ends, were correctly integrated; (ii) one-end-only integration events (upstream indicated as up or downstream indicated as down); and (iii) ectopic events (indicated as E when the cassette integrated into a different site), with respect to the total number of events (100%).

Based on the results presented above, we analyzed the cassette integration efficiency in yeast cells blocked at G1 phase with α-f. The results we obtained showed that the transformability with linear DNA molecules barely increased compared to an asynchronous population (Figure 1), but the translocation events were drastically reduced (Figure 4). These data, together with frequent ectopic integrations of SSI cassettes in G1 (Figure 5), suggest an enhanced plasticity of the genome, which would allow exogenous DNA integration via the non-homologous end-joining pathway (NHEJ), which is active throughout the cell cycle [23] and without the involvement of the Homologous Recombination System (HRS). It is in fact known that DSBs that occur during the G1 phase are mainly repaired through non-homologous end joining (NHEJ), whereas DSBs that are formed during the S and G2 phases are predominantly repaired by homologous recombination (HR) using the intact sister chromatid [21]. Furthermore, the above results in diploid cells confirmed a restriction of the HRS in DSB repair during G1 probably due to the lack of activity of Clb–CDK genes, which inhibits ssDNA resection, as proposed for haploid yeast cells [24].

3.5. The Enhancement of AD-Translocation Frequency during Synthesis Is Pol32-Dependent

Among all the experiments analyzed in this work throughout the cell cycle phases with the various constructs, BIT was strongly enhanced when cells, transformed with the AD cassette (Figure 1), were blocked with HU. The near three-fold increase in translocants recovered during S phase (Table S2C) correlated with a higher efficiency of transformation (Figure 1B; Table S2A.3) and with increased targeting at the *dur3* locus (Figure 5). This result for AD is peculiar since high translocation frequency with BIT cassettes is uncommon in wild-type strains as well as in mutants [9]. Therefore, we decided to investigate the mechanism responsible for this enhancement.

Since we previously surmised that the final step of BIT occurs through BIR [3], Pol32 could be responsible for the completion of the bridge at *dur3* and *ssu1* loci in the case of AD and SUSU translocation, respectively. To understand whether Pol32 could be involved in this regulation, we tested the Δ*POL32* San1 strain for the AD translocation proficiency in the presence and absence of HU. The results, reported in Figure 6, indicate that when the mutant is blocked during S phase, the efficiency of transformation was almost 65-fold higher than the that of the wild type (see the legend of Figure 6 for computation details).

The transformation frequency of the wild type with circular DNA (YCp50) was approximately 3-fold higher than that of the mutant, either with or without HU (see the table in Figure 6A). Therefore, the HU treatment does not affect the strain transformation efficiency per se and vice versa, after HU treatment, the increase in transformation frequency with a linear DNA was much more significant in the *pol32* mutant than in the wild type, as shown in Figure 6A. In this case, the sixty-eight G418-resistant transformants, which were collected in the S-phase-arrested mutant in three different transformations (Table S2), were checked by colony PCR. The distribution of the events (ectopic, translocated, *dur3*-integrated and *adh1*-integrated) in the asynchronous *pol32* mutant (Figure 6B) was similar to that found in the San1 wild type (Figure 4A) with a slight bias toward the translocation rate for the mutant. However, when the mutant was blocked in S phase with HU, the totality of the integrations occurred at the *adh1* locus (raw data are reported in Table S2).

Thus, when the cells were blocked in S phase, in the wild type the number of one-end integrations at *adh1* was decreased with respect to asynchronous cells (from 41% to 25%, see Figure 4A). In the mutant, the one-end integration in *adh1* reached 100%, which is a necessary event during synthesis to survive on G418 selection plates (Figure 6B). The survivors probably arose from the ectopic/microhomology integration of the second DNA end either within the same chromosome or in another chromosome (non-specific translocation). Some hypotheses on the molecular mechanisms that elicit these one-end integration events in the mutant are proposed in the next paragraph.

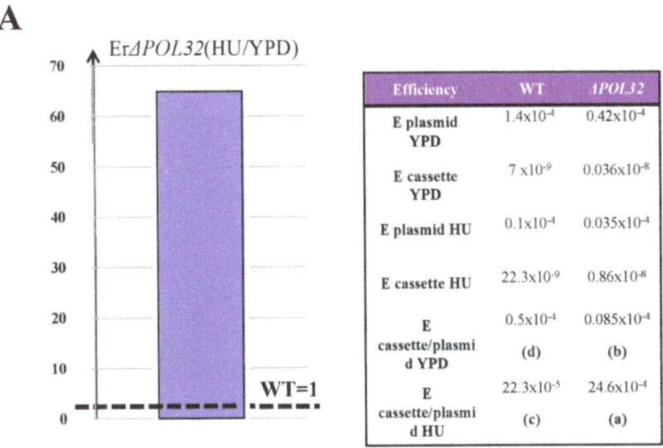

Efficiency	WT	$\Delta POL32$
E plasmid YPD	1.4×10^{-4}	0.42×10^{-4}
E cassette YPD	7×10^{-9}	0.036×10^{-8}
E plasmid HU	0.1×10^{-4}	0.035×10^{-4}
E cassette HU	22.3×10^{-9}	0.86×10^{-8}
E cassette/plasmid YPD	0.5×10^{-4} (d)	0.085×10^{-4} (b)
E cassette/plasmid HU	22.3×10^{-5} (c)	24.6×10^{-4} (a)

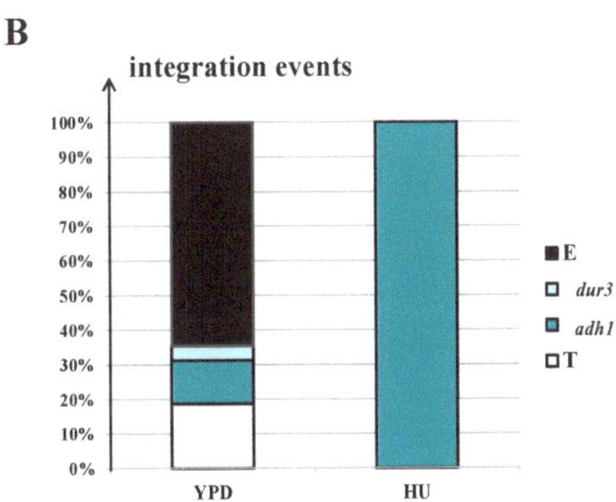

Figure 6. Analysis of the $\Delta POL32$ mutant. (**A**) The graphic on the left side was plotted using the data included in the table on the right. On the y-axis, the relative efficiency of the $\Delta POL32$ mutant respect to wild type (normalized to 1) is reported. The efficiency of transformation with the AD cassette of the mutant blocked in S phase (indicated as "a" in the table of Figure 6A) was divided by its own transformation efficiency with the same cassette in an asynchronous population (b, table in Figure 6A). Then, the efficiency of transformation with the AD cassette of the wild type blocked in S phase (c, table of Figure 6A) was divided by its own transformation efficiency in an asynchronous population (d, table of Figure 6A). The efficiencies with the linear cassettes were related to the transformability of each strain with the plasmid YCp50 in the same experimental conditions. Therefore, the bar reported in Figure 6A was obtained dividing a/b by c/d and represents the efficiency of transformability (Er) of the $\Delta POL32$ mutant with the AD cassette in S phase with respect to its own transformability in the asynchronous state and then normalized to the wild type. The raw data and statistical significance are reported in Table S2C. (**B**) The percentage distribution of the integration events of the AD cassette in the $\Delta POL32$ mutant. On the left: distribution of events in an asynchronous population of the mutant (YPD). On the right: the distribution of events when the mutant is blocked in S phase (HU). E: ectopic integration; T: translocants; *dur3*: integration at the *dur3* locus only; *adh1*: integration at the *adh1* locus only. Er: Relative Efficiency of transformability. The raw data for Figure 6B are reported in Table S2.

4. Discussion

Bridge-Induced Translocation is a molecular genetic tool to link together two heterologous or homologous chromosomes of a diploid yeast cell by exploiting the highly efficient HRS of *S. cerevisiae* é [1,25]. This system, developed while studying mitotic recombination hotspots [26], is based on the initial homologous DNA integration of one end of a double-stranded cassette in the chromosome harboring the homology, followed by the formation of a bridge with the second chromosome through BIR, with an efficiency of the repair that is inversely proportional to the distance of the second sequence homology from its telomere [9,17]. However, all of these previous observations were achieved in asynchronous cells without considering the checkpoint surveillance of the cell cycle transitions and the occurrence of distinct molecular pathways of DNA repair throughout the cell cycle.

Therefore, in this work, we synchronized diploid yeast cells after a block in G1, S and G2/M phase using α-f, HU and NOC, respectively. We then analyzed the efficiency and the distribution of genomic integration events following transformation with two different linear BIT DNA cassettes (Figure 2), one homologous to the XV–VIII chromosome set while the other homologous to the XVI–IX chromosome set, and compared the results with those obtained with the integration of SSI cassettes at the same genomic loci (Figure 3).

To quantify the efficiency of transformation—and hence of integration—with linear DNA, we first transformed the same strains in the same growing conditions with a circular centromeric plasmid, following previously published protocols [9]. We found fewer transformants in HU-synchronized cells with respect to unsynchronized cells (Table S2, Figure 6A). This phenomenon could be explained by a delayed kinetochore reassembly affecting replication after HU treatment, as was previously demonstrated in yeast [27].

Our data suggest that the two BIT cassettes (AD and SUSU) share a common behavior in S phase, represented by an increase in transformation efficiency (moderate for SUSU and substantial for AD, Table S2). These data were supported by a corresponding enhanced transformability during synthesis with the SSI cassettes (Figure 1A). However, in only two (*dur3*, *ssu1*) out of the four loci analyzed in this work, synchronization in S phase also increased the targeting, confirming previous experiments obtained with the *ade2* locus [7]. Nevertheless, as already documented, the efficiency of BIT [26] and a correct SSI targeting [28] strongly relied on the chosen chromosomal loci. For instance, targeting is favored especially in S phase if the locus overlaps a strong promoter, which could act as a hotspot for recombination. Indeed, it is very well known that in *S. cerevisiae*, rates of recombination are relatively low next to centromeres and telomeres whereas rates are higher in promoter-rich intergenic regions [29]. Broad remarks on the events distribution can be extrapolated from a thorough comparison between the two BIT cassettes. Transformation with SUSU, connecting together chromosomes IX and XVI, resulted in a homogeneous distribution of events throughout the whole cell cycle without peaks of translocations in any of the phases and with a slight bias toward DNA synthesis (Figure 4B). On the contrary, AD preferentially integrates during S phase (Figure 4A), tripling the frequency of translocation that occurs in asynchronous populations, probably due to increased targeting in S phase at the *dur3* locus (Figure 5A). We previously hypothesized that BIT occurs in two sequential steps accordingly to a proposed molecular model [3]. The efficiency of this phenomenon, which often generates a partial trisomy, is inversely proportional to the distance of the sequence homology from the telomeres and seems to be Pol32-independent [9]. Since AD integrates preferentially in S phase, we speculated whether Pol32 could modify this outcome in specific phases of the cell cycle. When both copies of *POL32* were deleted in asynchronous diploid cells, the efficiency of transformation with a BIT cassette decreased [9]. From 28 transformations, 48 Pol32 transformants were collected and analyzed (see Table S2 for details). All but one revealed stability of the translocated chromosome and absence of phenotypic defects. Transformant 26 showed a high mortality rate that was attributed to a verified rearrangement of the BIT-derived acentric fragment of chromosome XV [9]. We calculated in this work that the efficiency of transformation of the Pol32 mutant with the linear BIT AD cassette with respect to the wild-type strain was almost six-fold lower when compared to YCp50 (Figure 6A). However, the induced translocation was still possible and, paradoxically, the translocants even accounted for 20% of the total integration events (Figure 6B, Table S2). When the mutant was blocked in S phase with HU, the transformation efficiency with AD was only 2.6-fold lower than in the synchronized wild-type cells and similar results (2.9-fold) were obtained with the circular plasmid. Nevertheless, AD translocants were never recovered. An extensive analysis of all the 68 obtained transformants indicated that when the mutant was blocked during synthesis the only possible event is the integration at the *adh1* locus

(Figure 6B). Therefore, it seems that after the integration of the first end of the cassette through HRS, BIR cannot be completed and the repair perhaps switches to gene conversion (GC) after non-homologous tail removal [30]. The ensuing conclusion is that the BIT translocants obtained in the asynchronous Δ*POL32* mutant were never obtained during S phase, but were generated in other phases of the cell cycle. It has been already postulated that full or partial BIR is still possible in Δ*POL32* yeast cells although the experiments were performed only in asynchronous populations [31]. We can therefore infer from our data that when Pol32 activity is missing, cells can survive without BIR in S-phase, and in the other phases of the cell cycle the integration of the first DNA end and an inefficient or partial BIR repair of the second free end lead to massive cell death in the population because of the persistence of an unrepaired DSB. However, BIT translocation is still possible outside of S phase and accounts for the majority of the repair events. We speculated that these Pol32-independent repair pathways are either due to a traditional BIR preceding mitosis or are the result of half-crossover events similar to those previously described, leading to the loss of the template chromosome [31]. Our experiments support the first hypothesis. In fact, Δ*POL32* translocants revealed replication-derived partial trisomy, stability of the translocated chromosome and strong karyocinetic defects [9]. DNA sequence errors in the proximity of the break within the *dur3* locus ($\cong$1 Kb) were never observed in all the translocants obtained either with or without Pol32 despite the documented low proofreading efficiency of Polδ [31]. This result further confirms the hypothesis that BIR is not absent but reduced by 18-fold in the Δ*POL32* mutants [32]. Our results are in agreement with the evidence that Pol32 only affects the efficiency rather that the outcome of the BIR event [33], that Rad51-dependent BIR occurs efficiently in G2-arrested cells [34] and that Δ*POL32* mutants have a delay in G2/M of the cell cycle and DNA synthesis pausing [35,36]. We therefore propose that cell survival after DSBs originated by the BIT cassette is extremely rare but possible in the absence of the third subunit of polymerase δ. This result can shed light on the observations that mutations in POLD3, the human homologue of Pol32, increase genome instability [37] and predispose to colorectal cancer [38] and is a further demonstration of the value of yeast models for human diseases.

5. Conclusions

From this study we can postulate that BIT translocants recovered in the asynchronous Δ*POL32* mutant are the result of a programmed BIR, favored by a documented G2/M delay that is specific to this mutant [35,39] and confirmed in our strain by DAPI staining (Figure S3). The inferred conclusion is that the yeast cell actively regulates the repair pathway through the recombination execution checkpoint (REC), which signals if both ends of a DSB are engaged with the same template [40] and promotes BIR during a delayed G2/M phase associated with replication stress, safeguarding the cell's survival. Finally, since we already examined the CLS of AD and SUSU translocants and their apoptotic rate [2], this work is propaedeutic for investigating the efficiency of DNA transformation and integration not only in relation to the cell cycle but also with respect to the replicative age of the cells. The comparative study of CLS rate in the presence and the absence of *POL32* after BIT induction is currently under investigation.

Supplementary Materials: The following supporting information can be downloaded at https://www.mdpi.com/article/10.3390/biom13040614/s1, Figure S1: Arrest points of the *S. cerevisiae* mitotic cell cycle. Figure S2: Flow cytometry of translocant Susu5 compared to San1. Figure S3: The abnormal phenotype of the Pol32 mutant. Table S1: Primers used in this work. Table S2: Raw data of transformations.

Author Contributions: V.T. and C.V.B. contributed to the conception and design of the study. V.T. performed the experiments using the AD cassette (chromosomes XV and VIII), built the San1VTDMAT strain and used this strain for all the experiments related to the G1 arrest. B.R. performed the experiments using the SUSU construct (chromosomes XVI and IX) and contributed to the statistical analysis and data computing. J.S. contributed to the ROS data. B.R. and J.S. wrote sections of the manuscript. V.T. and C.V.B. collected the data and wrote the final version of the paper. All authors have read and agreed to the published version of the manuscript.

Funding: This research received no external funding.

Data Availability Statement: The authors acknowledge that the data presented in this study must be deposited and made publicly available in an acceptable repository, prior to publication. All the raw data are in the publicly accessible Table S1 and Table S2 (Supplementary Materials).

Acknowledgments: We thank Andrés Aguilera (Universidad de Sevilla, Spain) for the fruitful discussion and helpful suggestions. We also thank the Protein Structure and Bioinformatics Laboratory of ICGEB (Trieste, Italy) for having provided the α-factor peptide.

Conflicts of Interest: The authors declare that they have no conflicts of interest.

References

1. Tosato, V.; Waghmare, S.K.; Bruschi, C.V. Non-reciprocal chromosomal bridge-induced translocation (BIT) by targeted DNA integration in yeast. *Chromosoma* **2005**, *114*, 15–27. [CrossRef] [PubMed]
2. Sims, J.; Bruschi, C.V.; Bertin, C.; West, N.; Breitenbach, M.; Schroeder, S.; Eisenberg, T. High reactive oxygen species levels are detected at the end of the chronological life span of translocant yeast cells. *Mol. Genet. Genom.* **2016**, *291*, 423–435. [CrossRef] [PubMed]
3. Tosato, V.; Sims, J.; West, N.; Colombin, M.; Bruschi, C.V. Post-translocational adaptation drives evolution through genetic selection and transcriptional shift in *S. cerevisiae*. *Curr. Genet.* **2017**, *63*, 281–292. [CrossRef]
4. Nikitin, D.; Tosato, V.; Zavec, A.B.; Bruschi, C.V. Cellular and molecular effects of nonreciprocal chromosome translocation in *S. cerevisiae*. *Proc. Natl. Acad. Sci. USA* **2008**, *105*, 9703–9708. [CrossRef] [PubMed]
5. Bruschi, C.V.; Comer, A.R.; Howe, G.A. Specificity of DNA uptake during whole cell transformation of *S. cerevisiae*. *Yeast* **1987**, *3*, 131–137. [CrossRef]
6. Chaustova, L.; Zimkus, A. Study of cell wall permeability properties of synchronous *S. cerevisiae* cells in different phases of cell cycle. *Biologija* **2005**, *1*, 38–40.
7. Tsakraklides, V.; Brevnova, E.; Stephanopoulos, G.; Shaw, A.J. Improved gene targeting through cell cycle synchronization. *PLoS ONE* **2015**, *10*, e0133434. [CrossRef]
8. Prado, F. Genetic instability is prevented by Mrc1-dependent spatio-temporal separation of replicative and repair activities of homologous recombination. *BioEssays* **2014**, *36*, 451–462. [CrossRef]
9. Tosato, V.; Sidari, S.; Bruschi, C.V. Bridge-induced chromosome translocation in yeast relies upon a Rad/Rdh54 dependent, Pol32-independ pathway. *PLoS ONE* **2013**, *8*, e60926. [CrossRef]
10. Lydeard, J.R.; Jain, S.; Yamaguchi, M.; Haber, J.E. Break-induced replication and telomerase-independent telomere maintenance require Pol32. *Nature* **2007**, *448*, 820–823. [CrossRef]
11. Falbo, K.B.; Alabert, C.; Katou, Y.; Wu, S.; Han, J.; Wehr, T.; Xiao, J.; He, X.; Zhang, Z.; Shi, Y.; et al. Involvement of a chromatin remodeling complex in damage tolerance during DNA replication. *Nat. Struct. Mol. Biol.* **2009**, *16*, 1167–1172. [CrossRef]
12. Bruschi, C.V.; Howe, G.A. High frequency FLP-independent homologous DNA recombination of 2μ plasmid in the yeast *S. cerevisiae*. *Curr. Genet.* **1988**, *14*, 191–199. [CrossRef]
13. Lee, P.S.; Greenwell, P.W.; Dominska, M.; Gawel, M.; Hamilton, M.; Petes, T.D. A fine-structure map of spontaneous mitotic crossovers in the yeast *S. cerevisiae*. *PLoS Genet.* **2009**, *5*, e1000410. [CrossRef]
14. Storici, F.; Coglievina, M.; Bruschi, C.V. A 2-micron DNA-based marker recycling system for multiple gene disruption in the yeast *Saccharomyces cerevisiae*. *Yeast* **1999**, *15*, 271–283. [CrossRef]
15. Sambrook, J.; Fritsch, E.F.; Maniatis, T. Analysis of genomic DNA by Southern hybridization. In *Molecular Cloning: A Laboratory Manual*; Cold Spring Harbor Laboratory Press: Cold Spring Harbor, NY, USA, 1989.
16. Wach, A.; Brachat, A.; Pöhlmann, R.; Philippsen, P. New heterologous modules for classical or PCR-based gene disruptions in *Saccharomyces cerevisiae*. *Yeast* **1994**, *10*, 1793–1808. [CrossRef] [PubMed]
17. Rossi, B.; Noel, P.; Bruschi, C.V. Different aneuploidies arise from the same bridge-induced chromosomal translocation event in *Saccharomyces cerevisiae*. *Genetics* **2010**, *186*, 775–790. [CrossRef]
18. Lieberman, H.B. Cell Cycle Checkpoint Control Protocols. In *Methods in Molecular Biology*; Humana Press: Totowa, NJ, USA, 2004.
19. Rose, M.D.; Novick, P.; Thomas, J.H.; Botstein, D.; Fink, G.R. A *Saccharomyces cerevisiae* genomic plasmid bank based on a centromere-containing shuttle vector. *Gene* **1987**, *60*, 237–243. [CrossRef] [PubMed]
20. Nyberg, K.A.; Michelson, R.J.; Putnam, C.W.; Weinert, T.A. Toward maintaining the genome: DNA damage and replication checkpoints. *Annu. Rev. Genet.* **2002**, *36*, 617–656. [CrossRef]
21. Branzei, D.; Foiani, M. Regulation of DNA repair throughout the cell cycle. *Nat. Rev. Mol. Cell. Biol.* **2008**, *9*, 297–308. [CrossRef]
22. Symington, L.S. Mechanism and regulation of DNA end resection in Eukaryotes. *Crit. Rev. Biochem. Mol. Biol.* **2016**, *51*, 195–212. [CrossRef]

23. Moore, J.K.; Haber, J.E. Cell cycle and genetic requirements of two pathways of nonhomologous end-joining repair of double-strand breaks in *S. cerevisiae*. *Mol. Cell. Biol.* **1996**, *16*, 2164–2173. [CrossRef]
24. Aylon, Y.; Liefshitz, B.; Kupiec, M. The CDK regulates repair of double-strand breaks by homologous recombination during the cell cycle. *EMBO J.* **2004**, *23*, 4868–4875. [CrossRef]
25. Tosato, V.; Nicolini, C.; Bruschi, C.V. DNA bridging of yeast chromosomes VIII leads to near-reciprocal translocation and loss of heterozygosity with minor cellular defects. *Chromosoma* **2009**, *118*, 179–191. [CrossRef]
26. Tosato, V.; Bruschi, C.V. Per aspera ad astra: When harmful chromosomal translocation become a plus value in genetic evolution. Lessons from *S. cerevisiae*. *Microb. Cell.* **2015**, *2*, 363–375. [CrossRef]
27. Liu, H.; Liang, F.; Jin, F.; Wang, Y. The coordination of centromere replication, spindle formation and kinetochore microtubule interaction in budding yeast. *PLoS Genet.* **2008**, *4*, e1000262. [CrossRef]
28. Klinner, U.; Schäfer, B. Genetic aspects of targeted insertion mutagenesis in yeast. *FEMS Microbiol. Rev.* **2004**, *28*, 201–223. [CrossRef]
29. Gerton, J.L.; DeRisi, J.; Shroff, R.; Lichten, M.; Brown, P.O.; Petes, T.D. Global mapping of meiotic recombination hotspots and coldspots in the yeast *Saccharomyces cerevisiae*. *Proc. Natl. Acad. Sci. USA* **2000**, *97*, 11383–11390. [CrossRef] [PubMed]
30. Mehta, A.; Beach, A.; Haber, J.E. Homology requirements and competition between Gene Conversion and Break-Induced Replication during Double-Strand Break Repair. *Mol. Cell.* **2017**, *65*, 515–526. [CrossRef] [PubMed]
31. Deem, A.; Keszthelyi, A.; Blackgrove, T.; Vayl, A.; Coffey, B.; Mathur, R.; Chabes, A.; Malkova, A. Break-induced replication is highly inaccurate. *PLoS Biol.* **2011**, *9*, e1000594. [CrossRef] [PubMed]
32. Donnianni, R.A.; Symington, L.S. Break-induced replication occurs by conservative DNA synthesis. *Proc. Natl. Acad. Sci. USA* **2013**, *110*, 13475–13480. [CrossRef]
33. Ruiz, J.F.; Gómez-González, B.; Aguilera, A. Chromosomal translocations caused by either pol32-dependent or pol32-independent triparental break-induced replication. *Mol. Cell. Biol.* **2009**, *29*, 5441–5454. [CrossRef] [PubMed]
34. Malkova, A.; Naylor, M.L.; Yamaguchi, M.; Ira, G.; Haber, J.E. *RAD51*-dependent break-induced replication differs in kinetics and checkpoint responses from *RAD51*-mediated gene conversion. *Mol. Cell. Biol.* **2005**, *25*, 933–944. [CrossRef] [PubMed]
35. Karras, G.I.; Jentsch, S. The *RAD6* DNA damage tolerance pathway operates uncoupled from the replication fork and is functional behyond S phase. *Cell* **2010**, *141*, 255–267. [CrossRef] [PubMed]
36. Burgers, P.M.; Gerik, K.J. Structure and processivity of two forms of *Saccharomyces cerevisiae* DNA polymerase delta. *J. Biol. Chem.* **1998**, *273*, 19756–19762. [CrossRef]
37. Tumini, E.; Barroso, S.; Perez-Calero, C.; Aguilera, A. Roles of human POLD1 and POLD3 in genome stability. *Sci. Rep* **2016**, *6*, 38873. [CrossRef]
38. Dunlop, M.G.; Dobbins, S.E.; Farrington, S.M.; Jones, A.M.; Palles, C.; Whiffin, N.; Tenesa, A.; Spain, S.; Broderick, P.; Ooi, L.Y.; et al. Common variation near CDKN1A, POLD3 and SHROOM2 influences colorectal cancer risk. *Nat. Genet.* **2012**, *44*, 770–776. [CrossRef]
39. Huang, M.E.; Le Douarin, B.; Henry, C.; Galibert, F. The *Saccharomyces cerevisiae* protein YJR043C (Pol32) interacts with the catalytic subunit of DNA polymerase alpha and is required for cell cycle progression in G2/M. *Mol. Gen. Genet.* **1999**, *260*, 541–550. [CrossRef]
40. Jain, S.; Sugawara, N.; Lydeard, J.; Vaze, M.; Le Gac, N.T.; Haber, J.E. A recombination execution checkpoint regulates the choice of homologous recombination pathway during DNA double-strand break repair. *Genes Dev.* **2009**, *23*, 291–303. [CrossRef]

biomolecules

Article

Spontaneous Mutation Rates and Spectra of Respiratory-Deficient Yeast

Xinyu Tu [1,†], Fan Wang [1,†], Gianni Liti [2], Michael Breitenbach [3], Jia-Xing Yue [1,*] and Jing Li [1,*]

1   State Key Laboratory of Oncology in South China, Collaborative Innovation Center for Cancer Medicine, Guangdong Key Laboratory of Nasopharyngeal Carcinoma Diagnosis and Therapy, Sun Yat-sen University Cancer Center, Guangzhou 510060, China
2   IRCAN, INSERM, Université Côte d'Azur, 06107 Nice, France
3   Department of Biosciences, University of Salzburg, 5020 Salzburg, Austria
*   Correspondence: yuejiaxing@gmail.com (J.-X.Y.); lijing3@sysucc.org.cn (J.L.)
†   These authors contributed equally to the work.

Abstract: The yeast petite mutant was first discovered in the yeast *Saccharomyces cerevisiae*, which shows growth stress due to defects in genes encoding the respiratory chain. In a previous study, we described that deletion of the nuclear-encoded gene *MRPL25* leads to mitochondrial genome (mtDNA) loss and the petite phenotype, which can be rescued by acquiring *ATP3* mutations. The *mrpl25Δ* strain showed an elevated SNV (single nucleotide variant) rate, suggesting genome instability occurred during the crisis of mtDNA loss. However, the genome-wide mutation landscape and mutational signatures of mitochondrial dysfunction are unknown. In this study we profiled the mutation spectra in yeast strains with the genotype combination of *MRPL25* and *ATP3* in their wildtype and mutated status, along with the wildtype and cytoplasmic petite rho0 strains as controls. In addition to the previously described elevated SNV rate, we found the INDEL (insertion/deletion) rate also increased in the *mrpl25Δ* strain, reinforcing the occurrence of genome instability. Notably, although both are petites, the *mrpl25Δ* and rho0 strains exhibited different INDEL rates and transition/transversion ratios, suggesting differences in the mutational signatures underlying these two types of petites. Interestingly, the petite-related mutagenesis effect disappeared when *ATP3* suppressor mutations were acquired, suggesting a cost-effective mechanism for restoring both fitness and genome stability. Taken together, we present an unbiased genome-wide characterization of the mutation rates and spectra of yeast strains with respiratory deficiency, which provides valuable insights into the impact of respiratory deficiency on genome instability.

Keywords: respiratory deficiency; mtDNA loss; mutation rate; genome instability; *MRPL25*; *ATP3*

Citation: Tu, X.; Wang, F.; Liti, G.; Breitenbach, M.; Yue, J.-X.; Li, J. Spontaneous Mutation Rates and Spectra of Respiratory-Deficient Yeast. *Biomolecules* **2023**, *13*, 501. https://doi.org/10.3390/biom13030501

Academic Editor: Jürg Bähler

Received: 5 February 2023
Revised: 5 March 2023
Accepted: 7 March 2023
Published: 9 March 2023

1. Introduction

Mitochondria are the energy house of the eukaryotic cell, and generate adenosine triphosphate (ATP) to support cellular processes and functions. Mitochondrial dysfunction is associated with many disease-related defects, especially in brain and muscle tissue, where energy metabolism is highly active [1]. Most genes encoding for mitochondrial proteins reside in the nuclear genome, however, a core set of genes involved in respiration and oxidative phosphorylation are located in the mitochondrial genome (mtDNA) [2]. The mtDNA is conserved between humans and the unicellular model organism, the budding yeast *Saccharomyces cerevisiae*. Six out of eight protein-coding genes in the yeast mtDNA have homologs in the human mtDNA, including genes encoding for subunits I, II, and III of cytochrome c oxidase (*COX1*, *COX2*, *COX3*), apocytochrome b (*COB*) and subunits 6 and 8 of ATP synthase (*ATP6*, *ATP8*) [3]. Aside from protein coding genes, the numbers of tRNAs and rRNAs in the mtDNA are also similar. Moreover, the functions of mitochondrial genes involved in many cellular processes are also shared so that much of our current knowledge on mitochondrial dysfunction and associated human disorders comes from studies using *S. cerevisiae* [3–5]. The well-characterized genome and the rich cellular and molecular manipulation toolkits of the budding yeast makes it a key model to dissect and understand the functions of the mtDNA [3].

The mtDNA in *S. cerevisiae* lab strains (e.g., S288C) is not stable [6–9]. The absence (rho0) of mtDNA or the accumulation of mutations (rho⁻) in mtDNA can lead to a special phenotype, known as "petite". Such strains are phenotypically characterized by defects in the respiratory chain, slow growth in fermentable carbon sources (e.g., glucose), and the inability to grow in non-fermentable carbon sources (e.g., glycerol). In addition to the loss of or mutations in mtDNA (i.e., cytoplasmic petite), mutations occurring in the nuclear genome can also generate the petite phenotype (i.e., nuclear petite) if the nuclear genes required for respiratory growth or encoding mitochondrial proteins are disrupted. Moreover, mtDNA loss is lethal in the wildtype yeast species *Kluyveromyces lactis* (petite-negative) [10]. Thus, the petite phenotype reflects stresses that cells have to deal with when respiratory dysfunction occurs. The petite cells are able to evolve by acquiring suppressor mutations such as *ATP1, ATP2, ATP3* (respectively encoding α, β, and γ subunit of the F1 sector in F1F0 ATP synthase) and *SIT4* (encoding a phosphatase that regulates the subunit Atp2p of F1F0 ATP synthase) that alter the function of ATP synthase and restore growth [11–17].

In our previous paper, we described that deletion of the *MRPL25* gene (also known as *AFO1*; located on chromosome VII), which encodes the large subunit of mitochondrial ribosomes, can result in mtDNA loss and reveal the petite phenotype. The growth defects of the *mrpl25Δ* mutant can be rapidly rescued by acquiring the *atp3*^{G348T} suppressor mutation [18]. The energy charge and ATPase activity were almost the same between the *mrpl25Δ* mutant and the *mrpl25Δ atp3*^{G348T} double mutant, indicating that these factors are not causal for growth restoration [18]. Instead, using mutation accumulation experiments, we observed an elevated SNV rate in the *mrpl25Δ* strain, suggesting the potential occurrence of genome instability [18]. However, the genome-wide mutation landscape and mutational signatures of the *mrpl25Δ* strain remain to be characterized for a better understanding of its genome instability. In this study, we analyzed the whole genome sequencing data from our previous mutation accumulation lines (MALs) [18], including the MALs of yeast strains with different presence–absence combinations of the *mrpl25Δ* and *atp3*^{G348T} mutations as well as the wildtype and rho0 (cytoplasmic petite) control strains, to systematically characterize their respective mutation landscapes. We compared their mutation rates and spectra in terms of SNVs, INDELs, segmental and chromosomal copy number variants (CNVs and aneuploidies, respectively) and found both shared and background-specific mutational patterns for strains with respiratory dysfunction. Our findings shed light on the mutagenic effects brought about by respiratory dysfunction and provide insights into better understanding the cross-talks between nuclear and mitochondrial genome stability.

2. Materials and Methods

2.1. Strains

The *S. cerevisiae* strain C+ (*MAT alpha*) was treated with ethidium bromide to generate C+ rho0, in which the absence of mtDNA was validated by both DAPI staining and sequencing. The diploid strain JS760 was derived from a cross between C+ *mrpl25Δ* (this strain rapidly acquired *atp3*^{G348T}) and C+ *MAT a*. The four monosporic haploids JS760-6A, JS760-6B, JS760-6C and JS760-6D were isolated by micromanipulation of an ascus from JS760, which have the genotype combination of *MRPL25* and *ATP3* in their wildtype and mutated status. The genotype information of these strains was compiled in Table 1. The technical details of strain construction have been described in our previous paper [18].

Table 1. Basic information of yeast strains and their mutation accumulation lines.

Strain	Genotype	mtDNA Status	Number of Replicates	Time Interval between Bottlenecks (Day)	Total Number of Single-Cell Bottlenecks
JS760-6A	*MATa, MRPL25$^+$, atp3^{G348T}*	present, but tend to lose	4	4	60
JS760-6B	*MATa, mrpl25::NatMX, ATP3$^+$*	absent	4	4	60
JS760-6C	*MATalpha, MRPL25$^+$, ATP3$^+$*	present	4	2	120
JS760-6D	*MATalpha, mrpl25::NatMX, atp3^{G348T}*	absent	4	2	120
C+	*MATalpha, no auxotrophic markers*	present	4	2	120
C+ rho0	*MATalpha, no auxotrophic markers*	absent	4	4	60

2.2. Mutation Accumulation Lines (MALs)

The six strains described above (JS760-6A, JS760-6B, JS760-6C, JS760-6D, C+, and C+ rho0) were used for our mutation accumulation experiments [18]. Each strain was propagated on YPD plates (MP, 114001222, 2% peptone, 1% yeast extract, 2% glucose, 2% agar) with four replicates. To keep the number of cell divisions between bottlenecks consistent across different strains, the normal-growing strains (JS760-6C, JS760-6D, and C+) were plated for transfer every two days and the slow-growing strains (JS760-6A, JS760-6B, and C+ rho0) were transferred every four days. In total, we accomplished 120 bottlenecks for the normal growers and 60 bottlenecks for the slow growers. The total number of cell divisions throughout the entire mutation accumulation process was approximately 2520 in the normal-growing strains and 1260 in the slow-growing strains [18].

2.3. Whole-Genome Sequencing and Data Analysis

Genomic DNA was extracted from the six ancestors and 24 MAL end-point clones by the Yeast Master Pure kit (Epicenter, Madison, WI, USA). All samples were sequenced using the Illumina HiSeq 4000 PE150 platform by BGI Europe A/S (Copenhagen, Denmark) [18]. The average sequencing depth across all samples was 88×. The *S. cerevisiae* S288C genome sequence and annotation (Release R64-1-1) was used as reference genome for our downstream analysis. We applied the modules 01.Short_Read_Mapping, 02.Short_Read_SNP_IN DEL_Calling, and 04.Short_Read_CNV_Calling of our in-house genome analysis pipeline (https://github.com/yjx1217/Varathon, accessed on 10 August 2022) to investigate the genome-wide mutational landscape. For SNV and INDEL calling, both GATK4 [19] and freebayes [20] were used, with their outputs further intersected to reach a consensus. A series of filters were applied to remove SNV and INDEL calls that existed in ancestor strains, were located in complex regions (e.g., subtelomeres), and had poor mapping depths ($\leq$10). All SNVs and INDELs that passed these filters were visually checked in the Integrative Genomics Viewer (IGV) [21] and ambiguous ones were further discarded. The SNV mutational signatures were analyzed and fit to the COSMIC signatures using the R/Bioconductor MutationalPatterns package [22], in which a strict refitting function "fit_to_signatures_strict" and a cutoff of 0.004 were applied. Of note, we identified a potential cross-contamination between two replicates (replicate 1 and replicate 3) of the C+ strain MALs based on finding a high proportion of shared mutations. Therefore, we excluded the C+ replicate 3 MAL in our analysis. The copy number of the mtDNA was estimated by nuclear-normalized mapping depth. Statistical analysis in this work was carried out in R-4.2.2 (https://www.R-project.org/, accessed on 31 October 2022).

2.4. Variant Effect Prediction and GO Term Analysis

All IGV-verified de novo variants were fed into the Ensembl Variant Effect Predictor (VEP) [23] for variant effect prediction, during which functional impacts of all variants were scored (as "low", "moderate", or "high") and genic variants were assigned to their respective genes. All genic variants were used to perform Gene Ontology (GO) term enrichment analysis via the GO Term Finder module of Saccharomyces Genome Database

(SGD; https://www.yeastgenome.org, accessed on 19 September 2022). A p value cutoff of 0.01 was used for GO term enrichment analysis.

3. Results

3.1. Mutation Accumulation Experiment Setup

In order to systematically investigate the mutation rates and spectra of yeast strains with respiratory deficiency and the petite phenotype, we applied long-term random single-cell bottlenecks, also known as mutation accumulation lines (MALs), to a series of strains: the four monosporic strains (A, B, C, D) derived from the diploid strain JS760-6 with the genotype combination of *MRPL25* and *ATP3* in their wildtype and mutated status, as well as the positive control C+ and negative control C+ rho0 (cytoplasmic petite) (Table 1). The *MRPL25* gene encodes a protein of the large subunit of the mitochondrial ribosome, the deletion of which results in mtDNA loss, respiratory deficiency, and growth defects, such as the JS760-6B strain (*mrpl25Δ ATP3*$^+$) [24]. However, when accompanied with the suppressor mutation G348T in *ATP3*, the growth of the *mrpl25Δ* mutant can be restored, such as the JS760-6D strain (*mrpl25Δ atp3*^{G348T}) [17,18]. Strains with the suppressor mutation alone, such as the JS760-6A strain (*MRPL25*$^+$ *atp3*^{G348T}), rapidly revealed an enhanced number of petite colonies on the YPD plate, indicating a strong tendency towards losing mtDNA (which was confirmed by sequencing, as will be discussed further on). We performed mutation accumulation experiments with single-cell bottlenecks every two or four days, depending on the growth rate of the specific strain (See Section 2 for details) during 240 days continuous culturing, which accounted for approximately 2520 or 1260 generations of random mutation accumulation, respectively (Table 1).

3.2. Elevated Spontaneous SNV and INDEL Rates in the mrpl25Δ Mutant

The ancestors and end-time-point MALs of all six strains described in Table 1 were sent for whole genome sequencing to characterize their respective mutational profiles. Among these strains, the two petite strains, JS760-6B (*mrpl25Δ ATP3*$^+$) and C+ rho0, revealed the highest SNV rates (2.70×10^{-9} and 2.48×10^{-9} per base per generation, respectively), which are significantly higher when compared with the wildtype C+ control (1.15×10^{-9} per base per generation, t-test, $p < 0.05$) (Figure 1a, Tables S1 and S2). The more than 2-fold increases in genome-wide SNV rate revealed elevated genome instability associated with mtDNA loss. Such genome instability of petite strains was also reflected by the high INDEL rate of JS760-6B (4.27×10^{-10} per base per generation), although the C+ rho0 strain appeared to have a normal INDEL rate (Figure 1b, Tables S3 and S4). Interestingly, the double mutant JS760-6D (*mrpl25Δ atp3*^{G348T}) showed a comparable SNV rate (t-test, $p > 0.27$) but significantly higher INDEL rate (t-test, $p = 0.04$) relative to the C+ control, indicating that the genome instability triggered by *mrpl25Δ* can only be partially compensated by the *atp3*^{G348T} suppressor mutation. The INDEL rate of the double mutant JS760-6D (*mrpl25Δ atp3*^{G348T}) was on a par with that of JS760-6B (*mrpl25Δ ATP3*$^+$) (t-test, $p = 0.30$; Figure 1b and Table S4). Both of them showed higher INDEL rates than the other four strains without *mrpl25Δ*, suggesting that *mrpl25Δ* plays a role in the increase in INDEL mutation rate, although the underlying mechanism is unclear. This also explains the higher INDEL rates that we observed in JS760-6B (with *mrpl25Δ*) but not its C+ rho0 control (without *mrpl25Δ*).

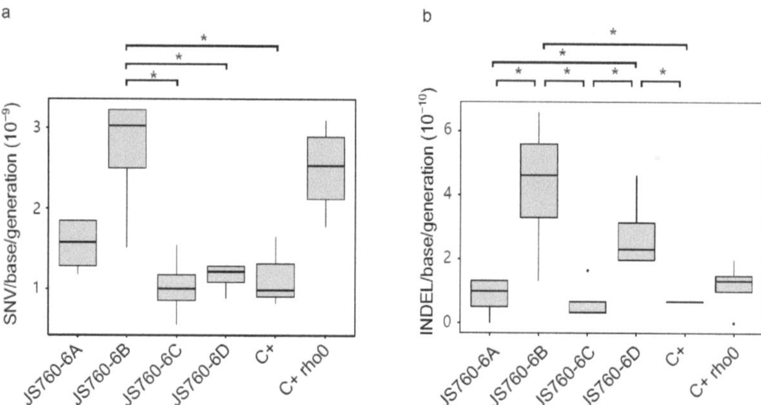

Figure 1. Mutation rates among the six strains. (**a**) SNV rate. (**b**) INDEL rate. The "*" symbols in the panels a and b indicate statistical significance with t test $p < 0.05$. The genotypes of the six strains are as follows: JS760-6A ($MRPL25^+$ $atp3^{G348T}$), JS760-6B ($mrpl25\Delta$ $ATP3^+$), JS760-6C ($MRPL25^+$ $ATP3^+$), JS760-6D ($mrpl25\Delta$ $atp3^{G348T}$), C+ (positive control), C+ rho0 (negative control).

3.3. Different Mutation Spectra Underlying the mrpl25Δ and rho0 Petite Strains

Next, we investigated the base substitution spectrum of SNVs (Figure 2a,b). Although the proportion of each individual substitution type does not vary significantly among the six strains (Figure 2a, chi-square test, $p = 0.90$), we observed notable differences when combining different substitution types for calculating the transition/transversion (Ts/Tv) ratio. Ts/Tv is expected to be 0.5 under the neutral model of evolution, but in reality Ts/Tv is often greater than 0.5 [6,25,26], which was also the case in this study. Notably, we observed a strong Ts/Tv bias in the C+ rho0 petite strain (mean Ts/Tv = 1.25; t-test, $p = 0.002$ compared with the C+ control; Table S5), which is largely driven by the high proportion of C > T transition (41%) (Figure 2a, b). Interestingly, such a strong Ts/Tv bias was not observed in the $mrpl25\Delta$ petite strain (Ts/Tv = 0.89, t-test, $p = 0.33$ compared with the C+ control; Table S5), reflecting different mutational signatures of these two types of petite strains. Yeast is known to have a guanine/cytosine (G/C) to adenine/thymine (A/T) mutational bias [6,27]. This was also observed in our experiments, with the number of G/C to A/T mutations relative to the number of A/T to G/C mutations ranging from 2.14 to 3.12 (Figure 2b and Table S6). The extent of such bias was similar among the six strains (t-test, $p > 0.05$). We further analyzed the SNVs based on the relative incidences of base substitution within a trinucleotide context and further fit the profile to the COSMIC (Catalogue of Somatic Mutations in Cancer) signatures that characterize mutational patterns derived from cancer (Figure 2c,d). The mutational signature of C+ recaptured what has been reported in the wildtype diploid yeast strain [28], suggesting the effectiveness of our signature identification pipeline. Using this approach, we found shared mutational signatures of polymerase eta (Pol η) activity in $mrpl25\Delta$ and rho0 strains (SBS9, single base substitutions signature, Figure 2d). Although the cancer signatures identified in humans are not completely applicable in yeast and Pol η is not the only error-prone polymerase in humans, our results provide empirical clues for future functional studies on the role of error-prone polymerases in driving the genome instability of strains with mitochondrial dysfunction. We also identified other signatures that were not completely shared between $mrpl25\Delta$ and rho0 strains. The etiology of these signatures was not clearly defined by COSMIC and remains to be further investigated.

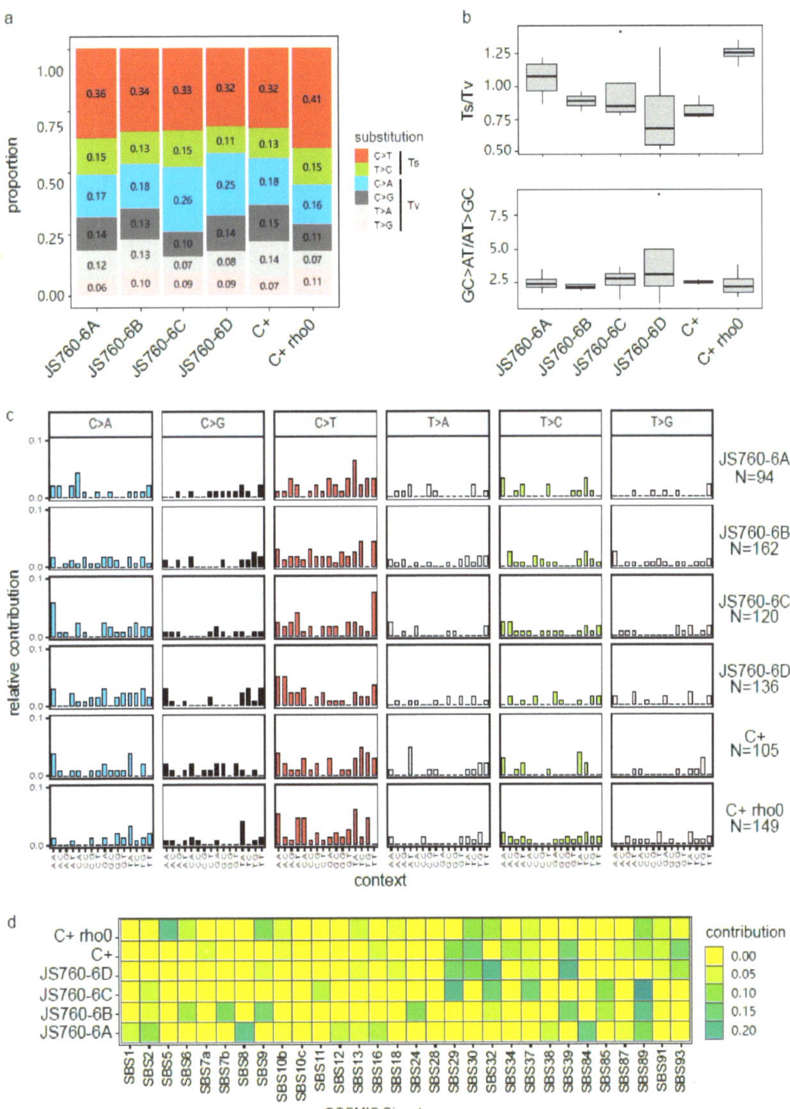

Figure 2. Mutation spectra comparison among the six strains. (**a**) Proportion of six types of substitutions. (**b**) Transition/Transversion ratio and G/C to A/T substitution bias. (**c**) SNV profiles within a trinucleotide context. The number of SNVs used for the analysis are 94, 162, 120, 136, 105, and 149 for JS760-6A, 6B, 6C, 6D, C+, and C+ rho, respectively. (**d**) The SNV profiles were fit to the COSMIC signature (https://cancer.sanger.ac.uk/cosmic/signatures, accessed on 13 December 2022) and the relative contributions of SBS (single base substitution) signatures are shown. The genotypes of the six strains are as follows: JS760-6A (*MRPL25*+ *atp3*^{G348T}), JS760-6B (*mrpl25*Δ *ATP3*+), JS760-6C (*MRPL25*+ *ATP3*+), JS760-6D (*mrpl25*Δ *atp3*^{G348T}), C+ (positive control), C+ rho0 (negative control).

3.4. No Sign of Selection during Mutation Accumulation

The biggest advantage of using a mutation accumulation experiment to study the mutational process lies in its non-selective nature, which allows for a random and unbiased accumulation of mutations. In this study, we analyzed the fractions of non-synonymous

SNVs and genic SNVs relative to total SNVs, which can be used as proxies to check if any sign of unintended selection was introduced during our experiment. In the model of neutral evolution, the fractions of non-synonymous SNVs and genic SNVs relative to total SNVs are expected to be 0.76 and 0.74, respectively [26]. By comparing to this neutral expectation, we found no sign of selection in any of the six strains, verifying the nature of random mutation accumulation in our experiments (Figure 3a,b, Fisher's exact test, $p > 0.37$). We also checked if there was any functional enrichment of genic SNVs found in our MALs, e.g., beneficial mutations in F1-ATPase would largely interrupt the neutrality of the mutation accumulation system. No such enrichment was identified except that the JS760-6A strain ($MRPL25^+$ $atp3^{G348T}$) showed an enrichment of mutated genes responsible for ATP-dependent activity and ATP-binding cassette transporter activity. In addition, we also examined the number of SNVs distributed among different chromosomes and found a strong linear correlation regarding chromosome lengths for all six strains (Figure 4, $R^2 > 0.61$, $p < 3.85 \times 10^{-4}$). This again reflects the randomness of mutations accumulated in our experiment.

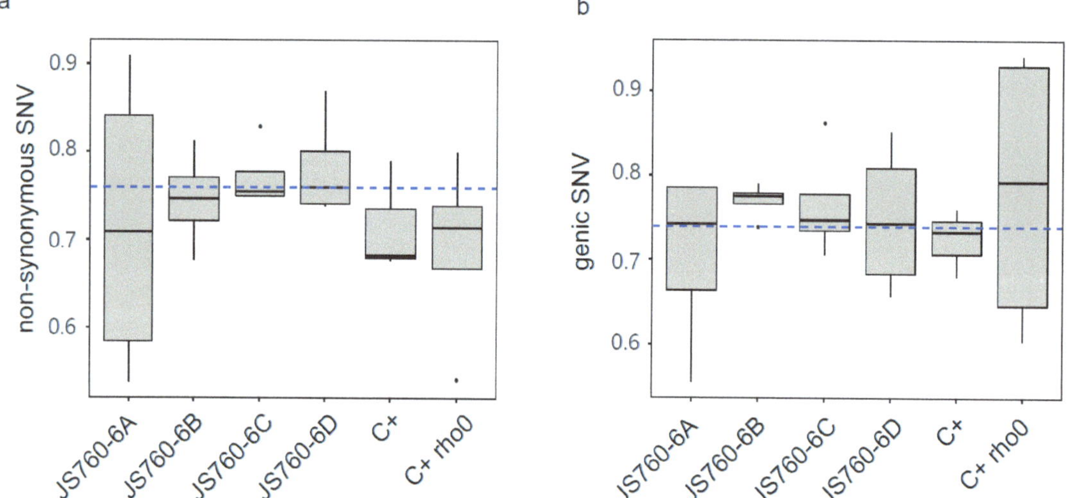

Figure 3. Non-synonymous and genic SNVs of the six strains. (**a**) Proportion of non-synonymous SNVs. (**b**) Proportion of genic SNVs. The blue lines in panels a and b represent the expected values for the proportion of non-synonymous SNVs and genic SNVs, respectively, under the neutral evolution model. The genotypes of the six strains are as follows: JS760-6A ($MRPL25^+$ $atp3^{G348T}$), JS760-6B ($mrpl25\Delta$ $ATP3^+$), JS760-6C ($MRPL25^+$ $ATP3^+$), JS760-6D ($mrpl25\Delta$ $atp3^{G348T}$), C+ (positive control), C+ rho0 (negative control).

3.5. The mtDNA Copy Number Variation and mtDNA Mutations

For the strains that initially had mtDNA (C+, JS760-6A and JS760-6C), we first analyzed the copy numbers of mtDNA across their MALs (Figure 5a). All MAL replicates of JS760-6A ($MRPL25^+$ $atp3^{G348T}$) lost mtDNA; this agrees with our observation that the petite phenotype appeared quickly in these lines soon after the first few transfers. Such rapid loss of mtDNA suggests a substantial mitochondrial genome instability in strains with the $atp3^{G348T}$ mutation. For wild type strains, the C+ control maintained mtDNA in all MAL replicates, while the JS760-6C strain ($MRPL25^+$ $ATP3^+$) lost its mtDNA in two MAL replicates and showed the petite phenotype accordingly. On average, the MALs that maintained the mtDNA had 14 mitochondrial genomes per haploid nuclear genome.

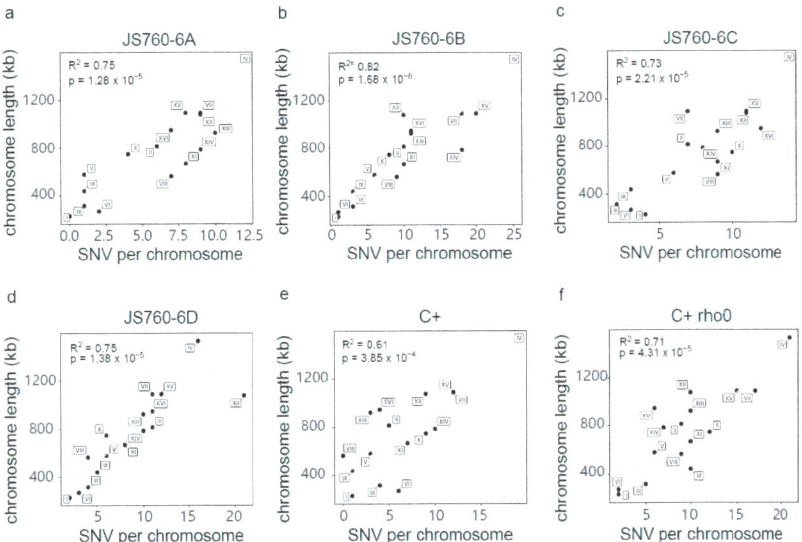

Figure 4. The number of SNVs positively correlates with chromosome lengths. The panels (**a**–**f**) show strong linear correlation between the number of SNVs identified on each chromosome for each strain (*x*-axis) and chromosome length (*y*-axis). The genotypes of the six strains are as below: JS760-6A (*MRPL25$^+$ atp3^{G348T}*), JS760-6B (*mrpl25Δ ATP3$^+$*), JS760-6C (*MRPL25$^+$ ATP3$^+$*), JS760-6D (*mrpl25Δ atp3^{G348T}*), C+ (positive control), C+ rho0 (negative control).

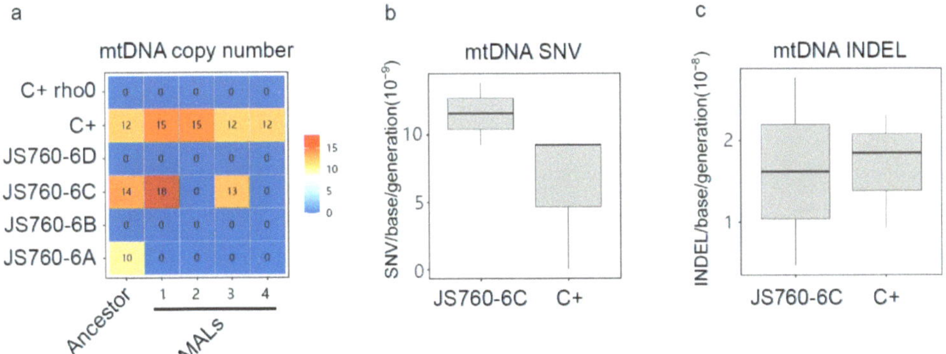

Figure 5. The copy number change and mutations of the mitochondrial genome. (**a**) The mitochondrial copy numbers of the six ancestor strains and their four MAL replicates. (**b**) The mitochondrial SNV rate. (**c**) The mitochondrial INDEL rate. The genotypes of the six strains are as follows: JS760-6A (*MRPL25$^+$ atp3^{G348T}*), JS760-6B (*mrpl25Δ ATP3$^+$*), JS760-6C (*MRPL25$^+$ ATP3$^+$*), JS760-6D (*mrpl25Δ atp3^{G348T}*), C+ (positive control), C+ rho0 (negative control).

Next, we examined the mitochondrial SNV and INDEL rates of these strains. We found the C+ and JS760-6C (*MRPL25$^+$ ATP3$^+$*) strains showed similar mutation rates in terms of both SNV and INDEL (*t*-test, $p > 0.26$), with an average mitochondrial SNV rate of 8.33×10^{-9} per base per generation (Figure 5b, Table S7) and mitochondrial INDEL rate of 1.67×10^{-8} per base per generation (Figure 5c, Table S7). Our estimates of the mitochondrial SNV rate were close to previous estimates using MALs of haploid yeast strains (12.2×10^{-9} by Lynch et al. [6]; 4.82×10^{-9} by Sharp et al. [27]), whereas our estimates of the mitochondrial INDEL rate were notably higher in comparison (7.48×10^{-9} by Lynch et al. [6]; 5.71×10^{-9} by Sharp et al. [27]). The background genetic differences of strains used across these three studies may have played a role in explaining this disparity as previously acknowledged [29].

3.6. Aneuploidies and Segmental Copy Number Variation

In addition to SNVs and INDELs, we also analyzed whole-chromosome (i.e., aneuploidy) and segmental copy number variants (CNVs) across all the MALs. We found the chromosome I (chrI) and XVI (chrXVI) aneuploidies existed in the JS760-6A ($MRPL25^+$ $atp3^{G348T}$) and JS760-6B ($mrpl25\Delta$ $ATP3^+$) ancestor strains (Figure 6a). This was unknown when we started the mutation accumulation experiment and was likely introduced during strain construction but before spore dissection given the 2:2 segregation ratios among JS760-6A, JS760-6B, JS760-6C, and JS760-6D. These aneuploidies may respond to $MRPL25$ deletion and the subsequent mitochondrial dysfunction, given it is known that aneuploidy can act as a transient or stable stage for survival or adaptation when cells are under stress [30–35]. During mutation accumulation, the chrI aneuploidy was kept to the end time point in all eight MALs, while the chrXVI aneuploidy was lost in three out of eight MALs. We also noticed that one JS760-6B MAL acquired two additional copies of chrI (a total of three copies of chrI) and one C+ MAL gained an extra copy of chrI. Aneuploidy has been reported to influence mutation rates [36]. In our system, we found that chrI and chrXVI aneuploidies do not affect our conclusion of the impact of $mrpl25\Delta$ on SNV and INDEL rates because both JS760-6A ($MRPL25^+$ $atp3^{G348T}$) and JS760-6B ($mrpl25\Delta$ $ATP3^+$) were chrI and chrXVI aneuploidies, but they showed distinct SNV and INDEL rates, indicating that aneuploidy was not the primary driver for genome instability. Moreover, the rho0 strain was euploid and showed a SNV rate as high as the JS760-6B strain ($mrpl25\Delta$ $ATP3^+$), suggesting mitochondrial dysfunction should have much larger effects on genome instability than aneuploidy. In addition to aneuploidy, we also analyzed segmental CNVs. Two such cases with convincing signals were presented (Figure 6b). The break points of both cases occurred in repeat regions such as long-terminal repeats (LTRs) and full-length Ty1 retrotransposable elements, which is consistent with a previous report on the strong association of genome rearrangement breakpoints and repetitive sequences in yeast genomes [37].

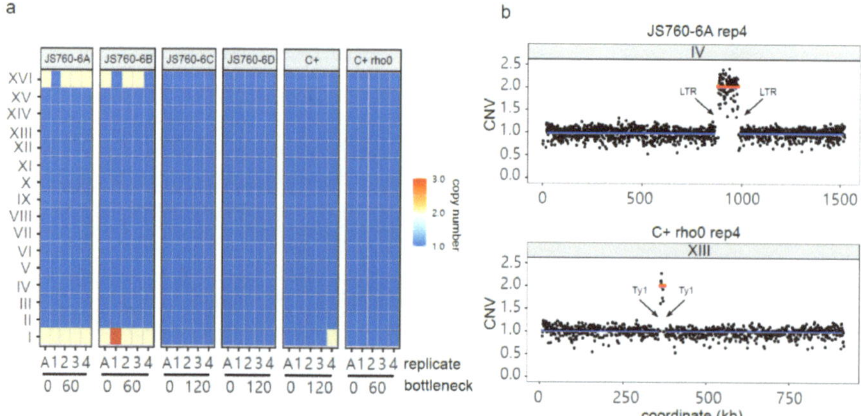

Figure 6. Aneuploidies and segmental copy number variants. (**a**) Chromosomal copy number changes of six ancestor stains (denoted with "A") and their four MAL replicates. (**b**) Two cases of segmental duplication with repeat-associated breakpoints (indicated with arrows). The blue lines represent normal copy number while the red lines represent segmental duplication. The genotypes of the six strains are as follows: JS760-6A ($MRPL25^+$ $atp3^{G348T}$), JS760-6B ($mrpl25\Delta$ $ATP3^+$), JS760-6C ($MRPL25^+$ $ATP3^+$), JS760-6D ($mrpl25\Delta$ $atp3^{G348T}$), C+ (positive control), C+ rho0 (negative control).

4. Discussion

The loss of mtDNA leads to a cellular crisis in yeast cells [38]. The most obvious phenotype reflecting this crisis is slow growth (petite) in YPD. On the quest for strategies to overcome such a growth defect, evolution commences and suppressor mutations such as those in *ATP3* can be acquired [17,18]. We sought to understand the mechanism of *ATP3* mutation suppression in terms of increased mitochondrial membrane potential caused by enhanced ATPase activity [38,39]. However, careful measurement did not reveal an increase in the JS760-6D strain (*mrpl25Δ atp3*^{G348T}) and thus we excluded this possibility [18]. Alternatively, one evolutionary strategy of quick adaptation is to induce genome instability and thereby increase the chance of gaining beneficial mutations. After acquiring such suppressor mutations, the growth advantage of the corresponding clone can outcompete its peers and rapidly take over the population. To test the mutagenetic effect of mtDNA loss, we applied the mutation accumulation approach that minimizes the effect of selection to allow almost all types of mutations to accumulate in an unbiased way. With this approach, we proved that the mtDNA loss itself can indeed promote genome instability.

Mitochondrial-dysfunction-induced genome instability has been previously reported [38,40–42]. For example, the mutation rate of petite cells was found to be elevated by estimating the LOH (loss of heterozygosity) rate at two loci [38]. Our work reinforced this observation by providing direct evidence for a genome-wide mutation rate increase in a set of yeast strains with mitochondrial dysfunction. Such stress-induced genome instability increases the chance of acquiring beneficial mutations in contingency scenarios [17,18,43]. Both *mrpl25Δ* and rho0 petite strains showed elevated SNV mutation rates. Moreover, their mutational signatures revealed the activity of error-prone DNA polymerase during replication. This resembles the stress-induced mutagenesis (SIM) observed in bacteria (e.g., via SOS pathway) [44], suggesting a potentially similar mechanism. Furthermore, we found once the petite cells acquired the *ATP3* suppressor mutation that conferred a growth advantage (e.g., JS760-6D: *mrpl25Δ atp3*^{G348T}), their mutation rates reversed to normal. Although the mechanism remains unclear, this observation hints that the mtDNA-loss-induced mutagenesis can be sophisticatedly regulated, which effectively reduces the evolutionary cost when growth disadvantage has been rescued and a high mutation rate is no longer needed. The mtDNA loss can lead to an insufficient supply of iron-sulfur cluster (ISC) and certain types of amino acids that are required for genome integrity, while the *ATP3* suppressor mutation can alleviate the perturbation in iron metabolism and restore the function of mitochondrial tricarboxylic acid cycle [17,38]. This could release signals to downregulate the mutagenesis activity. Although the mechanism of such mutation rate regulation needs further investigation, it is an effective strategy to deal with an acute cellular crisis. The *ATP3* mutant without *MRPL25* loss (e.g., JS760-6A: *MRPL25*$^+$ *atp3*^{G348T}) revealed a high tendency of losing mtDNA (4/4 MALs became petite) and moderately increased the SNV rate, indicating there are both genotypic and phenotypic costs of *ATP3* mutations. Taken together, the mechanisms underlying the mutagenesis regulation in petites could involve multiple molecular and cellular processes, which need further investigation.

5. Conclusions

We systematically examined the mutation rates and spectra of petite strains in the context of *MRPL25* and *ATP3* mutations by long-term mutation accumulation experiments. We found elevated mutation rates in strains with respiratory deficiency, which further revealed the mutational signatures of error-prone DNA polymerase activity probably contributing to the genome instability. We also found once the respiratory-deficient strain acquired a suppressor mutation to recover its fitness, its mutation rate returned to normal, hinting at delicate and effective molecular machinery regulating the mutation rate. Overall, our study provides a genome-wide mutational landscape for yeast cells with mitochondrial dysfunction, and deepens the understanding of respiratory deficiency and mtDNA loss on genome instability.

Supplementary Materials: The following supporting information can be downloaded at: https://www.mdpi.com/article/10.3390/biom13030501/s1. Table S1. List of SNVs accumulated in nuclear genome; Table S2. Nuclear SNV rates of all MALs; Table S3. List of INDELs accumulated in nuclear genome; Table S4. Nuclear INDEL rates of all MALs; Table S5. The transition/transversion (Ts/Tv) ratio of all MALs; Table S6. The G/C to A/T mutational bias; Table S7. List of SNVs and INDELs accumulated in the mitochondrial genome.

Author Contributions: Conceptualization, J.L. and J.-X.Y.; Funding acquisition, J.L., J.-X.Y., M.B. and G.L.; Investigation, X.T., F.W., J.L. and J.-X.Y.; Visualization, X.T., F.W., J.L. and J.-X.Y.; Writing-Original draft: J.L., J.-X.Y., X.T. and F.W.; Writing-Review and editing: J.L. and J.-X.Y. All authors have read and agreed to the published version of the manuscript.

Funding: This work is supported by the National Natural Science Foundation of China (32000395 to J.L. and 32070592 to J.-X.Y.), the Natural Science Foundation of Guangdong Province (2022A1515011873 to J.L. and 2022A1515010717 to J.-X.Y.), the Guangzhou Municipal Science and Technology Bureau (202102020938 to J.L.), the Guangdong Basic and Applied Basic Research Foundation (2019A1515110762 to J.-X.Y.), the Guangdong Pearl River Talents Program (2021QN02Y168 to J.L., 2019QN01Y183 to J.-X.Y.), the Austrian Science Fund FWF (P26713 to M.B.), and the Fondation pour la Recherche Médicale (EQU202003010413 to G.L.).

Institutional Review Board Statement: Not applicable.

Informed Consent Statement: Not applicable.

Data Availability Statement: The sequencing data of the ancestors and MALs are available in the SRA (Sequence Read Archive) database under the BioProject ID of PRJNA632985. The key raw data are uploaded to the Research Deposit public platform (www.researchdata.org.cn, accessed on 5 February 2023), with the approval RDD number of RDDB2022390844.

Acknowledgments: We would like to thank the reviewers for their constructive feedback for our manuscript.

Conflicts of Interest: The authors declare no conflict of interest.

References

1. Lin, M.T.; Beal, M.F. Mitochondrial Dysfunction and Oxidative Stress in Neurodegenerative Diseases. *Nature* **2006**, *443*, 787–795. [CrossRef]
2. Gray, M.W.; Lang, B.F.; Burger, G. Mitochondria of Protists. *Annu. Rev. Genet.* **2004**, *38*, 477–524. [CrossRef]
3. Malina, C.; Larsson, C.; Nielsen, J. Yeast Mitochondria: An Overview of Mitochondrial Biology and the Potential of Mitochondrial Systems Biology. *FEMS Yeast Res.* **2018**, *18*, foy040. [CrossRef]
4. Jacobs, H.T. Making Mitochondrial Mutants. *Trends Genet.* **2001**, *17*, 653–660. [CrossRef] [PubMed]
5. Shadel, G.S. Yeast as a Model for Human MtDNA Replication. *Am. J. Hum. Genet.* **1999**, *65*, 1230–1237. [CrossRef] [PubMed]
6. Lynch, M.; Sung, W.; Morris, K.; Coffey, N.; Landry, C.R.; Dopman, E.B.; Dickinson, W.J.; Okamoto, K.; Kulkarni, S.; Hartl, D.L.; et al. A Genome-Wide View of the Spectrum of Spontaneous Mutations in Yeast. *Proc. Natl. Acad. Sci. USA* **2008**, *105*, 9272–9277. [CrossRef]
7. Dimitrov, L.N.; Brem, R.B.; Kruglyak, L.; Gottschling, D.E. Polymorphisms in Multiple Genes Contribute to the Spontaneous Mitochondrial Genome Instability of Saccharomyces Cerevisiae S288C Strains. *Genetics* **2009**, *183*, 365–383. [CrossRef]
8. Stenberg, S.; Li, J.; Gjuvsland, A.B.; Persson, K.; Demitz-Helin, E.; González Peña, C.; Yue, J.-X.; Gilchrist, C.; Ärengård, T.; Ghiaci, P.; et al. Genetically Controlled MtDNA Deletions Prevent ROS Damage by Arresting Oxidative Phosphorylation. *eLife* **2022**, *11*, e76095. [CrossRef]
9. De Chiara, M.; Friedrich, A.; Barré, B.; Breitenbach, M.; Schacherer, J.; Liti, G. Discordant Evolution of Mitochondrial and Nuclear Yeast Genomes at Population Level. *BMC Biol.* **2020**, *18*, 49. [CrossRef]
10. Clark-Walker, G.D.; Chen, X.J. A Vital Function for Mitochondrial DNA in the Petite-Negative Yeast Kluyveromyces Lactis. *Molec. Gen. Genet.* **1996**, *252*, 746–750. [CrossRef]
11. Chen, X.J.; Clark-Walker, G.D. Mutations in MGI Genes Convert Kluyveromyces Lactis into a Petite-Positive Yeast. *Genetics* **1993**, *133*, 517–525. [CrossRef] [PubMed]
12. Chen, X.J.; Clark-Walker, G.D. Specific Mutations in Alpha- and Gamma-Subunits of F1-ATPase Affect Mitochondrial Genome Integrity in the Petite-Negative Yeast Kluyveromyces Lactis. *EMBO J.* **1995**, *14*, 3277–3286. [CrossRef]
13. Weber, E.R.; Rooks, R.S.; Shafer, K.S.; Chase, J.W.; Thorsness, P.E. Mutations in the Mitochondrial ATP Synthase Gamma Subunit Suppress a Slow-Growth Phenotype of Yme1 Yeast Lacking Mitochondrial DNA. *Genetics* **1995**, *140*, 435–442. [CrossRef]
14. Chen, X.J.; Clark-Walker, G.D. α and β Subunits of F1-ATPase Are Required for Survival of Petite Mutants in Saccharomyces Cerevisiae. *Mol. Gen. Genet.* **1999**, *262*, 898–908. [CrossRef]

15. Pereira, C.; Pereira, A.T.; Osório, H.; Moradas-Ferreira, P.; Costa, V. Sit4p-Mediated Dephosphorylation of Atp2p Regulates ATP Synthase Activity and Mitochondrial Function. *Biochim. Biophys. Acta (BBA) Bioenerg.* **2018**, *1859*, 591–601. [CrossRef]
16. Puddu, F.; Herzog, M.; Selivanova, A.; Wang, S.; Zhu, J.; Klein-Lavi, S.; Gordon, M.; Meirman, R.; Millan-Zambrano, G.; Ayestaran, I.; et al. Genome Architecture and Stability in the Saccharomyces Cerevisiae Knockout Collection. *Nature* **2019**, *573*, 416–420. [CrossRef]
17. Vowinckel, J.; Hartl, J.; Marx, H.; Kerick, M.; Runggatscher, K.; Keller, M.A.; Mülleder, M.; Day, J.; Weber, M.; Rinnerthaler, M.; et al. The Metabolic Growth Limitations of Petite Cells Lacking the Mitochondrial Genome. *Nat. Metab.* **2021**, *3*, 1521–1535. [CrossRef] [PubMed]
18. Li, J.; Rinnerthaler, M.; Hartl, J.; Weber, M.; Karl, T.; Breitenbach-Koller, H.; Mülleder, M.; Vowinckel, J.; Marx, H.; Sauer, M.; et al. Slow Growth and Increased Spontaneous Mutation Frequency in Respiratory Deficient Afo1- Yeast Suppressed by a Dominant Mutation in ATP3. *G3 Genes Genomes Genet.* **2020**, *10*, 4637–4648. [CrossRef]
19. McKenna, A.; Hanna, M.; Banks, E.; Sivachenko, A.; Cibulskis, K.; Kernytsky, A.; Garimella, K.; Altshuler, D.; Gabriel, S.; Daly, M.; et al. The Genome Analysis Toolkit: A MapReduce Framework for Analyzing next-Generation DNA Sequencing Data. *Genome Res.* **2010**, *20*, 1297–1303. [CrossRef]
20. Garrison, E.; Marth, G. Haplotype-Based Variant Detection from Short-Read Sequencing. *arXiv* **2012**, arXiv:1207.3907.
21. Robinson, J.T.; Thorvaldsdóttir, H.; Winckler, W.; Guttman, M.; Lander, E.S.; Getz, G.; Mesirov, J.P. Integrative Genomics Viewer. *Nat. Biotechnol.* **2011**, *29*, 24–26. [CrossRef]
22. Manders, F.; Brandsma, A.M.; de Kanter, J.; Verheul, M.; Oka, R.; van Roosmalen, M.J.; van der Roest, B.; van Hoeck, A.; Cuppen, E.; van Boxtel, R. MutationalPatterns: The One Stop Shop for the Analysis of Mutational Processes. *BMC Genom.* **2022**, *23*, 134. [CrossRef] [PubMed]
23. McLaren, W.; Gil, L.; Hunt, S.E.; Riat, H.S.; Ritchie, G.R.S.; Thormann, A.; Flicek, P.; Cunningham, F. The Ensembl Variant Effect Predictor. *Genome Biol.* **2016**, *17*, 122. [CrossRef] [PubMed]
24. Heeren, G.; Rinnerthaler, M.; Laun, P.; von Seyerl, P.; Kössler, S.; Klinger, H.; Jarolim, S.; Simon-Nobbe, B.; Hager, M.; Schüller, C.; et al. The Mitochondrial Ribosomal Protein of the Large Subunit, Afo1p, Determines Cellular Longevity through Mitochondrial Back-Signaling via TOR1. *Aging* **2009**, *1*, 622–636. [CrossRef] [PubMed]
25. Zhu, Y.O.; Siegal, M.L.; Hall, D.W.; Petrov, D.A. Precise Estimates of Mutation Rate and Spectrum in Yeast. *Proc. Natl. Acad. Sci. USA* **2014**, *111*, E2310–E2318. [CrossRef] [PubMed]
26. Liu, H.; Zhang, J. Yeast Spontaneous Mutation Rate and Spectrum Vary with Environment. *Curr. Biol.* **2019**, *29*, 1584–1591.e3. [CrossRef]
27. Sharp, N.P.; Sandell, L.; James, C.G.; Otto, S.P. The Genome-Wide Rate and Spectrum of Spontaneous Mutations Differ between Haploid and Diploid Yeast. *Proc. Natl. Acad. Sci. USA* **2018**, *115*, E5046–E5055. [CrossRef]
28. Loeillet, S.; Herzog, M.; Puddu, F.; Legoix, P.; Baulande, S.; Jackson, S.P.; Nicolas, A.G. Trajectory and Uniqueness of Mutational Signatures in Yeast Mutators. *Proc. Natl. Acad. Sci. USA* **2020**, *117*, 24947–24956. [CrossRef]
29. Jiang, P.; Ollodart, A.R.; Sudhesh, V.; Herr, A.J.; Dunham, M.J.; Harris, K. A Modified Fluctuation Assay Reveals a Natural Mutator Phenotype That Drives Mutation Spectrum Variation within Saccharomyces Cerevisiae. *eLife* **2021**, *10*, e68285. [CrossRef]
30. Kaya, A.; Mariotti, M.; Tyshkovskiy, A.; Zhou, X.; Hulke, M.L.; Ma, S.; Gerashchenko, M.V.; Koren, A.; Gladyshev, V.N. Molecular Signatures of Aneuploidy-Driven Adaptive Evolution. *Nat. Commun.* **2020**, *11*, 588. [CrossRef]
31. Gerstein, A.C.; Ono, J.; Lo, D.S.; Campbell, M.L.; Kuzmin, A.; Otto, S.P. Too Much of a Good Thing: The Unique and Repeated Paths Toward Copper Adaptation. *Genetics* **2015**, *199*, 555–571. [CrossRef]
32. Zhang, K.; Zheng, D.-Q.; Sui, Y.; Qi, L.; Petes, T.D. Genome-Wide Analysis of Genomic Alterations Induced by Oxidative DNA Damage in Yeast. *Nucleic Acids Res.* **2019**, *47*, 3521–3535. [CrossRef] [PubMed]
33. Chen, G.; Bradford, W.D.; Seidel, C.W.; Li, R. Hsp90 Stress Potentiates Rapid Cellular Adaptation through Induction of Aneuploidy. *Nature* **2012**, *482*, 246–250. [CrossRef] [PubMed]
34. Yona, A.H.; Manor, Y.S.; Herbst, R.H.; Romano, G.H.; Mitchell, A.; Kupiec, M.; Pilpel, Y.; Dahan, O. Chromosomal Duplication Is a Transient Evolutionary Solution to Stress. *Proc. Natl. Acad. Sci. USA* **2012**, *109*, 21010–21015. [CrossRef]
35. Liu, G.; Yong, M.Y.J.; Yurieva, M.; Srinivasan, K.G.; Liu, J.; Lim, J.S.Y.; Poidinger, M.; Wright, G.D.; Zolezzi, F.; Choi, H.; et al. Gene Essentiality Is a Quantitative Property Linked to Cellular Evolvability. *Cell* **2015**, *163*, 1388–1399. [CrossRef] [PubMed]
36. Sheltzer, J.M.; Blank, H.M.; Pfau, S.J.; Tange, Y.; George, B.M.; Humpton, T.J.; Brito, I.L.; Hiraoka, Y.; Niwa, O.; Amon, A. Aneuploidy Drives Genomic Instability in Yeast. *Science* **2011**, *333*, 1026–1030. [CrossRef]
37. Yue, J.-X.; Li, J.; Aigrain, L.; Hallin, J.; Persson, K.; Oliver, K.; Bergström, A.; Coupland, P.; Warringer, J.; Lagomarsino, M.C.; et al. Contrasting Evolutionary Genome Dynamics between Domesticated and Wild Yeasts. *Nat. Genet.* **2017**, *49*, 913–924. [CrossRef]
38. Veatch, J.R.; McMurray, M.A.; Nelson, Z.W.; Gottschling, D.E. Mitochondrial Dysfunction Leads to Nuclear Genome Instability via an Iron-Sulfur Cluster Defect. *Cell* **2009**, *137*, 1247–1258. [CrossRef]
39. Francis, B.R.; White, K.H.; Thorsness, P.E. Mutations in the Atp1p and Atp3p Subunits of Yeast ATP Synthase Differentially Affect Respiration and Fermentation in Saccharomyces Cerevisiae. *J. Bioenerg. Biomembr.* **2007**, *39*, 127–144. [CrossRef]
40. Dirick, L.; Bendris, W.; Loubiere, V.; Gostan, T.; Gueydon, E.; Schwob, E. Metabolic and Environmental Conditions Determine Nuclear Genomic Instability in Budding Yeast Lacking Mitochondrial DNA. *G3* **2013**, *4*, 411–423. [CrossRef]
41. Singh, K.K.; Rasmussen, A.K.; Rasmussen, L.J. Genome-Wide Analysis of Signal Transducers and Regulators of Mitochondrial Dysfunction in Saccharomyces Cerevisiae. *Ann. N. Y. Acad. Sci.* **2004**, *1011*, 284–298. [CrossRef]

42. Díaz de la Loza, M.D.C.; Gallardo, M.; García-Rubio, M.L.; Izquierdo, A.; Herrero, E.; Aguilera, A.; Wellinger, R.E. Zim17/Tim15 Links Mitochondrial Iron-Sulfur Cluster Biosynthesis to Nuclear Genome Stability. *Nucleic Acids Res.* **2011**, *39*, 6002–6015. [CrossRef]
43. Chen, X.J.; Clark-Walker, G.D. The Petite Mutation in Yeasts: 50 Years On. *Int. Rev. Cytol.* **2000**, *194*, 197–238. [CrossRef]
44. Erill, I.; Campoy, S.; Barbé, J. Aeons of Distress: An Evolutionary Perspective on the Bacterial SOS Response. *FEMS Microbiol. Rev.* **2007**, *31*, 637–656. [CrossRef]

biomolecules

Article

Quality Evaluation of *Ophiopogon japonicus* from Two Authentic Geographical Origins in China Based on Physicochemical and Pharmacological Properties of Their Polysaccharides

Zherui Chen [1,2,3], Baojie Zhu [1,2,3], Xin Peng [4,*], Shaoping Li [1,2,3,*] and Jing Zhao [1,2,3,*]

1 State Key Laboratory of Quality Research in Chinese Medicine, Institute of Chinese Medical Sciences, University of Macao, Macao SAR 999078, China
2 Joint Laboratory of Chinese Herbal Glycoengineering and Testing Technology, University of Macao, Macao SAR 999078, China
3 Macao Centre for Testing of Chinese Medicine, University of Macao, Macao SAR 999078, China
4 Ningbo Municipal Hospital of Traditional Chinese Medicine, Affiliated Hospital of Zhejiang Chinese Medical University, Ningbo 315000, China
* Correspondence: pengx@nit.zju.edu.cn (X.P.); spli@um.edu.mo or lishaoping@hotmail.com (S.L.); jingzhao@um.edu.mo (J.Z.)

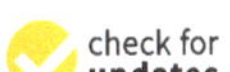

Citation: Chen, Z.; Zhu, B.; Peng, X.; Li, S.; Zhao, J. Quality Evaluation of *Ophiopogon japonicus* from Two Authentic Geographical Origins in China Based on Physicochemical and Pharmacological Properties of Their Polysaccharides. *Biomolecules* **2022**, *12*, 1491. https://doi.org/10.3390/biom12101491

Academic Editors: Mark Rinnerthaler and Markus Ralser

Received: 25 September 2022
Accepted: 12 October 2022
Published: 16 October 2022

Publisher's Note: MDPI stays neutral with regard to jurisdictional claims in published maps and institutional affiliations.

Abstract: *Ophiopogon japonicus* is widely used as a tonic herb in China. According to the origins, MaiDong of Chinese materia medica can be classified as *Zhe MaiDong* (*Ophiopogon japonicus* in Zhejiang), *Chuan MaiDong* (*Ophiopogon japonicus* in Sichuan), *Duanting Shan MaiDong* (*Liriope muscari*), and *Hubei MaiDong* (*Liriope spicata*). In terms of quality control, polysaccharides-based evaluations have not yet been conducted. In this study, microwave-assisted extraction (MAE) was used for the preparation of polysaccharides from 29 batches of *MaiDong*. HPSEC-MALLS-RID and HPAEC-PAD were employed to investigate their molecular parameters and compositional monosaccharides, respectively. The ability to scavenge ABTS radicals and immune promotion abilities, in terms of nitric oxide releasing and phagocytosis on RAW 264.7 macrophages, were also compared. The results showed that polysaccharides in different *MaiDong* varied in molecular parameters. All polysaccharides mainly contained fructose and glucose with small amounts of arabinose, mannose, galactose, and xylose. For polysaccharides of *Zhe MaiDong* and *Chuan MaiDong*, the molar ratio of Fru to Glc was roughly 15:1 and 14:1, respectively. *Zhe MaiDong* exhibited better antioxidant and immune promotion activity, and so did that of fibrous roots. The pharmacological activity, however, did not account for the variation in growth years. Finally, indicators for quality control based on multivariate statistical analysis included: yield, antioxidant activity, the content of fructose, and RI signal. It was concluded that *MaiDong*'s fibrous roots had similar components to the root, and their quality was not significantly affected by growth age. This may provide some guidance for the cultivation and use of *MaiDong*.

Keywords: *MaiDong* polysaccharides; quality control; physicochemical properties; pharmacological properties; multivariate statistical analysis

1. Introduction

The demand for traditional Chinese medicines (TCMs) continues to grow alongside their modernization and globalization. The contradiction between the supply and demand of TCMs could be effectively alleviated by artificial cultivation [1,2]. Studies have shown that TCMs from diverse origins often differ in chemical properties and pharmacological effects [3–5]. Authentic (*Daodi* in Chinese) medicinal materials, TCMs with regional characteristics grown in a specific ecological environment, are thought to be of higher quality, according to traditional Chinese medicine theory [6,7].

Ophiopogon japonicus (called *MaiDong* in Chinese) is a tonic of the same origin as medicine and food and has a long history of use in China. It is widely used in clinical practice because it has the effect of nourishing the Yin and moistening the lung. Modern pharmacological research has demonstrated the effects of *MaiDong*, which include antioxidation, anti-tumor, anti-inflammatory, and cardiovascular protective properties [4,8–12] Conventionally, Zhejiang (*Zhe MaiDong*) and Sichuan (*Chuan MaiDong*) provinces are two major authentic geographical origins of *O. japonicus* that originates in China. *Chuan MaiDong* can typically be harvested after growing for one year for medicinal purposes; however, *Zhe MaiDong* must grow for three years before harvest, resulting in greater cultivation expense. Therefore, their quality differences are the major concern in clinics [13]. To assess the quality of *Zhe MaiDong* and *Chuan MaiDong*, saponins, isoflavones, and other small molecules have been compared [11]. Furthermore, to compare *MaiDong* from various sources, the total homoisoflavonoid concentration [14], antioxidant activity [15], and anti-inflammatory activities of saponins [16] were evaluated. Additionally, the results indicated that those of *Zhe MaiDong* are usually higher than those of *Chuan MaiDong*.

Polysaccharides, with anti-glycemic, immunomodulatory, and gastrointestinal protective effects [4,12,17–20], are abundant in *MaiDong*. However, few studies have focused on the difference between *Zhe MaiDong* and *Chuan MaiDong* because of their complexity. In this study, physicochemical and pharmacological properties of polysaccharides from 29 batches of *Zhe MaiDong* and *Chuan MaiDong*, as well as *Duanting Shan MaiDong* (*Liriope muscari*) and *Hubei MaiDong* (*Liriope spicata*), were investigated and compared, which could be helpful to improve their quality control.

2. Materials and Methods

2.1. Materials and Chemicals

Raw materials of *O. japonicus*, *L. muscari* (named as *Shan MaiDong*), and *L. spicata* (named *Hubei MaiDong*) are listed in Table 1. Their species were identified by Professor Xin Peng, one of the authors of this paper. The samples were stored in the Institute of Chinese Medical Sciences, University of Macau.

2,2′-azino-bis (3-ethylbenzothiazoline-6-sulfonic acid) (ABTS) was purchased from International Laboratory (San Bruno, CA, USA). Potassium persulfate was purchased from Fluka (Selzer, Germany). Dulbecco's modified eagle medium (DMEM), fetal bovine serum (FBS), penicillin/streptomycin (P/S), and phosphate-buffered saline (PBS) were purchased from Gibco-Invitrogen (Paisley, UK). Cell counting kit 8 (CCK8) was purchased from MCE (MedChemExpress LLC, Monmouth Junction, NJ, USA). Griess reagent, fluorescein isothiocyanate-dextran (FITC-Dextran), and lipopolysaccharides (LPS) were purchased from Sigma-Aldrich (St. Louis, MO, USA). An endotoxin detection-specific Limulus test kit was purchased from Bioendo Technology (Xiamen, China). Trifluoroacetic acid (TFA) was purchased from Fisher Scientific (Thermo Fisher Scientific, Waltham, MA, USA). Nylon membrane filters (0.22/0.45 μm) were purchased from Millipore (Billerica, MA, USA). Deionized water was prepared using a Millipore MilliQ-Plus system (Millipore, Billerica, MA, USA). All the other reagents were of analytical grade.

2.2. Preparation of Polysaccharides

All samples with sticky nature were stored in a −80 °C refrigerator for 24 h, mixed with diatomaceous earth of equal amounts to avoid clogging during grounding, and sieved to a fine powder (250 μm ± 9.9 μm). Each sample (1.0 g) was immersed in 20.0 mL of 80% methanol, then refluxed in a Syncore parallel reactor (Büchi, Flawil, Switzerland) for 2 h at 65 °C. Subsequently, the extract solution was centrifuged at 4000 × *g* for 10 min (Allegra X-15 centrifuge; Beckman Coulter, Fullerton, CA, USA), and the supernatant was collected. The extracted residue was dried above a hot water bath (WB22 Memmert, Memmert Company, Schwabach, Germany). Then a microwave-assisted extractor (Multiwave 3000, Anton Paar GmbH, Graz, Austria) was used to extract polysaccharides from samples. The low molecular weight compounds of crude polysaccharides were removed using ultra centrifugal filters (molecular weight cut off 1 kDa, Pall Corporation, Port Washington, NY, USA) by centrifugation at 2500× *g*. After ultrafiltration, the fraction with a molecular weight of more than 1 kDa polysaccharides was obtained and stored in a 4 °C refrigerator for subsequent analysis after freeze-drying.

Table 1. The detailed information of 29 samples of *Maidong*.

Species	Sample ID	Origin	Plant part	Harvest Time	Growth Years
Ophiopogon japonicus	OJZ1 [1]	Zhejiang	Root	17 May 2020	3
	OJZ2	Zhejiang	Root	9 May 2020	3
	OJZ3	Zhejiang	Root	1 June 2020	3
	OJZ4	Zhejiang	Root	9 May 2020	3
	OJZ5	Zhejiang	Root	17 May 2020	3
	OJZ6	Zhejiang	Root	17 May 2021	3
	OJZ7	Zhejiang	Root	23 June 2021	3
	OJZ8	Zhejiang	Fibrous Roots	17 May 2020	3
	OJZ9	Zhejiang	Fibrous Roots	9 May 2020	3
	OJZ10	Zhejiang	Fibrous Roots	1 June 2020	3
	OJZ11	Zhejiang	Root	August 2020	3
	OJZ12	Zhejiang	Root	August 2020	2
	OJZ13	Zhejiang	Root	August 2020	2
	OJZ14	Zhejiang	Root	August 2020	2
	OJZ15	Zhejiang	Root	August 2020	2
	OJZ16	Zhejiang	Root	August 2020	1
	OJC1 [2]	Sichuan	Root	July 2020	1
	OJC2	Sichuan	Root	July 2020	1
	OJC3	Sichuan	Root	July 2020	1
	OJC4	Sichuan	Root	July 2020	1
	OJC5	Sichuan	Root	July 2020	1
	OJC6	Sichuan	Root	July 2020	1
	OJC7	Sichuan	Root	July 2020	1
	OJC8	Sichuan	Fibrous Roots	July 2020	1
	OJC9	Sichuan	Root	August 2021	1
	OJC10	Sichuan	Root	August 2021	1
Liriope muscari	LM1 [3]	Fujian	Root	August 2021	1
	LM2	Fujian	Root	August 2021	1
Liriope spicata	LS1 [4]	Hubei	Root	August 2021	1

[1] Polysaccharides from *O. japonicus* in Zhejiang; [2] Polysaccharides from *O. japonicus* in Sichuan; [3] Polysaccharides from *L. muscari*; [4] Polysaccharides from *L. spicata*.

2.3. HPSEC-MALLS-RID Analysis

HPSEC-MALLS-RID was employed to detect the fractions with molecular weight and content of polysaccharides according to our previous report [21]. HPSEC-MALLS-RID was composed of an Agilent 1260 series LC/DAD system (Agilent Technologies, Palo Alto, CA, USA), multi-angle light scattering detectors (MALLS, Wyatt Technology Co., Santa Barbara, CA, USA), and a refractometer (RID, Optilab rEX, Wyatt Technology Co.) in series at 35 °C. The chromatographic column was TOSOH gel columns TSKgel G2500 PWXL (300 mm × 7.8 mm), and the mobile phase was 0.9% NaCl solution. The flow rate was 0.5 mL/min with a 100 μL injection volume. ASTRA 7.3.2 software (Wyatt technology Co., CA, USA) was used to process the data.

2.4. HPAEC-PAD Analysis

HPAEC-PAD (Thermo Scientific™ Dionex™ ICS-5000⁺, Dionex, MA, USA) was used for the analysis of compositional monosaccharides according to a previous method with minor modification [22]. The sample solution was mixed with an equal volume of 2.0 mol/L TFA for complete acid hydrolysis at 80 °C for 3 h to gather complete acid hydrolysates (CAH). All the samples were filtered through a 0.45 μm membrane before analysis. The mobile phase consisted of 88% deionized water and 12% 10 mM NaOH, running for 22 min at a flow rate of 0.4 mL/min on a CarboPac PA200 (3 mm × 250 mm) (Thermo Scientific™ Dionex™ CarboPac™ PA200, Dionex, MA, USA) analytical column with a system temperature of 25 °C. Ten monosaccharide standards, including Fuc, Ara, Rha, Gal, Glc, Xyl, Man, Fru, GalA, and GlcA, were used to calculate the content of each monosaccharide in the samples.

2.5. ABTS Radical Scavenging Test

The assay of ABTS radical scavenging was slightly modified according to the method of Chen et al. [23]. In brief, 7 mmol/L ABTS aqueous solution and 2.5 mmol/L potassium persulfate aqueous solution were mixed in a ratio of 1:1, and then stood in dark for 12 h. This solution was diluted with deionized water to reach a 0.7 ± 0.05 absorbance value at 734 nm, and obtained the ABTS working solution. Sample solutions (0.5, 1, 2, 4, 8 mg/mL) were prepared. In a 96-well plate, 200 μL of ABTS working solution and 10 μL of the sample solution were added to each well, and the reaction was kept in the dark for 6 min. After the reaction, the absorbance was measured, and the ABTS clearance rate formula was as followed:

$$C(\%) = [1 - (A_1 - A_2)/A_0] \times 100 \tag{1}$$

where C is the clearance rate, A_0 is the control absorbance, A_1 is the sample absorbance, and A_2 is the background absorbance, which is to eliminate the interference of the tested solution.

2.6. Cell Culture

RAW 264.7 cells were purchased from American Type Culture Collection (ATCC, Rockville, MD, USA), and used for evaluating the effects of polysaccharides. Cells were cultured in DMEM supplemented with 10% FBS and 1% P/S at 37 °C in a humidified atmosphere of 5% CO_2.

2.7. Cytotoxicity Assay

RAW 264.7 cells (5×10^3 cells/well) were cultured in 96-well microplates overnight, then treated with LPS (0.4 μg/mL) and a series of concentrations of sample polysaccharides for 24 h, respectively. An equal volume of culture medium was used as blank control. Subsequently, the original culture medium was discarded and stained with 100 μL of culture medium containing 10% CCK8 for 2 h in dark. The absorbance values were read at 450 nm, and the cell viability was calculated as the ratio of absorbance values between the sample and vehicle control group.

2.8. Nitric Oxide (NO) Determination

RAW 264.7 cells (5×10^4 cells/well) were seeded in 96-well microplates overnight, and then cells were treated with a series of concentrations of sample polysaccharides and LPS (0.4 μg/mL) for 24 h, respectively. An equal volume of culture medium was used as vehicle control. Subsequently, 75 μL of supernatants were collected and mixed with an equal volume of modified Griess reagent at room temperature for 15 min. The absorbance was measured at 540 nm. NO production was expressed as the ratio of absorbance values between the sample and the LPS-treated group.

2.9. Phagocytic Activity Test

RAW 264.7 cells (5×10^4 cells/well) were cultured in 96-well plates overnight, and then incubated with culture medium, LPS (0.4 μg/mL), and a series of concentrations of samples polysaccharides for 18 h, respectively. Then, the cells were treated with FITC-dextran (0.1 mg/mL in culture medium) and incubated at 37 °C for an additional 1 h in dark. After incubation, the cells were collected with cold PBS after being washed three times. A BD Accuri™ C6 Cytometer (BD Biosciences, San Jose, CA, USA) was used to analyze. The percentage of phagocytosis was expressed as the ratio of phagocytic rate between treatment and control cells.

2.10. Determination of Endotoxin Contamination

An endotoxin detection-specific Limulus test kit was used for avoiding endotoxin contamination. The results indicated that endotoxin contamination in the tested sample could be excluded.

2.11. Statistical Analysis

GraphPad Prism 8.0.2 (Dotmatics, San Diego, CA, USA) was used to analyze and process the data. Data were presented as mean $\pm$ SEM from at least three independent experiments for each sample. Statistical significance between the experimental groups was determined by Student's t-test, and p values less than 0.05 were considered statistically significant. SPSS statistics 26 (IBM®, NY, USA) was used for Pearson correlation analysis. Origin Pro 2022b (OriginLab, MA, USA) was used for cluster analysis.

3. Results and Discussion

3.1. Optimization of Microwave-Assisted Extraction of Polysaccharides

Microwave-assisted extraction parameters were optimized by using sample OJZ7. Single-factor investigation was carried out on liquid-solid ratio, extraction power and time. All sample extracts were made up to 100 mL in volumetric flasks and filtered through a 0.45 μm filter membrane before analysis. HPSEC-MALLS-UV-RID profile of OJZ7 polysaccharides mainly had two peaks, Peak1 and Peak2, based on RI detection (Figure 1d). The effects of the three parameters on microwave-assisted extraction efficiency are also shown in Figure 1. The results indicated that microwave power had a great effect on polysaccharide extraction, and the contents of polysaccharides extracted at both 300 W and 100 W were similar, but the content rapidly decreased as the power increased to 500 W. The optimum extraction time and liquid-solid ratio were 15 min and 40, respectively (Figure 1b,c). From the optimization results of MAE, the polysaccharide content significantly decreased as the extraction power further increased, possibly due to the damage caused by the higher power on the polysaccharide structure (Figure 1a). The low polysaccharide content was seen at 10 min of extraction time, while as time increased, the poly-saccharides content increased at 15 min and decreased again at 20 min, which could be due to the excessive time that would lead to polysaccharide cleavage (Figure 1b). The polysaccharide content reached the maximum at a liquid-solid ratio of 40. This may be because as the liquid-solid ratio increases, more polysaccharides can be extracted. Then, due to solvent saturation, the increased liquid-solid ratio made it more difficult for microwave action on the samples (Figure 1c). Finally, to ensure stable extraction efficiency, we extracted OJZP, OJCP, LMP, and LSP, polysaccharides from 29 batches of samples under the power of 300 W, by using MAE with a liquid-solid ratio of 40 for 15 min.

3.2. Molecular Parameters of MaiDong Polysaccharides

All the sample polysaccharides were dissolved in HPSEC mobile phase at 2 mg/mL and filtered through a 0.45 μm filter membrane before analysis. HPSEC-MALLS-UV-RID chromatograms of representative samples are shown in Figure 2. Each was divided into three peaks based on MALLS, UV, and RI detection. Molecular parameters of the three peaks in all samples are listed in Table 2. The results indicated that in all samples, HPSEC profiles were very similar, but their molecular parameters, including Mw, could be variously different. Briefly, all samples had similar laser and UV signal profiles, but there were differences in the values, and the polysaccharides from different sources of *MaiDong* had some differences in the RID signal profiles; however, the RID signal profiles were similar when compared within the same source samples. Furthermore, all polysaccharide samples from fibrous roots had higher RID response values at Peak2 than those from the root. The Mw analysis revealed that OJZP, OJCP, and LMP had similar molecular weight distributions, with LSP having a larger molecular weight than LMP. Meanwhile, polysaccharides in OJZ8, OJZ9, and OJZ10 had higher molecular weights than other OJZP. However, polysaccharides in OJC8, extracted from the fibrous root, did not significantly differ in terms of molecular weight from other OJCP. Similarly, growth years had no consistent effect on polysaccharide molecular weight. The molecular weight of the polysaccharides, the RID signal, the laser signal, and the UV signal were all considered for inclusion in the quality control indicator, after combining the above information.

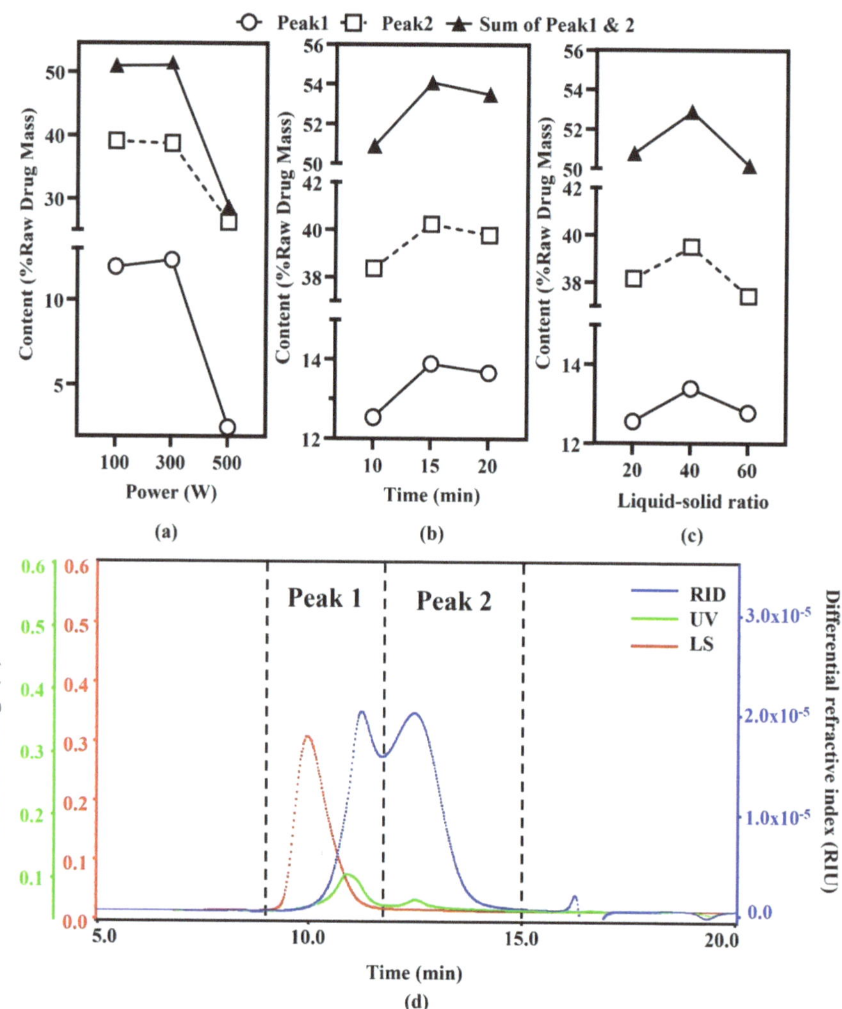

Figure 1. Effect of microwave power (**a**), extraction time (**b**), and liquid-solid ratio (**c**) on extraction efficiency of polysaccharides from *Ophiopogon japonicus* in Zhejiang based on (**d**) HPSEC-MALLS-RID chromatogram with peaks 1 and 2.

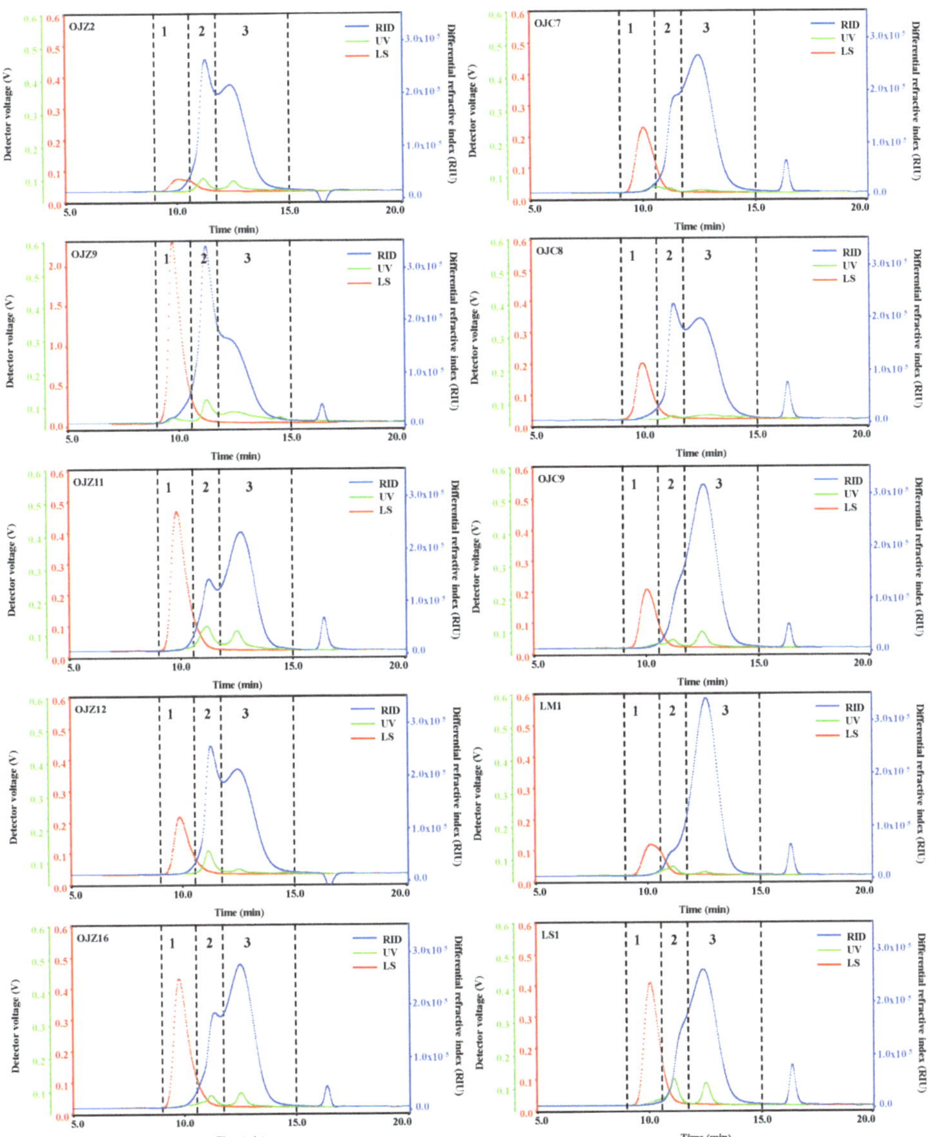

Figure 2. Representative HPSEC-MALLS-RID chromatograms with peaks 1, 2 and 3 of polysaccharides from different samples. The codes are the same as in Table 1.

Table 2. The molecular weight (Mw), polydispersity index (Mw/Mn), z-average radius of gyration (Rz), content of different peak (fraction), and yield of polysaccharides from *MaiDong* of different species or locations.

Sample ID	Peak 1				Peak 2				Peak 3				Recovery (%)
	Mw ($\times 10^6$ Da)	Mw/Mn	Rz (nm)	Content (%)	Mw ($\times 10^4$ Da)	Mw/Mn	Rz (nm)	Content (%)	Mw ($\times 10^3$ Da)	Mw/Mn	Rz (nm)	Content (%)	
OJZ1	7.2 (±2.8%)	2.9 (±3.3%)	96.5 (±1.9%)	0.7	6.0 (±1.1%)	3.4 (±1.6%)	25.3 (±6.3%)	32.3	6.0 (±1.8%)	1.0 (±2.5%)	25.6 (±10.1%)	40.2	7.3
OJZ2	6.7 (±2.3%)	2.7 (±3.3%)	93.4 (±1.6%)	0.9	6.8 (±0.9%)	3.7 (±1.4%)	25.1 (±5.1%)	35.1	6.0 (±1.6%)	1.0 (±2.2%)	23.1 (±10.5%)	42.2	7.4
OJZ3	12 (±2.8%)	3.8 (±3.1%)	86.0 (±1.9%)	0.6	6.5 (±1.3%)	3.2 (±1.9%)	32.0 (±4.6%)	32.0	7.9 (±1.9%)	1.0 (±2.6%)	39.3 (±4.8%)	40.5	6.9
OJZ4	6.8 (±2.2%)	3.1 (±2.5%)	77.7 (±1.7%)	0.6	5.9 (±1.1%)	3.2 (±1.7%)	24.2 (±7.1%)	32.7	6.2 (±1.9%)	1.0 (±2.6%)	23.2 (±12.6%)	36.8	6.0
OJZ5	61 (±7.6%)	5.6 (±7.7%)	158.1 (±2.3%)	0.6	13 (±1.1%)	4.3 (±2.1%)	41.4 (±2.5%)	26.3	9.1 (±2.9%)	1.1 (±4.0%)	60.4 (±3.3%)	47.4	7.9
OJZ6	170 (±9.5%)	4.9 (±9.8%)	201.4 (±1.4%)	1.1	43 (±3.4%)	2.7 (±5.8%)	85.6 (±2.2%)	30.2	77 (±4.1%)	1.2 (±5.9%)	147.7 (±1.2%)	47.8	8.1
OJZ7	48 (±6.0%)	4.2 (±6.2%)	141.3 (±1.9%)	1.2	24 (±1.4%)	4.0 (±2.7%)	42.4 (±2.6%)	32.0	22 (±3.7%)	1.1 (±5.1%)	81.4 (±2.2%)	41.2	8.6
OJZ8	190 (±13.1%)	8.1 (±13.3%)	204.1 (±1.5%)	2.2	44 (±4.1%)	2.1 (±7.0%)	102.2 (±2.0%)	40.4	220 (±5.3%)	1.2 (±7.5%)	223.2 (±1.2%)	32.8	10.3
OJZ9	180 (±12.4%)	5.3 (±12.9%)	209.8 (±1.4%)	4.0	71 (±5.4%)	2.0 (±8.6%)	116.2 (±1.8%)	47.6	390 (±6.4%)	1.1 (±9.0%)	229.7 (±1.1%)	34.3	7.2
OJZ10	170 (±11.8%)	7.3 (±12.0%)	192.5 (±1.7%)	1.7	31 (±3.1%)	2.6 (±5.6%)	83.5 (±1.8%)	37.3	92 (±4.8%)	1.2 (±6.8%)	169.8 (±1.2%)	34.7	7.3
OJZ11	120 (±12.3%)	8.0 (±12.4%)	184.7 (±2.5%)	1.0	32 (±1.9%)	3.2 (±4.6%)	55.9 (±2.1%)	23.1	28 (±5.2%)	1.2 (±7.4%)	123.2 (±1.7%)	49.9	8.2
OJZ12	52 (±9.0%)	5.8 (±9.2%)	150.3 (±2.8%)	0.8	10 (±1.5%)	3.9 (±2.4%)	42.9 (±2.9%)	33.5	9.9 (±3.3%)	1.1 (±4.5%)	61.9 (±3.5%)	44.7	8.0
OJZ13	130 (±8.6%)	4.7 (±8.8%)	145.7 (±2.3%)	0.7	21 (±1.4%)	4.4 (±2.6%)	51.6 (±1.8%)	30.3	18 (±3.2%)	1.1 (±4.4%)	86.5 (±1.7%)	53.2	7.0
OJZ14	340 (±26.8%)	4.5 (±27.0%)	232.5 (±3.5%)	0.4	18 (±2.3%)	4.2 (±4.4%)	65.9 (±1.8%)	29.2	25 (±5.2%)	1.3 (±7.1%)	120.4 (±1.6%)	45.4	6.8
OJZ15	100 (±17.9%)	10.3 (±18.0%)	208.7 (±2.7%)	1.1	13 (±2.4%)	3.0 (±4.6%)	58.0 (±2.4%)	39.9	28 (±5.1%)	1.2 (±7.3%)	119.6 (±1.8%)	46.9	7.4
OJZ16	87 (±11.3%)	9.1 (±11.4%)	185.4 (±2.0%)	1.3	20 (±2.3%)	3.4 (±5.3%)	57.3 (±2.4%)	29.8	23 (±6.3%)	1.2 (±8.7%)	112.6 (±2.2%)	55.1	4.4
Mean	105	5.6	160.5	1.2	22.8	3.3	56.8	33.2	60.5	1.1	103	43.3	7.4
OJC1	62 (±7.4%)	5.3 (±7.5%)	156.8 (±2.0%)	1.0	16 (±1.5%)	5.0 (±3.2%)	43.4 (±2.5%)	35.9	12 (±4.7%)	1.1 (±6.4%)	74.2 (±3.1%)	68.0	12.8
OJC2	44 (±4.3%)	4.1 (±4.5%)	139.6 (±1.4%)	0.8	15 (±1.2%)	6.6 (±2.4%)	35.5 (±2.9%)	32.8	9.1 (±3.3%)	1.1 (±4.3%)	54.6 (±3.8%)	60.0	12.9
OJC3	240 (±9.4%)	3.6 (±9.6%)	151.7 (±2.3%)	0.3	20 (±1.5%)	3.9 (±2.8%)	54.6 (±1.7%)	33.6	27 (±2.6%)	1.2 (±3.6%)	88.3 (±1.3%)	45.2	13.1
OJC4	65 (±8.9%)	5.1 (±9.0%)	156.1 (±2.6%)	0.8	22 (±1.4%)	6.3 (±2.7%)	40.5 (±2.8%)	26.1	11 (±3.2%)	1.1 (±4.2%)	66.6 (±2.6%)	55.9	13.7
OJC5	88 (±7.8%)	4.0 (±8.0%)	139.6 (±2.7%)	0.7	22 (±1.6%)	6.8 (±2.9%)	46.3 (±2.4%)	24.1	12 (±3.3%)	1.1 (±4.4%)	72.1 (±2.3%)	57.3	14.7
OJC6	110 (±12.7%)	7.6 (±12.9%)	161.9 (±2.9%)	0.7	13 (±2.0%)	3.9 (±3.7%)	53.9 (±2.3%)	34.3	17 (±4.1%)	1.2 (±5.6%)	89.0 (±2.1%)	56.2	15.3
OJC7	46 (±6.0%)	3.4 (±6.2%)	119.7 (±2.7%)	0.8	16 (±1.5%)	7.5 (±2.4%)	38.8 (±3.3%)	28.2	8.1 (±2.4%)	1.1 (±3.3%)	48.7 (±3.5%)	54.5	19.0
OJC8	41 (±8.8%)	4.8 (±8.9%)	194.1 (±2.3%)	1.3	11 (±2.2%)	3.2 (±4.5%)	59.1 (±2.1%)	31.0	18 (±5.2%)	1.1 (±7.3%)	97.2 (±2.3%)	45.1	13.6
OJC9	69 (±7.2%)	2.8 (±7.8%)	128.2 (±3.1%)	0.5	20 (±1.8%)	8.6 (±2.7%)	41.8 (±3.3%)	19.9	7.3 (±2.6%)	1.1 (±3.4%)	49.7 (±3.5%)	64.1	21.8
OJC10	100 (±9.3%)	4.3 (±9.6%)	142.6 (±3.7%)	0.3	19 (±1.9%)	6.5 (±3.0%)	45.8 (±3.0%)	20.5	8.0 (±2.8%)	1.1 (±3.8%)	51.2 (±3.5%)	67.4	18.7
Mean	86.5	4.5	149.0	0.7	13.4	5.8	46.0	28.6	12.9	1.1	69.2	57.4	15.6
LM1	37 (±5.3%)	2.4 (±5.7%)	116.1 (±2.3%)	0.5	29 (±1.5%)	9.5 (±2.5%)	35.4 (±3.9%)	14.4	6.8 (±2.6%)	1.1 (±3.4%)	44.3 (±4.4%)	64.0	15.5
LM2	36 (±3.3%)	1.7 (±3.7%)	88.0 (±2.0%)	0.3	17 (±1.4%)	7.7 (±2.2%)	36.3 (±3.6%)	20.6	7.0 (±2.1%)	1.1 (±2.9%)	39.6 (±4.7%)	65.6	14.6
Mean	36.5	2.0	102.1	0.4	23	8.6	35.8	17.5	6.9	1.1	41.9	64.8	15
LS1	150 (±9.9%)	2.4 (±10.1%)	139.4 (±3.4%)	0.4	36 (±1.5%)	8.4 (±2.5%)	49.1 (±2.1%)	22.4	17 (±2.6%)	1.1 (±3.7%)	78.1 (±1.8%)	50.0	14.6

The samples ID are the same as in Table 1.

3.3. Monosaccharide Composition

After complete acid hydrolysis, monosaccharide compositions of polysaccharides were determined by HPAEC-PAD. Table 3 shows their compositional monosaccharides molar ratios. From the monosaccharide composition results, all polysaccharides were mainly composed of Fru and Glc, with small amounts of Ara, Man, Gal, and Xyl. The molar ratio of Fru to Glc was approximately 15:1 in OJZP, 14:1 in OJCP, 11:1 in LMP, and 13:1 in LSP, with relatively similar results, implying that the different sources of polysaccharides in *MaiDong* have a consistent monosaccharide composition, but with minor differences in content, which was similar to other research [24,25]. Normally, the fibrous roots of *MaiDong* are removed for medicinal purposes, but the results of this study suggested that the polysaccharides in the fibrous roots may be complementary to those in the root, to some extent. At the same time, because the growth years did not result in significant changes in monosaccharide composition, it may be possible to shorten the growth years to save cultivation costs.

Table 3. Molar ratio of compositional monosaccharides of OJZP, OJCP, LMP, and LSP.

Samples	Fru	Glc	Ara	Man	Gal	Xyl
OJZ1	87.73	5.57	1.00	0.94	0.61	0.31
OJZ2	79.27	5.05	1.00	0.84	0.63	0.33
OJZ3	87.75	5.47	1.00	0.80	0.57	0.28
OJZ4	90.44	5.46	1.00	0.77	0.63	0.30
OJZ5	122.24	10.35	1.00	0.79	0.59	0.32
OJZ6	108.21	7.35	1.00	0.90	0.64	0.32
OJZ7	68.83	4.74	1.00	0.67	0.65	0.37
OJZ8	43.26	2.70	1.00	1.19	1.30	0.53
OJZ9	39.72	2.28	1.00	1.09	1.35	0.49
OJZ10	56.36	3.53	1.00	1.13	1.49	0.55
OJZ11	66.73	5.61	1.00	0.73	0.52	0.39
OJZ12	116.18	8.01	1.00	0.94	0.66	0.44
OJZ13	110.26	8.21	1.00	0.62	0.56	0.44
OJZ14	205.93	14.16	1.00	0.38	0.64	0.34
OJZ15	114.89	7.34	1.00	1.03	0.63	0.41
OJZ16	100.61	7.89	1.00	1.23	0.63	0.38
Mean	93.65	6.48	1.00	0.88	0.76	0.39
SD	39.48	2.97	0.00	0.23	0.31	0.08
RSD	42.16	45.78	0.00	25.76	41.48	21.44
OJC1	164.22	10.83	1.00	1.09	0.87	0.32
OJC2	192.41	13.76	1.00	1.61	1.03	0.41
OJC3	146.07	8.08	1.00	1.22	0.83	0.24
OJC4	152.66	10.54	1.00	0.94	2.26	0.35
OJC5	218.65	15.29	1.00	0.94	1.15	0.37
OJC6	164.62	11.54	1.00	1.57	0.98	0.37
OJC7	185.86	12.92	1.00	0.79	0.79	0.27
OJC8	126.33	9.82	1.00	1.77	0.61	0.46
OJC9	219.37	19.59	1.00	0.99	1.11	0.44
OJC10	180.42	15.96	1.00	1.22	1.19	0.48
Mean	175.10	12.83	1.00	1.21	1.08	0.37
SD	30.28	3.42	0.00	0.33	0.45	0.08
RSD	17.29	26.62	0.00	27.30	41.73	21.36
LM1	161.76	14.41	1.00	0.35	0.58	0.21
LM2	344.84	29.55	1.00	0.95	0.58	0.44
Mean	253.30	21.98	1.00	0.65	0.58	0.33
SD	129.50	10.71	0.00	0.42	0.00	0.16
RSD	51.13	48.73	0.00	65.28	0.00	50.03
LS1	236.06	17.31	1.00	0.89	0.64	0.35

3.4. ABTS Free Radical Scavenging Ability of MaiDong Polysaccharides

MaiDong was thought to have antioxidant properties [6]. Modern research has demonstrated that oxygen-containing radicals in the human body oxidize cell membranes, accelerate aging, and damage DNA [26]. ABTS radicals are a common form of oxygen-containing radical, and drugs' ability to scavenge them can, to some extent, reflect their antioxidant capacity [27]. In this study, we investigated and compared the ability of various polysaccharides from *MaiDong* to scavenge ABTS radicals. Figure 3 depicts the antioxidant capacity of all polysaccharides from *MaiDong*, and their 20% maximal inhibitory concentration (IC_{20}) for ABTS scavenging ability is listed in Table 4. All samples demonstrated the ability to scavenge ABTS radicals to varying degrees. According to Table 4, polysaccharides extracted from fibrous roots were more effective at ABTS scavenging than those extracted from the root. Polysaccharides from OJZ8, OJZ9, OJZ10, and OJC8 had IC_{20} values of 1.81, 1.87, 1.40, and 2.30 mg/mL, respectively, whereas the IC_{20} values of the polysaccharides extracted from the root were all greater than 3 mg/mL. The fibrous roots have a substantially lower molar ratio of Fru and Ara in *Zhe MaiDong* and *Chuan MaiDong* than that of their tuberous roots, which might contribute to the varied antioxidant activity. In general, the antioxidant capacity of OJZP was comparable to that of OJCP in terms of ABTS scavenging rate and IC_{20} value, and both were much higher than that of LMP and LSP, with LMP slightly higher than LSP. It is interesting to note that the antioxidant capacity of OJZP did not change consistently over the growth years, according to the results of the ABTS radical scavenging test (Figure 3 and Table 4). The ratio of various monosaccharides in the composition and how glycosidic bonds are connected, in our opinion, may be suspect for the variations in the antioxidant capacity of polysaccharides. Furthermore, different molecular weights and contents may have an effect on the results. Based on the antioxidant capacity of polysaccharides from *MaiDong*, the ABTS radical scavenging capacity should be used as an indicator for evaluating the quality of polysaccharides in different origins of *MaiDong*.

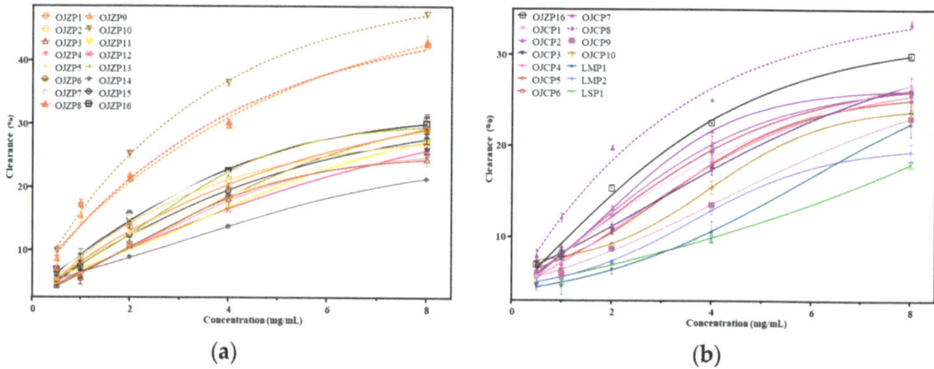

Figure 3. ABTS free radical scavenging activity of polysaccharides from *MaiDong* with different (**a**) and same (**b**) growth years. The codes are the same as in Table 1.

3.5. Immunostimulatory Activity of OJZP, OJCP, LMP and LSP

Macrophages play an indispensable role in the innate and adaptive immunity of the human body [28]. Studies have shown that high levels of NO are associated with immune responses during antitumor and antiviral processes, which can trigger cell proliferation, apoptosis, signal transduction, immune defense, and other physiological processes [29]. Phagocytosis is a basic cellular process that plays an important role in the immune system [30]. In this study, RAW 264.7 cells were treated with a series of concentrations of polysaccharides from selected samples and LSP, and their effects on NO production and phagocytic activity were investigated.

Table 4. ABTS free radical scavenging ability of *MaiDong* polysaccharides.

Sample	IC_{20} (mg/mL)
OJZP1	3.88
OJZP2	3.56
OJZP3	4.64
OJZP4	4.79
OJZP5	4.50
OJZP6	4.43
OJZP7	3.23
OJZP8	1.81
OJZP9	1.87
OJZP10	1.40
OJZP11	4.99
OJZP12	5.30
OJZP13	3.47
OJZP14	7.03
OJZP15	4.19
OJZP16	3.22
OJCP1	4.47
OJCP2	3.94
OJCP3	4.91
OJCP4	4.68
OJCP5	4.61
OJCP6	4.16
OJCP7	3.50
OJCP8	2.30
OJCP9	6.45
OJCP10	5.33
LMP1	7.10
LMP2	$>IC_{15}$: 4.73
LSP1	$>IC_{15}$: 6.66

We selected a series of sample concentrations to determine the effect of cell viability on the results of subsequent experiments, and finally chose concentrations of 5, 10, and 20 µg/mL. All samples in this series neither significantly promoted RAW 264.7 cell proliferation nor had cytotoxicity; the results are shown in Figure 4a. Effects of these polysaccharides on NO production of macrophages are shown in Figure 4b. The results of NO release showed that, among the polysaccharides extracted from the root of *O. japonicus* in Zhejiang with different growth years, all samples had a strong ability to promote NO release except OJZ13. Meanwhile, among the polysaccharides extracted from the root of *O. japonicus* in Sichuan, *L. muscari* and *L. spicata* with one-year-growth, not all samples promoted NO production, and showed no significant difference from the control group at the concentrations of 5, 10, and 20 µg/mL, except for OJC3. OJZ9 and OJC8 were both polysaccharides extracted from fibrous roots, and they were found to be more effective at promoting NO release than polysaccharides from the root. It has been reported that polysaccharides rich in β-1,4-D-Man*p* may help to activate the immune system by attaching to the mannose receptor on macrophages [31]. As a result, the varied monosaccharide composition ratios, particularly the variable Man molar ratios, are likely related to the immunological activities of OJZ9 and OJC8 on RAW 264.7 cells.

The results of phagocytosis experiments of polysaccharides from *MaiDong* were consistent with the results of NO release, reflecting the following overall trends (Figure 4c). First, the polysaccharides extracted from fibrous roots (OJZ9 and OJC8) had a stronger ability to promote FITC-Dextran phagocytosis by macrophages than that of the root. Second, OJZP had a stronger ability to promote phagocytosis than OJCP, LMP, and LSP. Third, LMP and LSP almost did not promote phagocytosis at the concentrations of 5, 10, and 20 µg/mL. Fourth, the macrophage-promoting ability of 3-year-old OJZ11 was stronger than that of 2-year-old OJZ12, but no significant difference from that of 1-year-old OJZP16, so it was not possible to determine whether the growth age had a significant effect on the phagocytosis of polysaccharides from *MaiDong*, and more samples are needed for further investigation; though *MaiDong* produced in Zhejiang should be collected at the third year. As can be seen in the typical phagocytosis flow charts (Supplementary Figure S1), the majority of the OJZP had a significant right shift compared to the control group, with the exception of the OJZ6 and OJZ13. This suggested that the stimulation of macrophages with OJZP could promote their phagocytosis to a certain extent. In contrast, as shown in the flow diagrams of samples from various origins, except for OJC3 and OJC8, the peaks of all samples showed no right shift when compared to the control group, which implied that the effect of OJCP, LMP, and LSP on enhancing the phagocytic ability of macrophages is not as significant as that of OJZP. In summary, the experimental results revealed that polysaccharides from fibrous roots promoted NO production (Figure 4b) and phagocytosis (Figure 4c) more effectively than those from the root. Additionally, OJZP promoted NO production more effectively than OJCP, and more strongly than LMP and LSP.

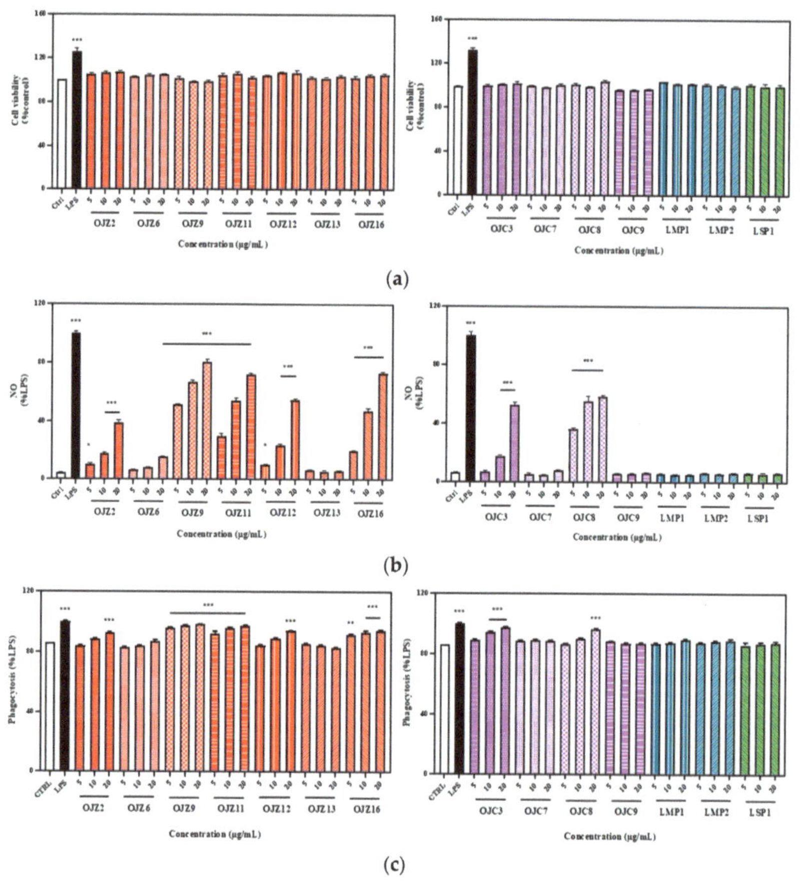

Figure 4. Effects of polysaccharides from representative samples of *MaiDong* on cell viability (**a**), NO production (**b**), and phagocytosis (**c**) of RAW 264.7 macrophages. All values were expressed as mean ± SEM of three independent experiments. * $p < 0.05$, ** $p < 0.01$, *** $p < 0.001$ vs control group, unmarked results indicate no significant difference vs control group. The codes are the same as in Table 1.

3.6. Multivariate Statistical Analysis for Polysaccharide Quality Control in MaiDong

3.6.1. Relationship of Chemicals and Activity

Scientific research has revealed that the biological activity of polysaccharides is frequently linked to their structure [32]. Using Pearson correlation analysis, we investigated the relationship of polysaccharide physicochemical properties with pharmacological activities, such as ABTS radical scavenging ability and immunological activity.

Pearson correlation analysis is used to reflect the linear correlation of variables, and the Pearson correlation coefficient is a measurement of the linear correlation between two sets of data [33]. A Pearson correlation coefficient that shows significance indicates that there is a correlation between the data. After that, the absolute value of the magnitude of the Pearson correlation coefficient reflects the closeness of the relationship between the data. A coefficient between 0.8 to 1 indicates a very close relationship; between 0.3 and 0.8 indicates a close relationship; and between 0 to 0.3 indicates a low correlation.

The correlation coefficients between the pharmacological activity and physicochemical properties of polysaccharides in *MaiDong* are recorded in Supplementary Table S1. In brief, Pearson correlation analysis was performed on 29 samples using ABTS radical scavenging ability as the reference, and 14 samples using immune activity as the reference. The comparison indicators were monosaccharide composition, Mw, RI, UV, and laser signal. In addition, the correlation values are calculated and recorded in Table 5. The correlation values ranged from 0 to 1. According to the table, the content of Man had the highest correlation value of 0.904 for antioxidant capacity among the 15 evaluated indicators, followed by the content of Xyl, with 0.896. Fru, the richest monosaccharide in *MaiDong* polysaccharides, had a significantly negative relationship to antioxidant and immunological activity of the polysaccharides. Meanwhile, the Mw at Peak1 and UV at Peak2 reflected no correlation with antioxidant activity. For immune activity, the content of Man also had the highest correlation value 0.817, followed by that of Ara at 0.713, while the UV signal at Peak2, and Mw at both Peak1 and Peak2, did not show a correlation with immune activity.

Table 5. Pearson correlation analysis between activities and physicochemical characters.

Rank	Antioxidant Activity		Immune Activity	
	Indicator	Correlation Coefficient	Indicator	Correlation Coefficient
1	Man	0.904 **	Man	0.817 **
2	Xyl	0.896 **	Ara	0.713 **
3	Gal	0.863 **	Xyl	0.688 **
4	Ara	0.832 **	RI-Peak3	−0.657 *
5	RI-Peak1	0.815 **	RI-Peak2	0.626 *
6	Fru	−0.800 **	RI-Peak1	0.616 *
7	LS	0.753 **	Fru	−0.598 *
8	Mw-Peak3	0.708 **	Gal	0.567 *
9	RI-Peak2	0.686 **	UV-Peak3	0.509
10	RI-Peak3	−0.609 **	LS	0.491
11	UV-Peak3	0.515 **	Mw-Peak3	0.442
12	Glc	−0.510 **	Glc	−0.439
13	Mw-Peak2	0.507 **	UV-Peak2	0.341
14	Mw-Peak1	0.238	Mw-Peak1	0.229
15	UV-Peak2	0.151	Mw-Peak2	0.207

$* p < 0.05, ** p < 0.01$.

3.6.2. Quality Marker Optimization

In addition to the molecular parameters, monosaccharide composition, and pharmacological activity of polysaccharides, the yield of polysaccharides is also an important factor to evaluate their quality [34]. The higher yield of polysaccharide means that the cost of planting can be reduced, which is beneficial to the marketization of *MaiDong*. In conclusion, seven indicators were chosen for polysaccharide quality assessment: yield, Mw, RI, laser, UV, monosaccharide composition, and ABTS scavenging activity.

Cluster analysis, by considering the seven indicators together, results in a spectrum plot as shown in Figure 5a. Such a clustering method can distinguish fibrous roots from the root and distinguish OJZP from OJCP. However, it does not distinguish OJCP from LMP and LSP, nor can it distinguish OJZP by growth years. Considering that, in general, polysaccharides do not have UV signals, we excluded the UV indicator. The results after optimization of the indexes are shown in Figure 5b. The results showed that by using the indicators of yield, ABTS clearance capacity, and the content of fructose, RID signal intensity could most effectively distinguish *O. japonicus* in Zhejiang from other *MaiDong* and could distinguish the difference between fibrous roots and root. *L. muscari* and *L. spicata* were also distinguished from *O. japonicus* in Sichuan. However, the growing years did not show an influence on the results. MaiDong produced in Sichuan usually can be collected in the first year, but *MaiDong* produced in Zhejiang should be collected in the third year. The difference might be mainly derived from the climate, and cultivation conditions should also be carefully investigated in the future. Nevertheless, chemical variation [11,14–16] has been noticed and their clinical efficacy should be carefully compared.

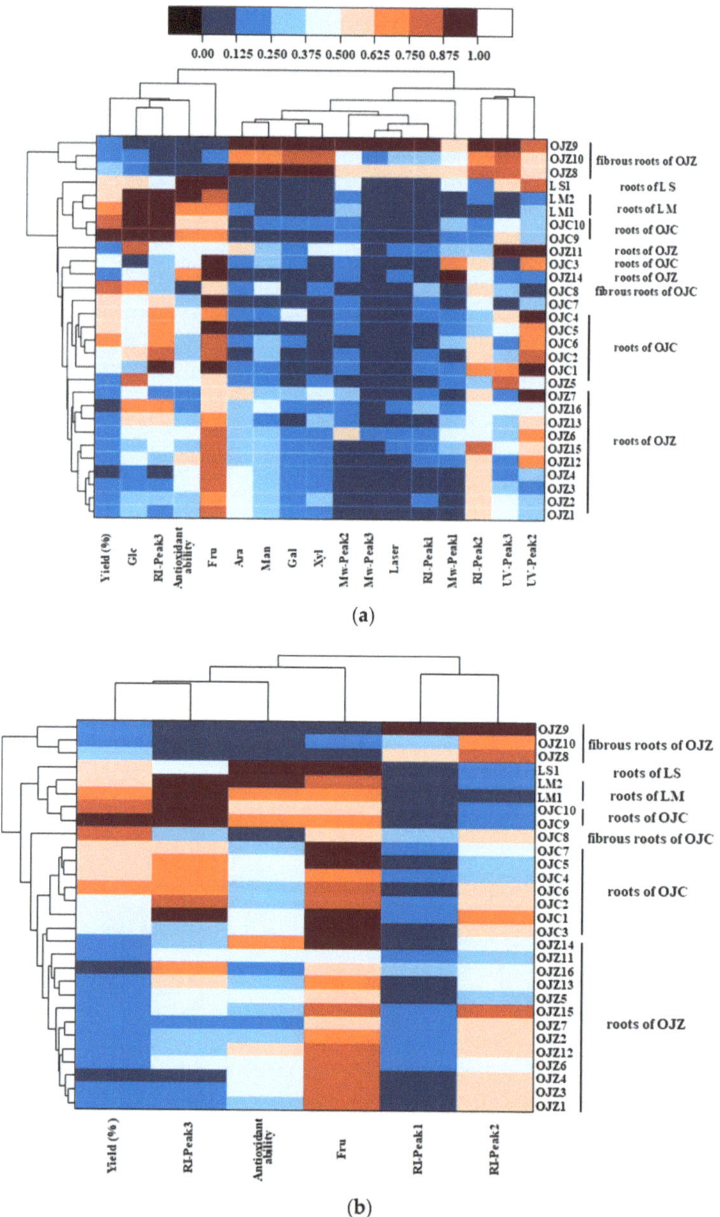

(a)

(b)

Figure 5. Hierarchical clustering and heat-map of 29 batches of *MaiDong* polysaccharides based on (**a**) all investigated indicators or (**b**) optimized indicators.

Through the Pearson correlation analysis, we found that Man, Ara, and Xyl showed significantly high correlation values for antioxidant capacity and immune activity, while the signal of UV and Mw showed low correlation values. It has been shown that the structure of mannose backbone or side chain modifications in polysaccharides has a relevant link to their mediated biological activities, such as immunomodulatory activities [35]. This may be one of the reasons why the Man content reflects a strong correlation in this analysis. As for how these physicochemical property indicators affect pharmacological activity, we believe that further in-depth exploration is needed through subsequent experiments.

Finally, through the cluster analysis (Figure 5), the yield, content of Fru, ABTS scavenging activity, and the RI signal were used as indicators for clustering. Through the results, a clear distinction can be made between the OJZ, OJC, LS, and LM, and the fibrous roots and the root. This may provide an innovative, simplified, and prompt strategy for the quality control of *MaiDong*. However, most likely due to the nature of the samples themselves, the properties of polysaccharides in *L. muscari*, *L. spicate*, and *O. japonicus* in Sichuan were closer.

4. Conclusions

In this study, the physicochemical properties and pharmacological activities of different sources of *MaiDong* were comprehensively compared, and the results showed that different *MaiDong* had similar molecular parameters and compositional monosaccharides. *O. japonicus* in Zhejiang had better antioxidant and immune-promoting activities than *O. japonicus* in Sichuan. There was no effect of growing age on the pharmacological activity of *O. japonicus* in Zhejiang. The fibrous root had better pharmacological activity than the root. The polysaccharides yield, content of Fru, antioxidant activity, and RI signal can be employed as quality indicators of *MaiDong*, which is helpful to the cultivation, selection, use, and quality control of *MaiDong*.

Supplementary Materials: The following supporting information can be downloaded at: https://www.mdpi.com/article/10.3390/biom12101491/s1, Figure S1: Typical flow cytometric profiles of phagocytosis of culture medium, LPS at 0.4 µg/mL, OJZP, OJZP, LMP and LSP at 20 µg/mL treated RAW 264.7 macrophages; Table S1: The Pearson correlation coefficients between the antioxidant/immune activity and physicochemical properties of polysaccharides in *MaiDong*.

Author Contributions: Conceptualization, S.L., J.Z. and X.P.; methodology, S.L. and J.Z.; investigation, Z.C. and B.Z.; validation, Z.C.; software, Z.C. and B.Z.; formal analysis, Z.C.; resources, S.L., J.Z. and X.P.; writing—original draft preparation, Z.C.; writing—review and editing, S.L., J.Z. and Z.C.; supervision, S.L. and J.Z.; project administration, S.L., X.P. and J.Z.; funding acquisition, S.L., X.P. and J.Z. All authors have read and agreed to the published version of the manuscript.

Funding: This research was partially funded by grants from the Science and Technology Development Fund, Macau SAR (File no. 0017/2019/AKP), the Key-Area Research and Development Program of Guangdong Province (File no. 2020B1111110006) and the University of Macau (File no. MYRG2018-00083-ICMS/MYRG2019-00128-ICMS/CPG2021-00009-ICMS) and Ningbo City Science and Technology Innovation 2025 Major Research Project (2019B10008).

Institutional Review Board Statement: Not applicable.

Informed Consent Statement: Not applicable.

Data Availability Statement: Data will be provided upon reasonable request.

Conflicts of Interest: The authors declare no conflict of interest.

References

1. Wang, H.Y.; Kang, C.Z.; Zhang, W.J.; Zhou, L.Y.; Wan, X.F.; Lyu, C.G.; Huang, L.Q.; Liu, D.H.; Guo, L.P. Land use strategy of ecological agriculture of Chinese materia medica in future development. *CJCMM* **2020**, *45*, 1990–1995. [CrossRef]
2. Li, X.; Chen, Y.; Lai, Y.; Yang, Q.; Hu, H.; Wang, Y. Sustainable utilization of traditional chinese medicine resources: Systematic evaluation on different production modes. *Evid.-Based Complement. Altern. Med.* **2015**, *2015*, 218901. [CrossRef] [PubMed]

3.   Lu, Y.; Yao, G.; Wang, X.; Zhang, Y.; Zhao, J.; Yu, Y.-J.; Wang, H. Chemometric discrimination of the geographical origin of licorice in China by untargeted metabolomics. *Food Chem.* **2022**, *380*, 132235. [CrossRef] [PubMed]

4.   Dong, F.; Lin, J.; You, J.; Ji, J.; Xu, X.; Zhang, L.; Jin, Y.; Du, S. A chemometric modeling-free near infrared barcode strategy for smart authentication and geographical origin discrimination of Chinese ginseng. *SAA Spectrochim. Acta Part A* **2020**, *226*, 117555. [CrossRef]

5.   Wang, P.; Yu, Z. Species authentication and geographical origin discrimination of herbal medicines by near infrared spectroscopy: A review. *J. Pharm. Anal.* **2015**, *5*, 277–284. [CrossRef]

6.   Yang, X.; Tian, X.; Zhou, Y.; Liu, Y.; Li, X.; Lu, T.; Yu, C.; He, L. Evidence-based study to compare daodi traditional Chinese medicinal material and non-daodi traditional Chinese medicinal material. *Evid.-Based Complement. Altern. Med.* **2018**, *6763130*. [CrossRef]

7.   Zhao, Z.; Guo, P.; Brand, E. The formation of daodi medicinal materials. *J. Ethnopharmacol.* **2012**, *140*, 476–481. [CrossRef]

8.   Chen, M.H.; Chen, X.J.; Wang, M.; Lin, L.G.; Wang, Y.T. *Ophiopogon japonicus*—A phytochemical, ethnomedicinal and pharmacological review. *J. Ethnopharmacol.* **2016**, *181*, 193–213. [CrossRef]

9.   Lei, F.; Weckerle, C.S.; Heinrich, M. Liriopogons (Genera *Ophiopogon* and *Liriope*, Asparagaceae): A Critical Review of the Phytochemical and Pharmacological Research. *Front. Pharmacol.* **2021**, *12*, 769929. [CrossRef]

10.   Li, Z.; Wu, Y.Y.; Yu, B.X. Methylophiopogonanone A, an *Ophiopogon* homoisoflavonoid, alleviates high-fat diet-induced hyperlipidemia: Assessment of its potential mechanism. *Braz. J. Med. Biol. Res.* **2020**, *53*, 10. [CrossRef]

11.   Zhou, Y.F.; Wang, L.L.; Chen, L.C.; Liu, T.B.; Sha, R.Y.; Mao, J.W. Enrichment and separation of steroidal saponins from the fibrous roots of *Ophiopogon japonicus* using macroporous adsorption resins. *RSC Adv.* **2019**, *9*, 6689–6698. [CrossRef]

12.   Fang, J.; Wang, X.; Lu, M.; He, X.; Yang, X. Recent advances in polysaccharides from *Ophiopogon japonicus* and *Liriope spicata* var. *prolifera*. *Int. J. Biol. Macromol.* **2018**, *114*, 1257–1266. [CrossRef]

13.   Jiao, Y. Fragment-Based Drug Discovery Library Based on Ligusticum Chuanxiong, a Traditional Chinese Medicine. Master's Thesis, Griffith University, Queensland, Australia, 2020. [CrossRef]

14.   Ge, Y.; Chen, X.; Godevac, D.; Bueno, P.C.P.; Salome Abarca, L.F.; Jang, Y.P.; Wang, M.; Choi, Y.H. Metabolic Profiling of Saponin-Rich *Ophiopogon japonicus* Roots Based on 1H NMR and HPTLC Platforms. *Planta Med.* **2019**, *85*, 917–924. [CrossRef]

15.   Lin, Y.; Zhu, D.; Qi, J.; Qin, M.; Yu, B. Characterization of homoisoflavonoids in different cultivation regions of *Ophiopogon japonicus* and related antioxidant activity. *J. Pharm. Biomed. Anal.* **2010**, *52*, 757–762. [CrossRef]

16.   Zhao, M.; Xu, W.F.; Shen, H.Y.; Shen, P.Q.; Zhang, J.; Wang, D.D.; Xu, H.; Wang, H.; Yan, T.T.; Wang, L.; et al. Comparison of bioactive components and pharmacological activities of *Ophiopogon japonicas* extracts from different geographical origins. *J. Pharm. Biomed. Anal.* **2017**, *138*, 134–141. [CrossRef]

17.   Lin, C.; Kuo, T.C.; Lin, J.C.; Ho, Y.C.; Mi, F.L. Delivery of polysaccharides from *Ophiopogon japonicus* (OJPs) using OJPs chitosan/whey protein co-assembled nanoparticles to treat defective intestinal epithelial tight junction barrier. *Int. J. Biol. Macromol.* **2020**, *160*, 558–570. [CrossRef]

18.   Sun, W.J.; Hu, W.J.; Meng, K.; Yang, L.M.; Zhang, W.M.; Song, X.P.; Qu, X.H.; Zhang, Y.Y.; Ma, L.; Fan, Y.P. Activation of macrophages by the ophiopogon polysaccharide liposome from the root tuber of *Ophiopogon japonicus. Int. J. Biol. Macromol.* **2016**, *91*, 918–925. [CrossRef]

19.   Wang, H.Y.; Guo, L.X.; Hu, W.H.; Peng, Z.T.; Wang, C.; Chen, Z.C.; Liu, E.Y.L.; Dong, T.T.X.; Wang, T.J.; Tsim, K.W.K. Polysaccharide from tuberous roots of *Ophiopogon japonicus* regulates gut microbiota and its metabolites during alleviation of high-fat diet-induced type-2 diabetes in mice. *J. Funct. Food* **2019**, *63*, 10. [CrossRef]

20.   Zhu, N.; Lv, X.C.; Wang, Y.Y.; Li, J.L.; Liu, Y.M.; Lu, W.F.; Yang, L.C.; Zhao, J.; Wang, F.J.; Zhang, L.S.W. Comparison of immunoregulatory effects of polysaccharides from three natural herbs and cellular uptake in dendritic cells. *Int. J. Biol. Macromol.* **2016**, *93*, 940–951. [CrossRef]

21.   Zhu, B.J.; Yan, Z.Y.; Hong, L.; Li, S.P.; Zhao, J. Quality evaluation of *Salvia miltiorrhiza* from different geographical origins in China based on qualitative and quantitative saccharide mapping and chemometrics. *J. Pharm. Biomed. Anal.* **2020**, *191*, 113583. [CrossRef]

22.   Wu, D.T.; Cheong, K.L.; Wang, L.Y.; Lv, G.P.; Ju, Y.J.; Feng, K.; Zhao, J.; Li, S.P. Characterization and discrimination of polysaccharides from different species of *Cordyceps* using saccharide mapping based on PACE and HPTLC. *Carbohydr. Polym.* **2014**, *103*, 100–109. [CrossRef] [PubMed]

23.   Chen, Z.R.; Zhu, B.J.; Chen, Z.X.; Cao, W.; Wang, J.Q.; Li, S.P.; Zhao, J. Effects of steam on polysaccharides from *Polygonatum cyrtonema* based on saccharide mapping analysis and pharmacological activity assays. *Chin. Med.* **2022**, *17*, 97. [CrossRef] [PubMed]

24.   Gu, D.; Huang, L.L.; Chen, X.; Wu, Q.H.; Ding, K. Structural characterization of a galactan from *Ophiopogon japonicus* and anti-pancreatic cancer activity of its acetylated derivative. *Int. J. Biol. Macromol.* **2018**, *113*, 907–915. [CrossRef] [PubMed]

25.   Gong, Y.; Zhang, J.; Gao, F.; Zhou, J.; Xiang, Z.; Zhou, C.; Wan, L.; Chen, J. Structure features and in vitro hypoglycemic activities of polysaccharides from different species of Maidong. *Carbohydr. Polym.* **2017**, *173*, 215–222. [CrossRef]

26.   Li, X.; Ouyang, X.; Cai, R.; Chen, D. 3′, 8″-Dimerization enhances the antioxidant capacity of flavonoids: Evidence from acacetin and isoginkgetin. *Molecules* **2019**, *24*, 2039. [CrossRef]

27.   Re, R.; Pellegrini, N.; Proteggente, A.; Pannala, A.; Yang, M.; Rice-Evans, C. Antioxidant activity applying an improved ABTS radical cation decolorization assay. *Free. Radic. Biol. Med.* **1999**, *26*, 1231–1237. [CrossRef]

28. Hao, N.B.; Lü, M.H.; Fan, Y.H.; Cao, Y.L.; Zhang, Z.R.; Yang, S.M. Macrophages in tumor microenvironments and the progression of tumors. *Clin. Dev. Immunol.* **2012**, *2012*, 11. [CrossRef]
29. Boscá, L.; Zeini, M.; Través, P.G.; Hortelano, S. Nitric oxide and cell viability in inflammatory cells: A role for NO in macrophage function and fate. *Toxicology* **2005**, *208*, 249–258. [CrossRef]
30. Greenberg, S.; Grinstein, S. Phagocytosis and innate immunity. *Curr. Opin. Immunol.* **2002**, *14*, 136–145. [CrossRef]
31. Leung, M.; Liu, C.; Zhu, L.; Hui, Y.; Yu, B.; Fung, K.J.G. Chemical and biological characterization of a polysaccharide biological response modifier from Aloe vera L. var. chinensis (Haw.) Berg. *Glycobiology* **2004**, *14*, 501–510. [CrossRef]
32. Huang, G.L.; Chen, X.; Huang, H.L. Chemical Modifications and Biological Activities of Polysaccharides. *Curr. Drug Targets* **2016**, *17*, 1799–1803. [CrossRef]
33. Hauke, J.; Kossowski, T. Comparison of Values of Pearson's and Spearman's Correlation Coefficients on the Same Sets of Data. *Quageo* **2011**, *30*, 87–93. [CrossRef]
34. Yan, Y.L.; Yu, C.H.; Jing, C.; Li, X.X.; Wei, W.; Li, S.Q. Ultrasonic-assisted extraction optimized by response surface methodology, chemical composition and antioxidant activity of polysaccharides from *Tremella mesenterica*. *Carbohydr. Polym.* **2011**, *83*, 217–224. [CrossRef]
35. Zhang, S.; Lin, Z.; Wang, D.; Xu, X.; Song, C.; Sun, L.; Mayo, K.H.; Zhao, Z.; Zhou, Y. Galactofuranose side chains in galactomannans from *Penicillium* spp. modulate galectin-8-mediated bioactivity. *Carbohydr. Polym.* **2022**, *292*, 119677. [CrossRef] [PubMed]

MDPI AG

Grosspeteranlage 5

4052 Basel

Switzerland

Tel.: +41 61 683 77 34

Biomolecules Editorial Office

E-mail: biomolecules@mdpi.com

www.mdpi.com/journal/biomolecules